# DEVONIAN–EARLY MISSISSIPPIAN CARBONATES OF THE WESTERN CANADA SEDIMENTARY BASIN: A SEQUENCE STRATIGRAPHIC FRAMEWORK



BY

JACK WENDTE

Institute of Sedimentary and Petroleum Geology,
Geological Survey of Canada
3303 33rd Street N.W.
Calgary, Alberta T2L 2A7

FRANK A. STOAKES

Stoakes Campbell Geoconsulting Ltd.
500, 839 - 5th Avenue S.W.
Calgary, Alberta T2P 3C8

CLARENCE V. CAMPBELL

Stoakes Campbell Geoconsulting Ltd.
500, 839 - 5th Avenue S.W.
Calgary, Alberta T2P 3C8



SEPM Short Course No. 28
Calgary, June 20-21, 1992

ISBN #0-918985-96-X

Additional Copies of this publication may be ordered from SEPM.
Send your order to:

SEPM
Post Office Box 4756
Tulsa, OK 74159-0756
U.S.A.


Printed in the United States of America

# PREFACE

The Devonian and early Mississippian strata in the Western Canada Sedimentary Basin include a wide diversity of shallow-water carbonate and basin filling carbonate, shale and evaporite facies. Of these, large Devonian platform-reef complexes are the most spectacular. They occur in magnificent exposures in the Front Ranges of the Canadian Rockies and in the subsurface of Alberta. In the subsurface, these complexes pool many of the largest oil and gas accumulations in Western Canada.

This short course is intended to provide a summary of Devonian and early Mississippian deposits in the subsurface of the Western Canada Sedimentary Basin. One of our major goals is to present the evolution of these strata in a sequence-stratigraphic context. The role of sea-level, tectonic and depositional controls on "stacking" and facies patterns are considered. A second major goal is to relate the occurrence of hydrocarbon pools to this sequence-stratigraphic framework.

This manual provides the texts and some of the illustrations for talks given during the first day of the short course. Chapter 1 presents an overview of the Devonian System in Western Canada. Chapter 2 summarizes Devonian cyclicity in Western Canada. Chapters 3, 4, and 5 discuss the basic depositional components or building blocks of the Devonian successions - carbonate platform, isolate reef and basin-fill deposits. Chapters 6 through 10 consider the evolution of the major Devonian sequences. Chapter 11 provides a discussion of early Mississippian strata in the Peace River Arch area. Chapter 12 summarizes the major aspects of the course.

We hope that these discussions stimulate further research on many of the unanswered questions regarding Devonian and early Mississippian depositional systems in Western Canada.

*Jack Wendte* *Frank Stoakes* *Clarence Campbell*

DEDICATION OF SEPM SHORT COURSE NO. 28
TO DR. PERRY GLAISTER

This short course manual is dedicated to Dr. R. Perry Glaister by way of thanks for his support and encouragement over the years. Perry is a stratigrapher, sedimentologist and petroleum geologist who has contributed through his work to our understanding of the Western Canada Sedimentary Basin. Born in Calgary, he joined Imperial Oil Ltd (later Esso Resources Canada Ltd) in 1952 and but for an educational leave of absence to obtain his PhD under Dr. L.L. Sloss at Northwestern University, spent his entire professional career with this organization.

Since then, Perry has contributed numerous papers and articles to the literature of the AAPG and the ASPG/CSPG. He has been an active participant at core conferences, seminars and field trips and was awarded the ASPG Medal of Merit (with co-authors W.R. Maiklem and D.G. Bebout) for his paper "Classification of Anhydrites - a Practical Approach". He became an honorary member of the CSPG in 1985.

Although Perry's interests span both siliciclastics and carbonates, he is perhaps best known as co-editor of the first edition of the "Atlas of the Geological History of Western Canada" a formidable undertaking at that time and one that has remained a standard reference ever since.

In was under his supervision as Head of the Facies group at ERCL that two of us (JCW, FAS) first came in contact with Perry. As a supervisor he provided an open work environment with enough direction to keep us focused. For a while we even thought that we'd weaned him away from siliciclastics as he argued sucessfully with upper management for the establishment of an in-house modern carbonate course to Florida and the Bahamas. During subsequent trips to modern carbonate environments Perry rekindled his carbonate ideas at the same time as proving to be a formidable shrimp predator!

Since retiring from ERCL Perry has continued his involvement in training, started at Esso, giving siliciclastic courses for SCG with one of us (CVC). Finally during one course, given in Pakistan, Perry was able to find someone who still believed in environmental interpretation from grain size analysis!

In conclusion then we would like to thank Perry for his patience, support and encouragement over the years. His dedication to the pursuit of the scientific answer, combined with his amiable and approachable disposition means that he will always be "Sir" Perry Glaister to us.

## ACKNOWLEDGEMENTS


This manual represents the culmination of thought and effort over a long period of looking at and puzzling over Devonian rocks in the Western Canada Sedimentary Basin. We regard it not as a conclusion but rather a milestone in our understanding of this basin system. On a broader scale it has helped us to understand the true nature of and controls on cyclicity and it's application to exploration and exploitation. None of this would have been possible without the existence of the ERCB and it's mandate to collect and make available subsurface data from Alberta. In particular the friendly and co-operative staff of the ERCB under the direction of Art Shephard made our task a pleasant one at the Core Research Centre.

First and foremost however we would like to thank our former and present employers for the opportunity to pursue lines of investigation. Two of us (FAS, JCW) thank Esso Resources Canada Ltd (ERCL) in whose employ we first began to build and test our ideas of the basin's evolution. The list of co-workers is too long to mention but special thanks must go to Perry Glaister in whose group we initiated our study (see dedication).

Our employers changed and one of us (JCW) more directly applied our work to the exploration effort with Canadian Hunter Exploration Ltd. A recent change to the Institute of Sedimentary and Petroleum Geology (a division of the Geological Survey of Canada) allowed him to return to more "research" avenues of investigation.

The consulting business was the route that two of us (FAS, CVC) pursued initially with RPI Canada Inc and since 1986 with Stoakes Campbell Geoconsulting Ltd (SCG). This afforded us the opportunity to see a wide stratigraphic and geographic array of rocks to confirm and temper our ideas. We have our client base to thank for their support in this endeavour.

More specifically certain individuals have helped us directly in assembling this manual. Unpublished sections and some diagrams were supplied by Iain Muir (Esso) and Myles Bosman (CanHunter), drafting by Al Stewart (consultant), typing by Wilma Ell (ISPG), and the SCG staff assissted in all aspects (Drue Robinson, Jana King, Barb McAuliffe and Sandra Goucher).

Finally Ben Robertson (Chevron) first suggested that we make our material a short course and Tricia Auberle at the SEPM office patiently listened to our delays and excuses.

Two of us (FAS, CVC) wish to acknowledge the effort of Jack Wendte in helping to bring about the completion of this manual. He tirelessly completed his sections in addition to molding and editing many of the others. Thanks Jacques!

TABLE OF CONTENTS

CHAPTER 1

# OVERVIEW OF THE DEVONIAN OF THE WESTERN CANADA SEDIMENTARY BASIN

J.C. WENDTE

## INTRODUCTION

Devonian strata of the Western Canada Sedimentary Basin (WCSB) occur in a widespread belt extending from the Arctic Ocean and the Alaska-Yukon border on the north to the southeast through the Northwest Territories, northeast British Columbia, and Alberta into Saskatchewan and southwest Manitoba (Figure 1). This short course manual discusses the temporal and spatial evolution of Devonian carbonates and associated deposits in the WCSB, along with a discussion of early Mississippian strata. The initial 10 chapters of this manual, including this introductory chapter, consider the evolution of Devonian deposits; the 11th chapter deals with the latest Devonian-earliest Mississippian sequence. The 12th and final chapter provides a summary of the entire course.

The Devonian of the WCSB is, of course, best known for the occurrence of reefs. These reefs occur in multiple phases of sedimentation and vary according to depositional setting, evolution, related facies associations, areal extent, thickness, fauna and morphology. Many of these reefs are limited in distribution to the subsurface; others are magnificently exposed in the Rocky Mountains. A summary of all of these reefs is provided by Moore (1988a).

The occurrence of Devonian reefs in the WCSB subsurface is of prime significance to the economy of Western Canada, especially that in Alberta. Since the initial discovery of oil in Leduc reefs of Late Devonian age in 1947, the exploration and exploitation of these complexes has provided the backbone of the hydrocarbon industry in Western Canada. Even though large hydrocarbon reserves have been discovered in younger Paleozoic and Mesozoic rocks, the Devonian System still ranks as the major oil producing interval in Western Canada, accounting for 55 percent of recoverable conventional crude oil reserves and about 27 percent of recoverable gas reserves (Podruski and others, 1988 and Reinson and others, 1992). Recent estimates of the potential reserves in Devonian strata remaining undiscovered point to continued exploration emphasis in these units. The Geological Survey of Canada estimates 23 percent of Devonian conventional oil reserves and 56 percent of Devonian gas reserves remain to be discovered (Podruski and others, 1988 and Reinson and others, 1992).

According to Burrowes and Krause (1987), approximately 36,500 wells drilled in Alberta penetrated rocks of Devonian age. The wireline logs, cores, cuttings, test and completion results from these wells and others in British Columbia, Saskatchewan and the Northwest Territories are available to the public no later than one

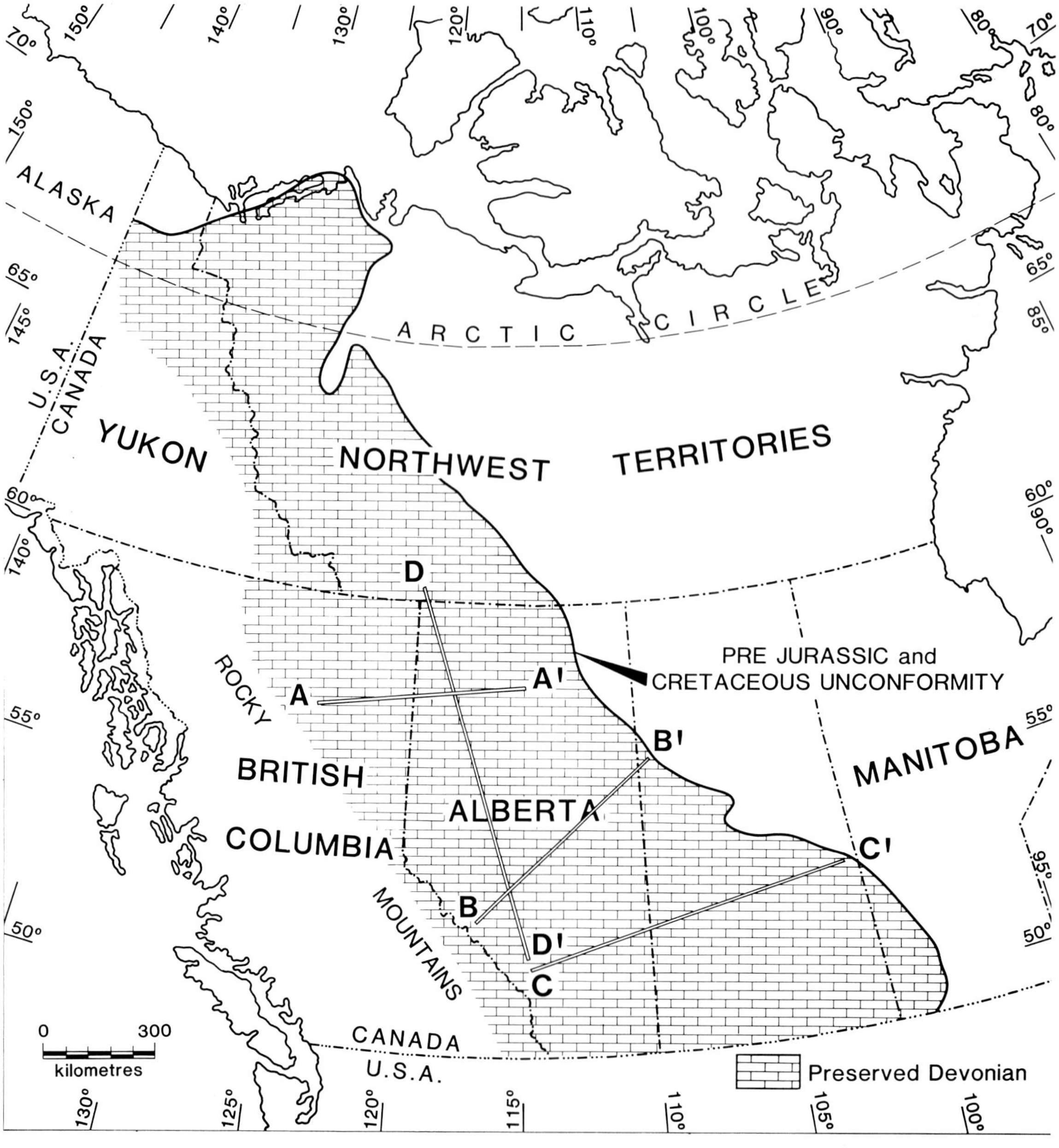


Figure 1. Index map of western Canada showing the limits of preserved Devonian sediments and the lines of section illustrated in Figures 10 and 11 (after Moore, 1989).

year following rig release. This subsurface information, along with ideas gained from examining surface exposures, provides an unparalleled opportunity to explore the genesis and evolution of reefs and associated facies. In addition, the accessibility of the superb subsurface information allows geologists to characterize the nature and occurrence of hydrocarbon pools perhaps in more detail than anywhere else.

Devonian strata of the WCSB include a diverse variety of depositional facies that change through time. Basic facies belts include platform carbonates and evaporites, reefal carbonates, and basin-filling shales, limestones and evaporites. Although significant aspects of these general facies vary from one megasequence to another, recurring lateral and vertical facies relationships exist.

This short course focusses on the subsurface occurrence of Devonian carbonates and associated facies in an area extending from the Middle Devonian Slave Point barrier in northeastern British Columbia and the southern Northwest Territories to those deposits in central Alberta. Stratigraphically, our discussion is mainly limited to the interval above the Middle Devonian Chinchaga anhydrite. This selected stratigraphic interval and area encompasses deposits that contain the vast majority of hydrocarbons in Devonian strata of the WCSB. These deposits include the well known Middle Devonian Keg River and Swan Hills reef complexes and the Upper Devonian Leduc and Nisku reef complexes. Stratigraphic nomenclature used in this course corresponds to that of the Alberta-northeast British Columbia subsurface as employed by practising petroleum geologists in Calgary. The stratigraphy and ages of the lithostratigraphic units are shown in Figure 2. Participants desiring information either from outside the area or interval of focus are referred to the recent Devonian synthesis of the western interior of Canada by Moore (1988b and 1989) and that of the eastern Canadian Cordillera by Morrow and Geldsetzer (1988).

One major goal of this short course is to provide an overview of Middle and Upper Devonian strata of the WCSB in a sequence stratigraphic context. Emphasis is placed on discussing the evolution of carbonate platforms and reefs as well as basin-fill shales and limestones. The stacking patterns of shallow-water facies are considered relative to basin position. The significance and applicability of sea level, tectonic and depositional controls on carbonate stacking patterns and cyclicity, carbonate platform evolution and reef inception, and basin-filling mechanisms are evaluated.

A second major goal of this short course is to elucidate the occurrence of hydrocarbons. The major hydrocarbon occurrences are characterized with respect to reservoir, source and seal. Maturation and migration patterns for the major stratigraphic intervals are presented to relate the exploration potential of various plays basinwide.

## GLOBAL AND REGIONAL SETTING

The Western Canada Sedimentary Basin is a simple northeasterly trending wedge of sedimentary rocks more than 6 km thick, extending

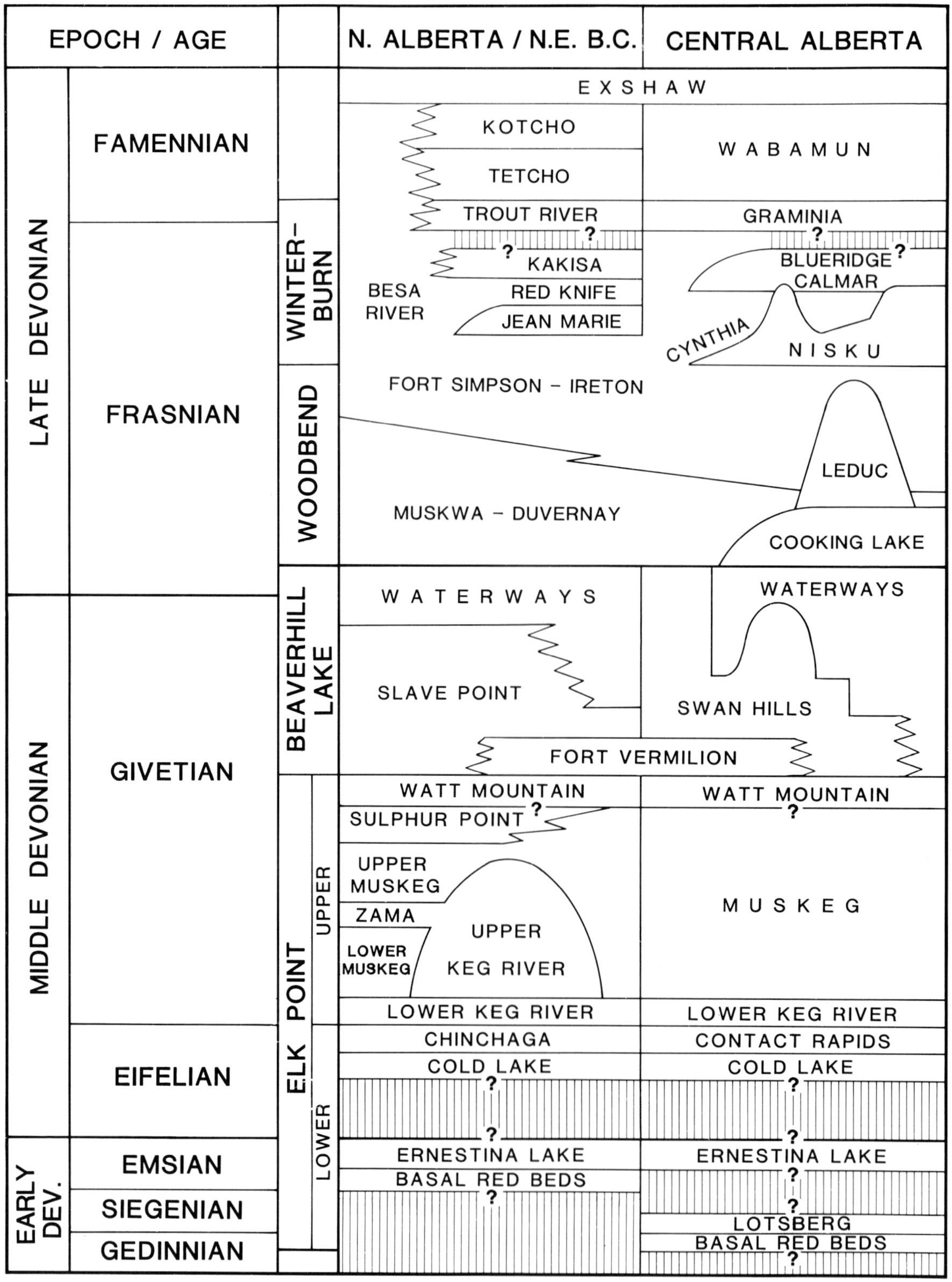


Figure 2. Stratigraphic nomenclature and correlations of Devonian strata in Alberta and northeastern British Columbia.

from the Canadian Shield southwest to the Cordilleran foreland thrust belt. Figure 3 shows an isopach of Phanerozoic rocks of the WCSB. This sedimentary wedge is underlain by Archaean and Proterozoic crystalline basement and consists of sedimentary rocks of mid-Proterozoic to Paleocene age. The evolution of this basin represents a continuum of tectonic phases varying from rift and passive margin in the late Precambrian and Cambrian to foreland basin in Middle Jurassic to Early Cretaceous and again from Late Cretaceous to Paleocene. Structural loading and the western down-tilting of the WCSB during these two phases resulted in the deep burial of Devonian strata along the western side of the WCSB. This deep burial led to maturation of organic-rich Devonian source intervals. Hydrocarbons migrated updip to the east and northeast.

During Devonian time, the WCSB was situated along the western margin of the Laurussia Continent. The Laurussia Continent formed when the North America-Greenland craton converged, collided and was joined to the northwestern European craton in latest Silurian time (Ziegler, 1988). The suture joining the two cratons is known as the Caledonian Fold Belt.

The position of the WCSB relative to the paleo-equator on published plate tectonic reconstructions is highly variable. Scotese and others (1985) show reconstructions relying heavily on paleomagnetic data but supplemented by geological evidence. They show the paleo-equator in Early Devonian time (late Gedinian-early Siegenian) passing through northern Alberta in a northeast-southwest direction, tilted approximately 30 degrees in a counter-clockwise direction from the position of the present equator. In their reconstructions, the Laurussia Continent subsequently rotates a few degrees in a clockwise manner and moves northward. By Late Devonian time (late Frasnian-Famennian), their model places the WCSB entirely in the northern hemisphere.

Witzke and Heckel (1988) provide reconstructions relying more heavily on paleoclimatic, paleobiogeographic and tectonic considerations and less on paleomagnetic data. Their reconstructions of the Laurussia Continent in Early, Middle and Late Devonian times are shown in Figure 4. Unlike other models, their reconstructions place virtually all of the WCSB in the southern hemisphere throughout most of Devonian time. Their models depict a paleo-equator striking in a northeast-southwest direction and passing through the Canadian Arctic Archipelago in Middle and Late Devonian time. The WCSB during the same interval falls in the arid, easterly trade wind belt. Their reconstructions indicate a northerly movement of the Laurussia Continent during Devonian time.

The model of Witzke and Heckel (1988) is compatible with the depositional aspects of a composite WCSB-Canadian Arctic Archipelago system. Studies of upper Middle and Upper Devonian terrigenous strata in the Canadian Arctic Archipelago indicate a hot and wet climate and support deposition in an equatorial belt (Embry, 1988 and Personal Communication). In a compatible manner, the following aspects of Middle and Upper Devonian strata in the WCSB support deposition in a trade wind belt: (1) the extensive occurrence of evaporites, indicating a dry, tropical climate and (2) indicators of northeast to southwest water circulation,

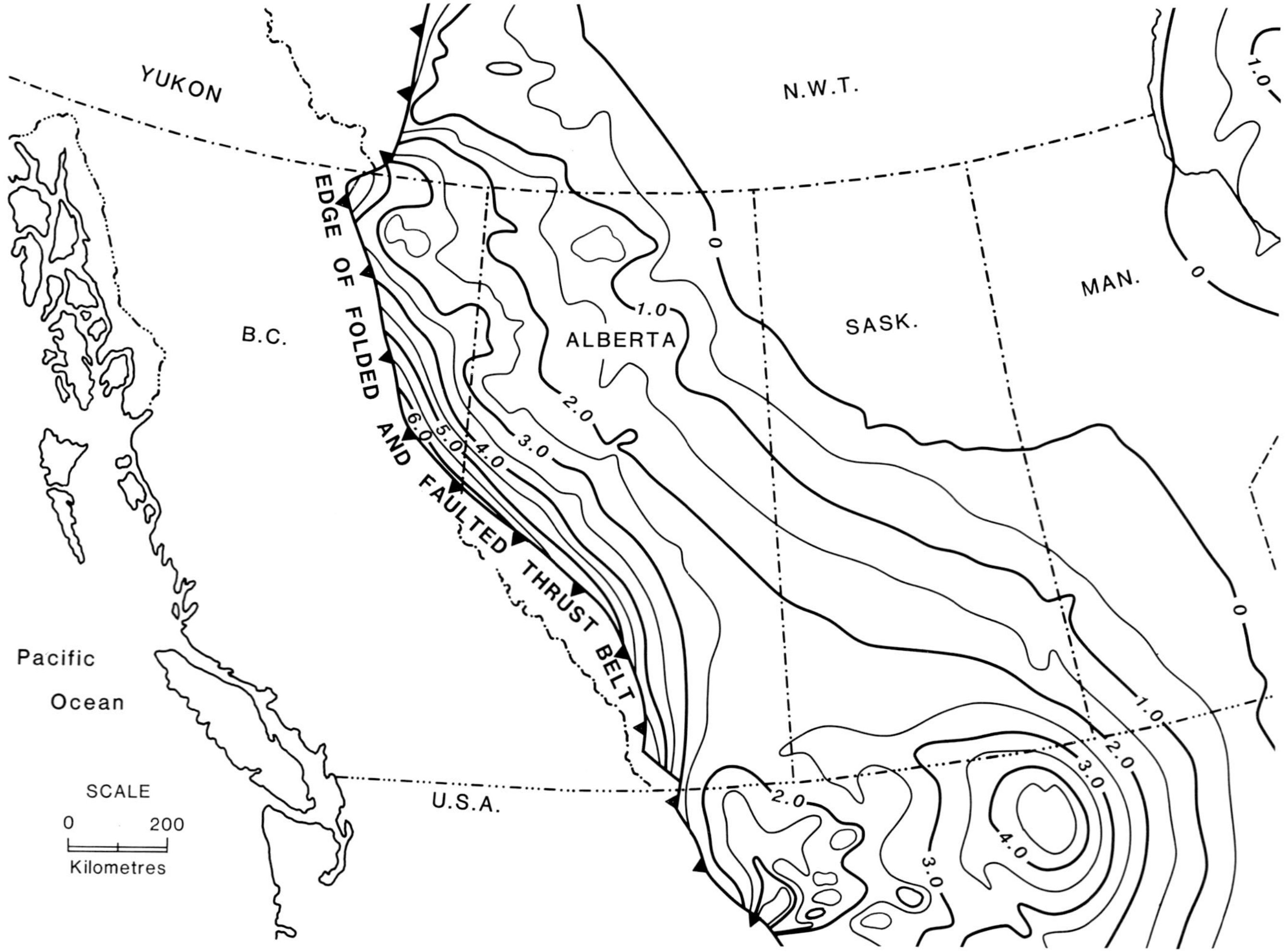


Figure 3. Total preserved thickness of Phanerozoic rocks east of the folded and faulted rocks of the Cordilleran foreland thrust belt. Contours are in kilometres (after Porter, Price and McCrossan, 1982).

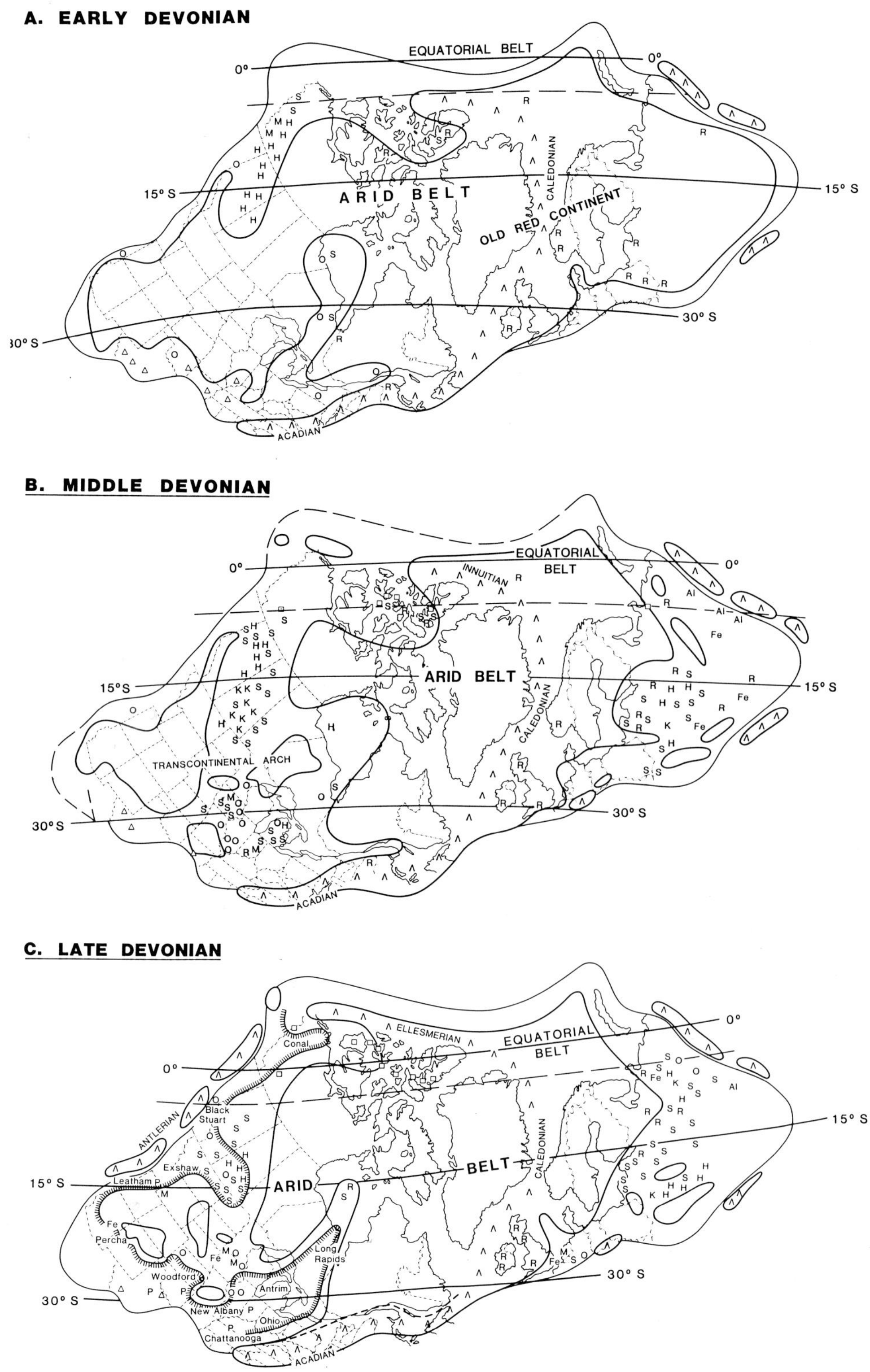


Figure 4. Devonian paleographic reconstructions of the Laurussia Continent (after Witzke and Heckel, 1988).

consistent with an easterly trade wind belt parallel to the paleo-equator.

## PRE DEVONIAN GEOLOGY

Carbonate deposition is especially sensitive to topographic features that provide areas of nucleation and influence water circulation. Many of the features that influenced Devonian deposition in the WCSB formed during Precambrian or early Paleozoic time. The pursuant discussion provides a summary of the geology of the WCSB prior to Devonian time.

### *Precambrian*

Precambrian continental crust is thought to have extended approximately 200 km west of the present limit of the WCSB (Monger, 1989). Ross and Stephenson (1989) concluded that the present configuration of the Precambrian crystalline basement in Alberta was assembled between 2.0 Ga and 1.8 Ga and involved plate collisions similar in character and gravity to present day subduction zones and collisional orogens. On the basis of aeromagnetic and gravity data, and U-Pb, zircon and monzonite geochronology, they divided the Alberta basement into a series of Archaean and Proterozoic tectonic domains. The distribution of these domains is illustrated in Figure 5. They recognized three crustal discontinuities. The Snowbird tectonic zone in central Alberta includes the Rimbey Arc. As noted by Ross and Stephenson (1989), the position of this arc is coincident with that of the Late Devonian (Frasnian) Leduc reef trend. The Great Slave Shear zone incorporates the Hay River Fault zone of northern Alberta. The Vulcan Low in southern Alberta is coincident with the Precambrian Southern Alberta Aulacogen.

Proterozoic and lower Paleozoic strata show a pronounced westward thickening from the pericratonic platform to the miogeocline as illustrated in Figure 6. Two sequences of Proterozoic strata overlie the crystalline basement and are confined to the structurally deformed and thrusted part of the southern Canadian Cordillera. Both sequences are totally lacking on the autochthonous platform of the WCSB. The older assemblage corresponds to the Purcell Supergroup and its equivalent in the northwest United States, the Belt Supergroup. The Purcell-Belt succession is up to 20 km thick and is considered to have accumulated between 1570 and 1200 Ga (Ross and others, 1989). The younger sequence is named the Windermere Supergroup. This succession is up to 9 km thick and is of late Proterozoic age (780-570 Ga) (Ross and others, 1989). Interpreted tectonic settings for these two assemblages have ranged from rifted to passive continental margin.

### *Early Cambrian to Mid-Ordovician*

Rocks of these ages correspond to the Sauk Sequence of Sloss (1963). The onset of Cambrian sedimentation immediately followed the rifting phase that gave birth to the North America craton,

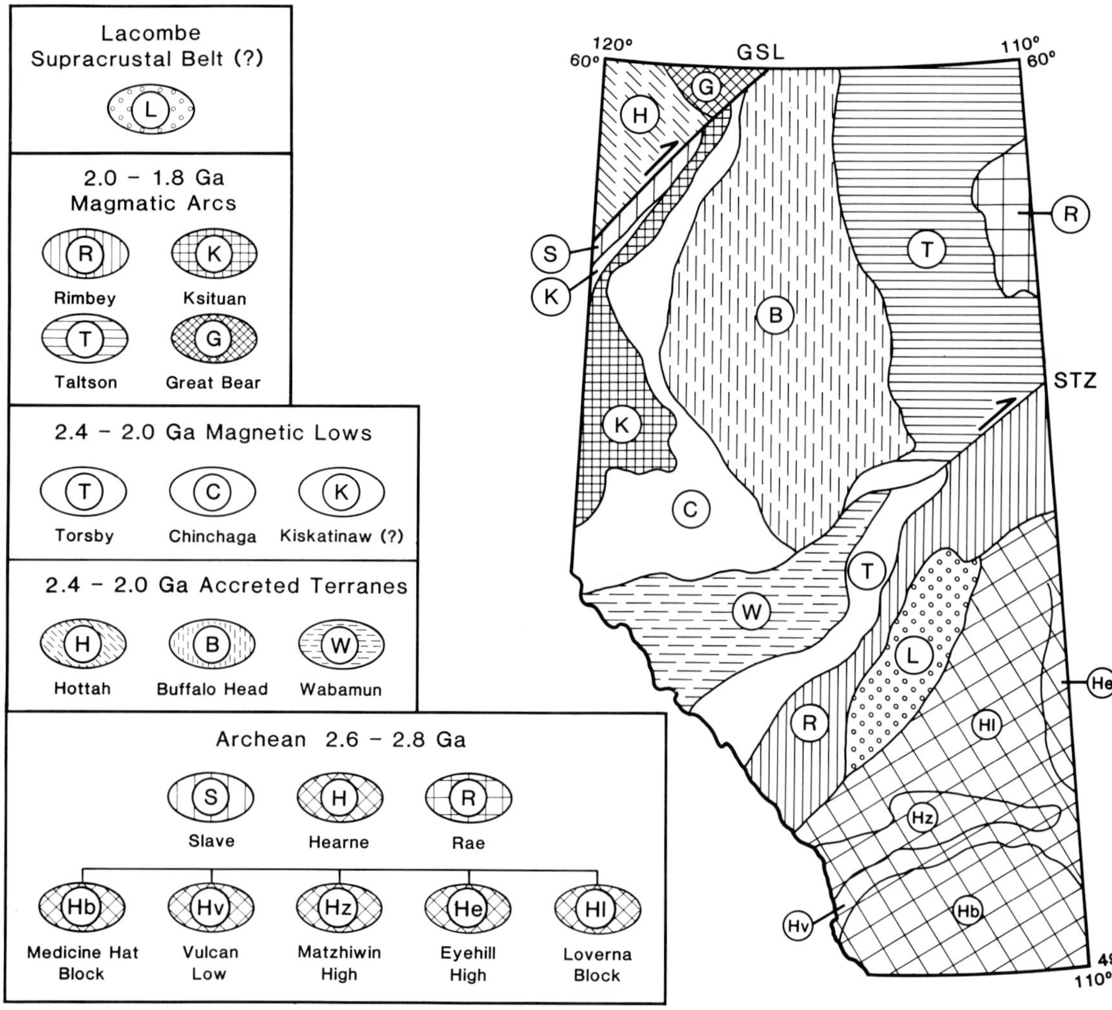


Figure 5. Map of tectonic domains postulated in the basement of Alberta. The outline of the domains corresponds to aeromagnetic boundaries. Ages for each domain are based on U/Pb zircon and monazite geochronology (after Ross and Stephenson, 1989).

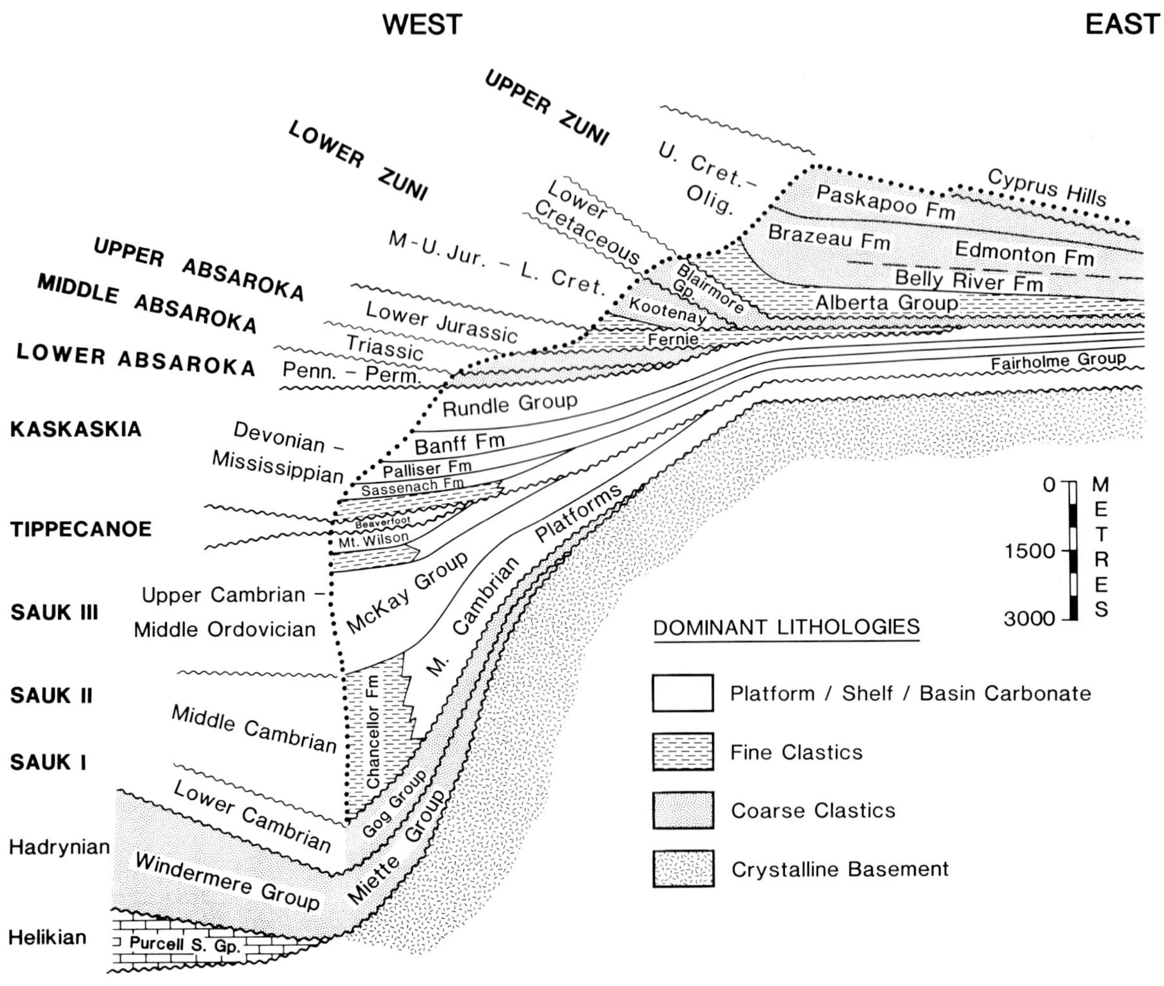


Figure 6. Generalized stratigraphic cross section of the westward thickening, Western Canada Sedimentary Basin miogeocline-platform and foreland basin succession, indicating the principle tectono-stratigraphic sequences and their bounding unconformities, time-stratigraphic and lithostratigraphic units (after Ricketts, 1989 as modified from Porter, Price and McCrossan, 1982).

postulated by Bond and Kominz (1984) to be about 575 to 600 Ma. The Sauk Sequence is essentially a simple northeasterly tapering cratonic wedge that was deposited along a passive continental margin (Porter and others, 1982 and Aitken, 1989). It shows only limited evidence of the influence of cratonic arches and basins (Porter and others, 1982 and Aitken, 1989). Aitken (1989) identifies two sedimentary domains: (1) a miogeoclinal assemblage consisting of a thick wedge of sedimentary and minor volcanic rocks deposited on a newly rifted margin of the western North America craton, and (2) an interior platform comprising a much thinner cover of purely sedimentary rocks deposited inland from the passive margin and directly upon crystalline basement. A regional eastward bevelling accompanied a uniform tilting of the craton during emergence prior to the deposition of the Tippecanoe Sequence.

### *Middle Ordovician to Late Silurian*

This assemblage comprises the Tippecanoe Sloss Sequence and reflects a continuation of depositional styles from the underlying Sauk Sequence. Deposits are dominantly carbonate that accumulated along the passive western margin of North America and over a wide area of the craton. Tippecanoe sediments in the WCSB are only preserved in the Cordillera miogeocline and in the Williston Basin of the Canadian plains. These successions are separated by a broad area in Alberta where Tippecanoe rocks were stripped during emergence and erosion prior to the Middle Devonian (Osadetz and Haidl, 1989). A Devonian subcrop map is shown in Figure 7.

## TECTONIC AND PHYSIOGRAPHIC ELEMENTS

The prominent Devonian physiographic and tectonic elements of the WCSB and the Canadian Arctic Archipelago are shown in Figure 8. These features influenced depositional patterns and resulting lithofacies. The origins of many of these features are not well understood.

Shallow-water carbonate and evaporite deposition was mainly confined to broad positive areas. There areas include the Porcupine Platform in the Yukon, the peri-cratonic MacDonald-Mackenzie-Peel Shelf of northeastern British Columbia and the Northwest Territories, and certain areas in the Alberta "basin".

The Porcupine Platform and the MacDonald-Mackenzie-Peel Shelf were the locus of shallow-water carbonate and evaporite sedimentation through early Middle Devonian time (Morrow and Geldsetzer, 1988). In late Middle Devonian time these areas were drowned and subsequently became sites of dominantly deep-water shale deposition for the remainder of the Devonian. The focus of shallow-water carbonate and evaporite deposition subsequently shifted to the Alberta "basin".

The area west of the MacDonald-Mackenzie Shelf area was a lower lying region farther removed from the Precambrian Shield. It was a pericratonic basin throughout Early and Middle Devonian time. Deposits in this basin are mainly deep-water shales. A continuation of this deep-water seaway, the Richardson Trough, separated the Mackenzie-Peel Shelf from the Porcupine Platform on

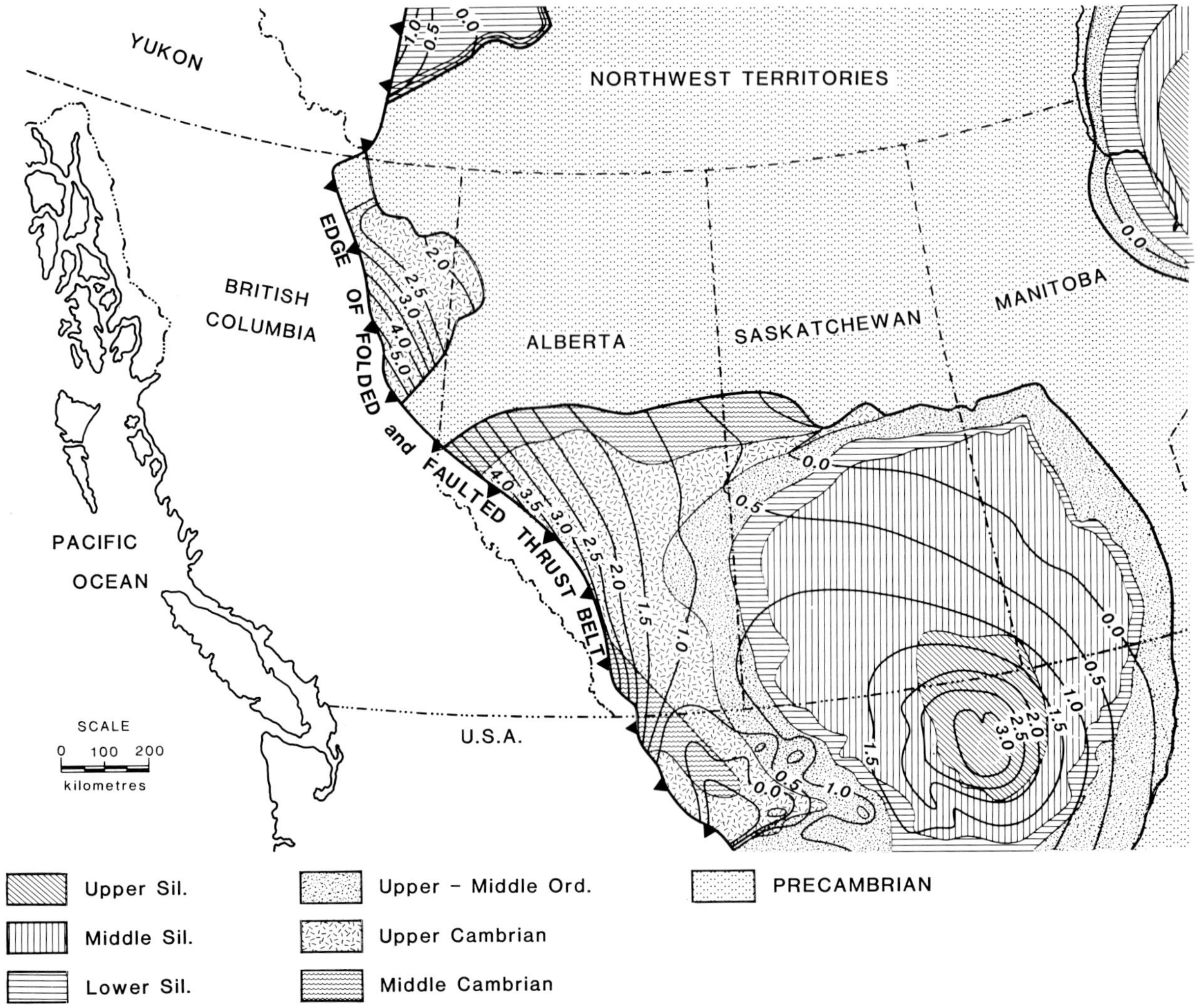


Figure 7. Structure and paleogeology of the pre-Devonian erosion surface (after Porter, Price and McCrossan, 1982).

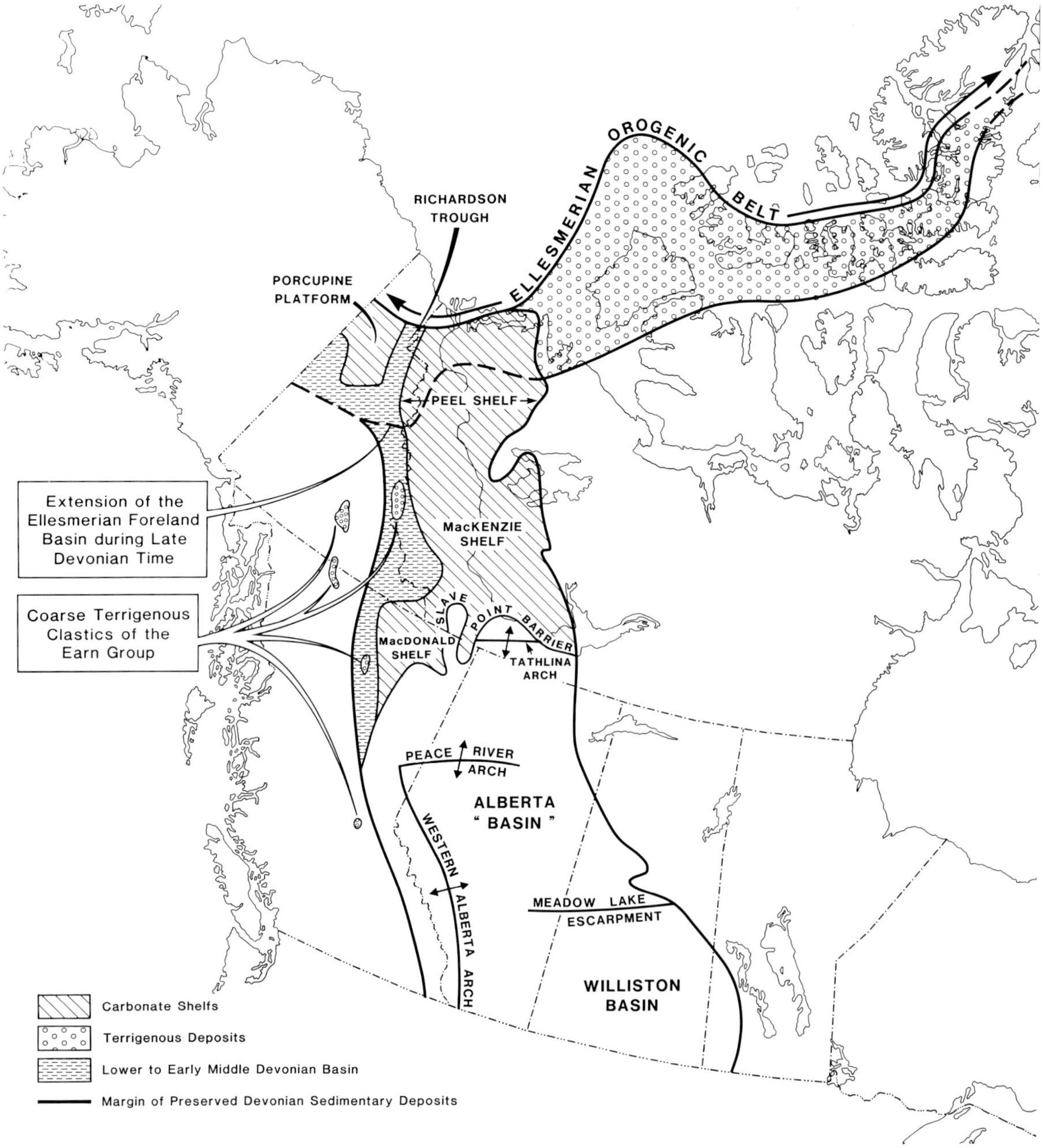


Figure 8. Physiographic and tectonic elements of the Western Canada Sedimentary Basin. Compiled from Embry (1988), Gordey (1988), Moore (1989), Morrow and Geldsetzer (1988), Muir (1988), O'Connell and others (1990) and Pugh (1983).

the west. This trough remained a deep-water seaway throughout the Devonian. It may have developed as a Middle Cambrian aulocogen (Gabriels and Yorath, 1989).

During the Late Devonian highlands arose in the present day Cordilleran area of the Yukon and British Columbia. These highlands provided the source for the deposition of nearby coarse terrigenous clastics and possibly shales that filled in widespread areas of the WCSB. Two geographically separate and fundamentally different tectonic regimes seem responsible for these uplifts.

In the northern Yukon uplift involved compression and granitic intrusion in Late Devonian (Frasnian) to early Mississippian time (Gordey, 1988). The uplifted landmass corresponds to the Borovian Highland of Morrow and Geldsetzer (1988) and is considered to represent the southwest terminus of the Ellesmerian Fold Belt of the Canadian Arctic Archipelago (Morrow and Geldsetzer, 1988 and Embry, 1988 and Personal Communication). According to Embry (1988), Middle and Upper Devonian siliciclastic strata of the Canadian Arctic Archipelago were deposited in a foreland basin which developed in front of the southwestwards advancing Ellesmerian Orogeny. Embry attributed the orogeny to the oblique, southwestwards directed, collision of an unknown Arctic plate with the northern margin of the Laurussia Continent. As a consequence of the oblique collision, the focus of uplift and subsequent foreland basin deposition migrated to the southwest. Gordey (1988) attributed the thick and widespread basin-filling deposition of the Late Devonian Imperial sequence in the Northwest Territories to this northerly source. An earlier phase of this orogeny may account for the source of the Middle Devonian Hare Indian basin-filling sequence in the Northwest Territories. As shown by Muir (1988), lithofacies patterns in this unit indicate a northeasterly source.

In central Yukon and British Columbia, five widely scattered areas of coarse terrigenous deposits including chert pebble conglomerates document the presence of at least small local uplifts (Gordey, 1988). These highlands have been interpreted by Morrow and Geldsetzer (1988) to be joined and are collectively referred to as the Selwyn Highland. The extent of these uplifts remains unknown. The coarse terrigenous sediments derived from these landmasses are part of the Upper Devonian-Mississippian Earn Group. Gordey (1988) attributed the uplift of these landmasses to a non-compressional regime, either as a consequence of regional extension or strike-slip faulting. The timing of these uplifts is coincident with the Antler Orogeny in the western United States.

Although both of these uplifts are along the border of, or are outside the WCSB, they are potentially significant to understanding the source of the Frasnian Ireton and Fort Simpson shales in the Alberta "basin". Both these shales clinoform to the west. This offlap direction would tend to preclude a significant source area to the west. Despite being distant from the Alberta "basin", the Ellesmerian Fold Belt appears to be the most likely major source of the Ireton and Fort Simpson shales. The manner of delivery of sediments from this orogenic belt into the Alberta "basin" is unknown. One possibility is transport from southerly directed marine currents over the drowned Peel and MacKenzie shelf areas.

Another possibility is fluvial transport of sediments over the Canadian Shield and debouching of the load into the "basin" in northeast Alberta.

The Alberta basin was bordered in part by the Tathlina Arch on the north and the Western Alberta Arch on the west. Both of these features appear to have been passive topographic elements with a history of gradual onlap and burial (O'Connell and others, 1990). Both features influenced Devonian sedimentation patterns by localizing reef growth. The Tathlina Arch was covered by Middle Devonian time (Givetian) and the Western Alberta Arch was ultimately buried in Late Devonian time (earliest Frasnian).

The most pronounced tectonic feature in the Alberta basin is the Peace River Arch. This feature is a major structure in which granitic Precambrian rocks are uplifted 800 to 1000 m above the regional basement elevation (Cant, 1988). The Arch is 140 km wide and extends about 400 km cratonward from the Cordillera in an east to northeast direction. The Arch was an active feature throughout at least the Middle Devonian. During this time, the Peace River Arch influenced subsidence and sedimentation patterns, and provided a source of terrigenous sediment.

The origin of the Arch is unknown. The axis of the Arch trends at a high angle to basement structure (Ross and Stephenson, 1989). Its origin is clearly not the result of the uplift of a single or discrete Precambrian structure (O'Connell and others, 1990).

The timing of the initial uplift of the Peace River Arch is unclear because of the lack of lower Paleozoic rocks in the area. McMechan (1990) suggests that the arch was uplifted initially in the late Proterozoic based on the nature of the contact between Cambrian and latest Proterozoic rocks exposed in the Canadian Cordillera. Lithofacies patterns in Cambrian, Ordovician and Silurian sequences indicate an emergent arch (Aitken, 1989 and Norford, 1990). By mid-Devonian, the Arch was emergent and the Elk Point Group onlapped its lower flanks. At this time, the first documented episode of normal faulting occurred on the Arch (Cant, 1988). The Arch continued to become onlapped and buried. By the end of the Devonian, only the highest portions of the Arch remained emergent. During the Mississippian the Arch was completely buried.

The Meadow Lake Escarpment in central Alberta and Saskatchewan is a north-facing paleotopographic feature about 250 m high (Porter and others, 1982). It represents the northern erosional limit of Ordovician and Silurian rocks of what is now the Williston Basin. Truncation occurred during the pre-Middle Devonian unconformity when the protective cover of Tippecanoe rocks was breached, permitting deep erosion of the less resistant siliciclastic rocks of the underlying Sauk Sequence (Porter and others, 1982).

## INTRODUCTION TO DEVONIAN SEQUENCES IN THE ALBERTA BASIN

The subdivision of lithologic successions into depositional sequences provides a useful means for introducing the stratigraphy and facies distributions in sedimentary basins. Moore (1988a, 1988b and 1989) synthesized Devonian successions of the western interior of Canada. He recognized major sequences that he correlated from the Yukon and the Northwest Territories, through British Columbia and Alberta, into Saskatchewan and Manitoba. In general, each major sequence oversteps the preceding one toward the craton and thus comprises an overall onlapping episode. In the Alberta Basin, most of Moore's major sequences consist of a lower phase with widespread shallow-marine carbonates and an upper basin-filling stage. The definition of these sequences generally conforms with or incorporates multiple stages of the authors (Stoakes and Wendte, 1987). Moore recognized five major sequences (Figure 9).

*Major Sequence 1 (Late Silurian to Siegenian)*

This major sequence represents the initial Devonian flooding of the Alberta "basin" and is restricted in occurrence to a low-lying sub-basin southeast of the Peace River Arch, northeast of the Western Alberta Arch and northwest of the Meadow Lake Escarpment. This succession contains two thick redbed-halite sequences of the Lotsberg Formation.

*Major Sequence 2 (Emsian to Middle Eifelian)*

Marine waters inundated the area between the Tathlina and Peace River arches, joining this area with the sub-basin to the southeast of the Peace River Arch. Facies include restricted carbonates, siliciclastics and anhydrites of the Ernestina Lake Formation, Cold Lake salts, and restricted carbonates and anhydrites of the lower Chinchaga Formation.

*Major Sequence 3 (Middle Eifelian to Middle Givetian)*

Open-marine seas inundated the Alberta Basin and flooded the Meadow Lake Escarpment. This incursion linked the Alberta Basin with the Williston Basin, establishing a huge interior basin (the Elk Point Basin). A barrier-reef fringed platform nucleated on the Tathlina Arch. Behind this barrier platform, isolate reefs of the Upper Keg River Formation grew on widespread, open-marine carbonates of the Lower Keg River Formation. Restriction of the Elk Point Basin led to termination of reef growth and to the deposition of basin-filling evaporites of the Muskeg Formation.

*Major Sequence 4 (Middle Givetian to Frasnian)*

An open-marine sea again inundated the Alberta Basin. During this interval, three major episodes of reefing occurred: Swan Hills, Leduc and Nisku Zeta Lake. The basin gradually became filled during late Frasnian time. Shallow-marine and coastal plain deposits of the Upper Winterburn cap this sequence.

*Major Sequence 5 (Famennian)*

Ramp carbonates and restricted platform evaporites of the Wabamun Formation were deposited over virtually all of the Alberta

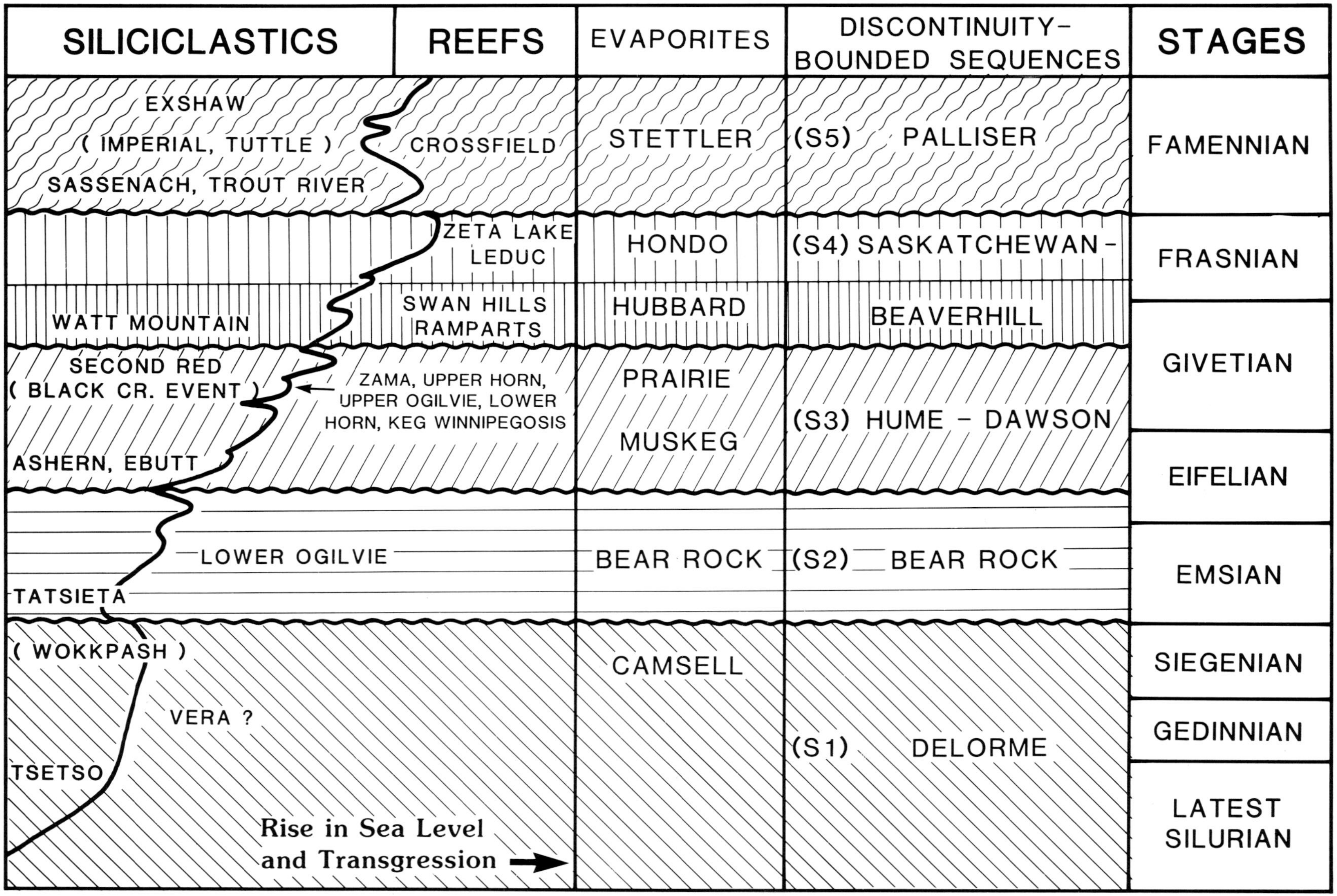


Figure 9. Devonian sequences of the Western Canada Sedimentary Basin. The sequences show an onlapping style of deposition (after Moore, 1988b).

Basin. These deposits gradually shale out in British Columbia toward the west and northwest. Deposition was halted by a major drowning.

Figures 10 and 11 reproduce four of Moore's cross sections and show the distribution of facies in each of his major sequences. All these cross sections depict the gross stratigraphic relationships between carbonate platforms and reefs, and basin-filling shales, limestones and evaporites. Cross sections B-B' (Figure 10) and D-D' (Figure 11) demonstrate the onlap and thinning of strata onto the Western Alberta Arch and Peace River Arch, respectively.

## SUMMARY OF HYDROCARBON EXPLORATION

Hydrocarbon exploration in Devonian strata in the WCSB occurred in major pulses following the initial discovery in each of the major plays. The sequence of plays reflects the progression of geological understanding and ease of seismic recognition. The majority of hydrocarbons are pooled in isolate reefs and secondarily along carbonate platform margins. Significant pools occur in the Elk Point, Beaverhill Lake and the Woodbend sequences. Geological Survey of Canada estimates of in-place and recoverable reserves for all the major sequences are listed in Table 1.

| | OIL ($10^6 m^3$) | | GAS ($10^6 m^3$) | |
|---|---|---|---|---|
| | In-Place | Initial Recoverable | In-Place | Initial Marketable |
| Elk Point | | | | |
| Siliciclastics | 318 | 130 | 25665 | 10516 |
| Carbonates | 384 | 145 | 170576 | 118580 |
| Beaverhill Lake | 1030 | 419 | 524176 | 247211 |
| Woodbend | 767 | 461 | 599645 | 292264 |
| Winterburn | 327 | 164 | 94654 | 48192 |
| Wabamun | 28 | 5 | 178440 | 75848 |
| TOTAL | 2854 | 1324 | 1593156 | 792611 |

Table 1. Hydrocarbon reserves in the WCSB. (after Podruski and others, 1988 and Reinson and others, 1992).

Discovery of oil at Leduc-Woodbend in 1947 initiated an intensive search for other Leduc reefs in central Alberta. Despite the primitive state of seismic technology, the strong impedance contrast between the reefs and the surrounding Ireton Shale enabled seismic to be used as an effective exploration tool. By the mid-1950's, most of the large pools in Leduc reefs, discovered to date, had been found. These discoveries include Redwater, the second largest Devonian oil pool with OOIP of 207 X $10^6 m^3$ (1302 MMB), Bonnie Glen (125 X $10^6 m^3$ or 786 MMB OOIP), Wizard Lake (62 X $10^6 m^3$ or 391 MMB OOIP) and Golden Spike (49.6 X $10^6 m^3$ or 312 MMB OOIP) in addition to the initial discovery at Leduc (61.2 X $10^6 m^3$ or 385 MMB OOIP).

The discovery of Virginia Hills in 1956 spurred a second major round of reef exploration. Lack of velocity contrast between these

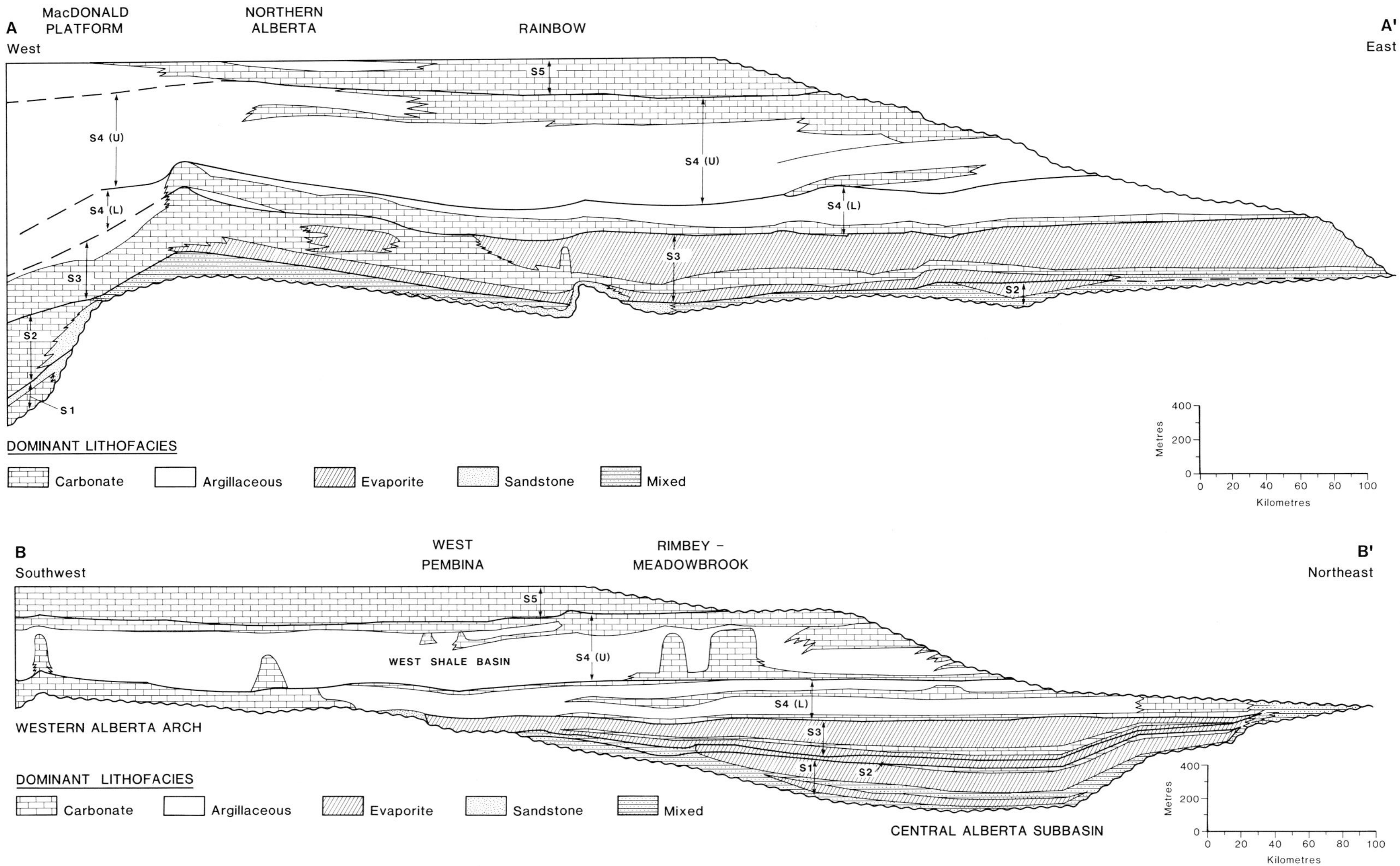


Figure 10. Cross sections illustrating the distribution of lithologies in Devonian sequences of the Western Canada Sedimentary Basin. Cross section A-A' crosses northern Alberta and intersects the Keg River and Slave Point barrier reefs. Cross section B-B' crosses central Alberta and extends westward onto the Western Alberta Arch. Transects of section shown on Figure 1 (modified from Moore, 1989).

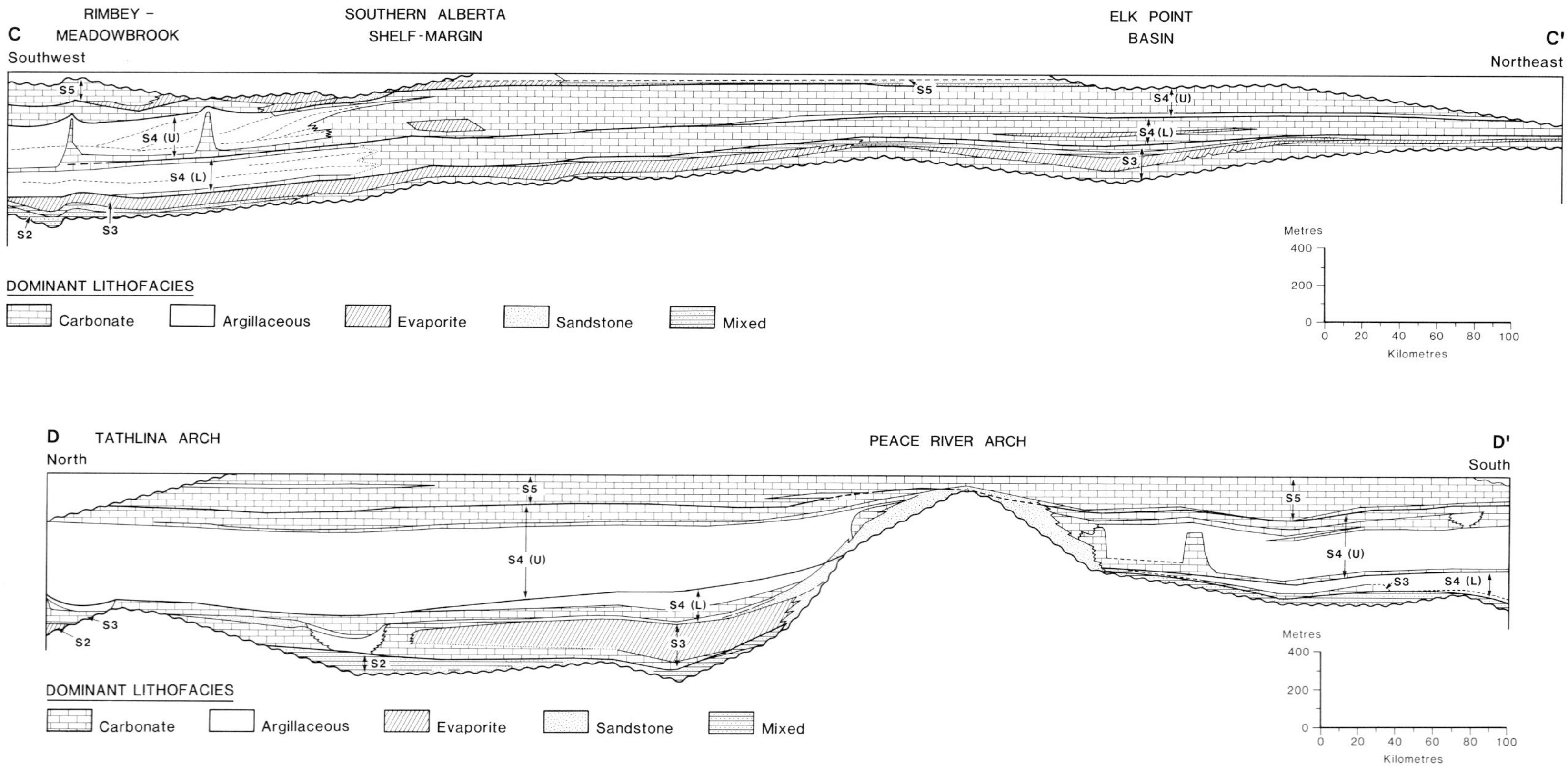


Figure 11. Cross sections illustrating the distribution of lithologies in Devonian sequences of the Western Canada Sedimentary Basin. Cross section C-C' extends from the Elk Point Basin of Saskatchewan into southern Alberta. Cross section D-D' crosses the Peace River and Tathlina Arches. Transects of section shown on Figure 1 (modified from Moore, 1989).

Beaverhill Lake Swan Hills reefs and the enclosing Waterways basinal limestones, and the lack of appreciable drape made seismic ineffective for this play. Nevertheless, all the major fields found to date were discovered within a few years. These pools include Swan Hills (approximately 221 x $10^6m^3$ or 1392 MMB OOIP), Swan Hills South (144 X $10^6m^3$ or 906 MMB OOIP), Judy Creek (130 X $10^6m^3$ or 818 MMB OOIP) and Virginia Hills (75.4 X $10^6m^3$ or 474 MMB OOIP) in addition to the platform-margin Kaybob South pool, the largest Devonian gas accumulation with 104,424 X $10^6m^3$ OGIP (3.7 TCF). Collectively, more oil is pooled in these reefs than in any other Devonian reef system in the WCSB.

The discovery at Rainbow in 1965 demonstrated that reefs could form in evaporite-filled basins and pool oil. Because of the strong impedance contrast between these Keg River reefs and the surrounding salts, seismic proved to be extremely effective. Within two years of the initial discovery, the nine largest Keg River pools found to date had been discovered. These reefs are smaller in areal extent and consequently have lower reserves than either the Leduc or Swan Hills reefs. The Rainbow B pool, the largest Keg River pool, has in-place reserves of 43.8 x $10^6m^3$ (276 MMB). Approximately three-fourths of all Keg River reef pools in northern Alberta have in-place reserves less than .5 x $10^6m^3$ (3.15 MMB). Collectively, however, these reefs contain significant oil reserves because of their great number. Four hundred and thirty nine pools had been discovered by 1987 (Podruski and others, 1988).

The discovery of Nisku Zeta Lake reefs at West Pembina in central Alberta in 1977 revitalized Devonian reef exploration in the WCSB. These buildups were the first Devonian reefs in the WCSB to be found on a shale-limestone clinoform. Rapid industry response, using 1200 percent CDP seismic surveys, turned this into a mature play within a short period. Most of the pools found to date were discovered within 4 years of the initial West Pembina find. Individually, these reefs contain small reserves when compared to Leduc or Swan Hills complexes. The largest pool is the Brazeau River A with in-place reserves of 5.3 X $10^6m^3$ (33MMB OOIP). Forty-five pools had been discovered up to 1988 (Podruski and others, 1988). Collectively, the hydrocarbon reserves are significant but considerably less than the three previous reef plays.

The reversion of deep rights sparked another wave of Devonian exploration from 1980 to 1985 (Podruski and others, 1988). Instead of focusing in on one play, exploration in these and later years was broad based. Discoveries were made in all the major Devonian sequences. But no new impact plays were initiated. The most significant discovery following West Pembina was Shell's find at Caroline in west-central Alberta in 1985. This pool occurs along a Swan Hills platform margin and has in-place reserves of 61,150 X $10^6m^3$ (2.2 TCF) of gas and 39.75 X $10^6m^3$ (250 MMB) of condensate.

Perhaps stimulus generated in this short course will lead to new ideas and discoveries.

## REFERENCES


AITKEN, J.D., 1989, The Sauk Sequence: Cambrian to Lower Ordovician Miogeocline and Platform, *in* Ricketts, B.D., ed., Western Canada Sedimentary Basin - A Case History: Canadian Society of Petroleum Geologists, Special Publication no. 30, p. 105-120.

BOND, G.C., AND KOMINZ, M.A., 1984, Construction of tectonic subsidence curves for the early Paleozoic miogeocline, southern Canadian Rocky Mountains: Implication for subsidence mechanisms, age of break-up and crustal thinning: Geological Society of America Bulletin, v. 95, p. 155-173.

BURROWES, O.G., AND KRAUSE, F.F., 1987, Overview of the Devonian System: Subsurface Western Canada Basin, *in* Burrowes, O.G., and Krause, F.F., eds., Devonian Lithofacies and Reservoir Styles in Alberta: 13th Canadian Society of Petroleum Geologists Core Conference and Second International Symposium on the Devonian System, p. 1-20.

CANT, D.J., 1988, Regional structure and development of the Peace River Arch, Alberta: A Paleogene faulted-rift system?: Bulletin of Canadian Petroleum Geology, v. 36, no. 3, p. 284-295.

EMBRY, A.F., 1988, Middle-Upper Devonian sedimentation in the Canadian Arctic Islands, *in* McMillan, N.J., Embry, A.F., and Glass, D.J., eds., Devonian of the World: Canadian Society of Petroleum Geologists, Second International Symposium on the Devonian System, Memoir 14, v. 2, p. 15-28.

GABRIELSE, H., AND YORATH, C.J., 1989, DNAG #4. The Cordillera Orogeny in Canada: Geoscience Canada, v. 2, p. 67-83.

GORDEY, S.P., 1988, Devono-Mississippian clastic sedimentation and tectonics in the Canadian Cordilleran Miogeocline, *in* McMillan, N.J., Embry, A.F., and Glass, D.J., eds., Devonian of the World: Canadian Society of Petroleum Geologists, Second International Symposium of the Devonian System, Memoir 14, v. 2, p. 1-14.

McMECHAN, M.E., 1990, Upper Proterozoic to Middle Cambrian History of the Peace River Arch region: evidence from the Rocky Mountains: Bulletin of Canadian Petroleum Geology, v. 38A, p. 36-44.

MONGER, J.W.H., 1989, Overview of Cordilleran Geology, *in* Ricketts, B.D., ed., Western Canada Sedimentary Basin - A Case History: Canadian Society of Petroleum Geologists, Special Publication no. 30, p. 9-32.

MOORE, P.F., 1988a, Devonian reefs in Canada and some adjacent areas, *in* Geldsetzer, H.H.J., James, N.P., and Tebbutt, G.E., eds., Reefs Canada and Adjacent Areas: Canadian Society of Petroleum Geologists, Memoir 13, p. 367-390.

MOORE, P.F., 1988b, Devonian Geohistory of the Western Interior of Canada, *in* McMillan, N.J., Embry, A.F., and Glass, D.J., (eds.), Devonian System of the World: Canadian Society of Petroleum Geologists, Second International Symposium on the Devonian System, Memoir 14, v. 1, p. 67-84.

MOORE, P.F., 1989, The Kaskasia Sequence: Reefs, platforms and foredeeps - The Lower Kaskasia Sequence - Devonian, *in* Ricketts, B.D., ed., Western Canada Sedimentary Basin - A Case History: Canadian Society of Petroleum Geologists, Special Publication no. 30, p. 139-164.

MORROW, D.W., AND GELDSETZER, H.H.J., 1988, Devonian of the Eastern Canadian Cordillera, *in* McMillan, N.J., Embry, A.F., and Glass, D.J., eds., Devonian of the World: Canadian Society of Petroleum Geologists, Second International Symposium on the Devonian System, Memoir 14, v. 1., p. 85-122.

MUIR, I.D., 1988, Devonian Hare Indian and Ramparts Formation, Mackenzie Mountains, N.W.T.: Basin-Fill, platform and reef development: Unpublished PhD Dissertation, University of Ottawa, 593 p.

NORFORD, B.S., 1990, Ordovician and Silurian stratigraphy, paleogeography and depositional history in the Peace River Arch area, Alberta and British Columbia: Bulletin of Canadian Petroleum Geology, v. 36, no. 3, p. 284-295.

O'CONNELL, S.C., DIX, G.R., AND BARCLAY, J.E., 1990, The origin, history, and regional structural development of the Peace River Arch, Western Canada: Bulletin of Canadian Petroleum Geology, v. 38A, p. 4-24.

OSADETZ, K.G., AND HAIDL, F.M., 1989, Tippecanoe Sequence: Middle Ordovician to Lowest Devonian-Vestiges of a Great Epeiric Sea, *in* Ricketts, B.D., ed., Western Canada Sedimentary Basin - A Case History: Canadian Society of Petroleum Geologists, Special Publication no. 30, p. 121-138.

PODRUSKI, J.A., BARCLAY, J.E., HAMBLIN, A.D., LEE, P.J., OSADETZ, K.G., PROCTOR, P.M., AND TAYLOR, W.C., 1988, Part 1: Resource Endowment of Conventional Oil Resources of Western Canada (Light and Medium): Geological Survey of Canada, Paper 87-26, p. 1-126.

PORTER, J.W., PRICE, R.A., AND McCROSSAN, R.G., 1982, The Western Canada Sedimentary Basin: Philosophical Transactions of Royal Society of London, v. 305, p. 169-192.

PUGH, D.C., 1983, Pre-Mesozoic Geology in the subsurface of Peel River map area, Yukon Territory and District of Mackenzie: Geological Survey of Canada, Memoir 401, 61p.

REINSON, G.E., LEE, P.J., WARTERS, W., OSADETZ, K.G., BELL, L.L., PRICE, P.R., TROLLOPE, F., AND BARCLAY, J.E., 1992, Devonian gas resources of Western Canada Sedimentary Basin -- Play definition and resource assessment: Geological Survey of Canada Paper (in press).

ROSS, G., MCMECHAN, M.E. AND HEIN, F.J., 1989, Proterozoic History: the birth of the miogeocline, *in* Ricketts, B.D., ed., Western Canada Sedimentary Basin - A Case History: Canadian Society of Petroleum Geologists, Special Publication no. 30, p. 79-104.

ROSS, G., AND STEPHENSON, R.A., 1989, Crystalline basement: the Foundation of Western Canada Sedimentary Basin, *in* Ricketts, B.D., ed., Western Canada Sedimentary Basin - A Case History: Canadian Society of Petroleum Geologists, Special Publication no. 30, p. 33-46.

SCOTESE, C.R., VAN DER VOO, R., AND BARRETT, S.F., 1985, Silurian and Devonian base maps: Philosophical Transactions Royal Society of London, v. 309, p.57-77.

SLOSS, L.L., 1963, Sequences in cratonic interior of North America: Geological Society of America Bulletin, v. 74, p.93-114.

STOAKES, F.A., AND WENDTE, J.C., 1987, The Woodbend Group, *in* Burrowes, O.G., and Krause, F.F., eds., Devonian Lithofacies and Reservoir Styles in Alberta: 13th Canadian Society of Petroleum Geologists Core Conference and Second International Symposium on the Devonian System, p. 153-170.

WITZKE, B.J., AND HECKEL, P.H., 1988, Paleoclimatic indicators and inferred Devonian Paleolatitudes of Euramerica, *in* McMillan, N.J., Embry, A.F., and Glass, D.J., eds., Devonian of the World: Canadian Society of Petroleum Geologists, Second International Symposium on the Devonian System, Memoir 14, v. 1, p. 49-68.

ZIEGLER, P.A., 1988, Laurussia - the Old Red Continent, *in* McMillan, N.J., Embry, A.F., and Glass, D.J., eds., Devonian of the World: Canadian Society of Petroleum Geologists, Second International Symposium on the Devonian System, Memoir 14, v. 1, p. 15-48.

CHAPTER 2

# CYCLICITY OF DEVONIAN STRATA IN THE WESTERN CANADA SEDIMENTARY BASIN

J.C. WENDTE

## INTRODUCTION

One of the goals of science is to create order out of apparent disorder. Inherent in this pursuit is the need to classify and subdivide naturally occurring phenomena in order to better understand their origin. Stratigraphers since at least the early 20th century have sought to divide lithologic successions into genetic sequences. In regard or application to carbonate strata, stratigraphers have approached this goal from two general perspectives.

One group has emphasized the significance of the areal distribution of strata and the nature of their bounding surfaces. Blackwelder in 1909 expounded on a scheme showing the areal distribution and ages of unconformity-bounded, Phanerozoic sequences of a part of the central United States. Sloss (1963) widened the scope of this endeavour and provided a division of Phanerozoic strata over the cratonic interior of North America into six unconformity-bounded sequences. Explicit in both schemes was the recognition that unconformities are of a long duration toward the centre of the craton and disappear into conformable successions along cratonic margins. The utility of Sloss Sequences has gained wide acceptance and has been used as a means for discussing the Phanerozoic evolution of sedimentary basins, for example the Western Canada Sedimentary Basin (Ricketts, 1988).

A more recent advance in this aspect has been the development of the concept of seismic stratigraphy by Vail and his Exxon colleagues (Vail and others, 1977). Using a large seismic data base, these investigators recognized and delineated various stratal patterns. From these studies they identified sequences that they interpreted were bounded by unconformities or their correlative conformities. They attributed the origin of these sequences to global or eustatic changes of sea level. The *Depositional Sequences* of Vail and his colleagues are on the order from 1 to 2 million years in duration.

Carbonate sedimentologists and stratigraphers during the same period investigated the lithologic and environmental aspects of both Holocene and ancient Phanerozoic deposits. A common recurrent theme in these investigations was the recognition, from core and outcrop studies, of upward-shoaling sequences. These sequences typically have a sharp to thinly transitional base and top, and consist of an asymmetric shoaling-upward succession of facies. Studies of the Holocene from both the Bahamas (Shinn, Ginsburg and Lloyd, 1969) and the Persian Gulf (Kinsman, 1969) documented a coastal progradational origin for these sequences. Prior to the development of seismic stratigraphy, investigators like Wilson (1967a and 1967b), Purser (1972) and Roehl (1967) also identified asymmetric upward-shoaling sequences in ancient carbonates. These sequences occur in a repetitive or cyclic manner. In this course, an upward-shoaling succession of facies comprises either a sequence or a cycle. The terms cycle and sequence are used interchangeably. The durations of the sequences or cycles identified from these studies were generally, but not always, of significantly less time than the sequences recognized from seismic stratigraphic studies. The sequences identified by Wilson (1967a and 1967b) and Roehl (1967) correspond to durations from tens to hundreds of thousands of years. Purser (1972), however, identified major shoaling-upward sequences from Middle Jurassic strata in France with durations from 5 to 10 million years, a span of time greater than those of Vail's *Depositional Sequences*.

In the last 10 years numerous investigators have attempted to link the larger, seismically defined *Depositional Sequences* of Vail and his Exxon colleagues with the generally shorter duration, upward-shoaling sequences identified by sedimentologists. The most widely known of these models is the "Exxon" evolutionary Systems Tract Model based on pioneering stratigraphic modelling by Jervey (1988). These systems tract models (Posamentier and others, 1988 and Posamentier and Vail, 1988) relate the stacking pattern of siliciclastic strata to changes of eustatic sea level, but also recognize the importance of subsidence rates and sediment supply. Sarg (1988) presented a companion paper on carbonate sequence stratigraphy but differed from the previous papers in that he stated that he believed eustatic changes to be <u>the</u> major control on carbonate stacking patterns and distribution of lithofacies. The significance of other potential controls was downplayed. His model has come under criticism from many carbonate sedimentologists. Most of these criticisms have focused on the remarkable similarity between his model for carbonates with that of Posamentier and others (1988) for siliciclastics (Schlager, 1989; Handford and Loucks, 1990 and Wendte and others, 1990). Are the controls so similar? Some points put forth question whether eustasy is the only major control on carbonate stacking patterns (Wendte and others, 1990) and what happens to rimmed carbonate shelfs when sea level lowers (Schlager, 1989). These questions provide food for thought as we discuss the evolution of Devonian sequences in the WCSB.

The following sections are designed to introduce the nature of Devonian cyclicity in the WCSB.

## GENERAL MODEL OF DEVONIAN CYCLICITY

Figure 1 shows a composite schematic cross section that depicts the cyclicity and gross facies of Devonian strata in the WCSB. The cross section is intended to illustrate the evolutionary aspects of the key elements of Devonian stratigraphy. The occurrence of these elements is from widely spaced areas. The transect representing the section below the base of the Watt Mountain Formation is from northern Alberta, extending from the

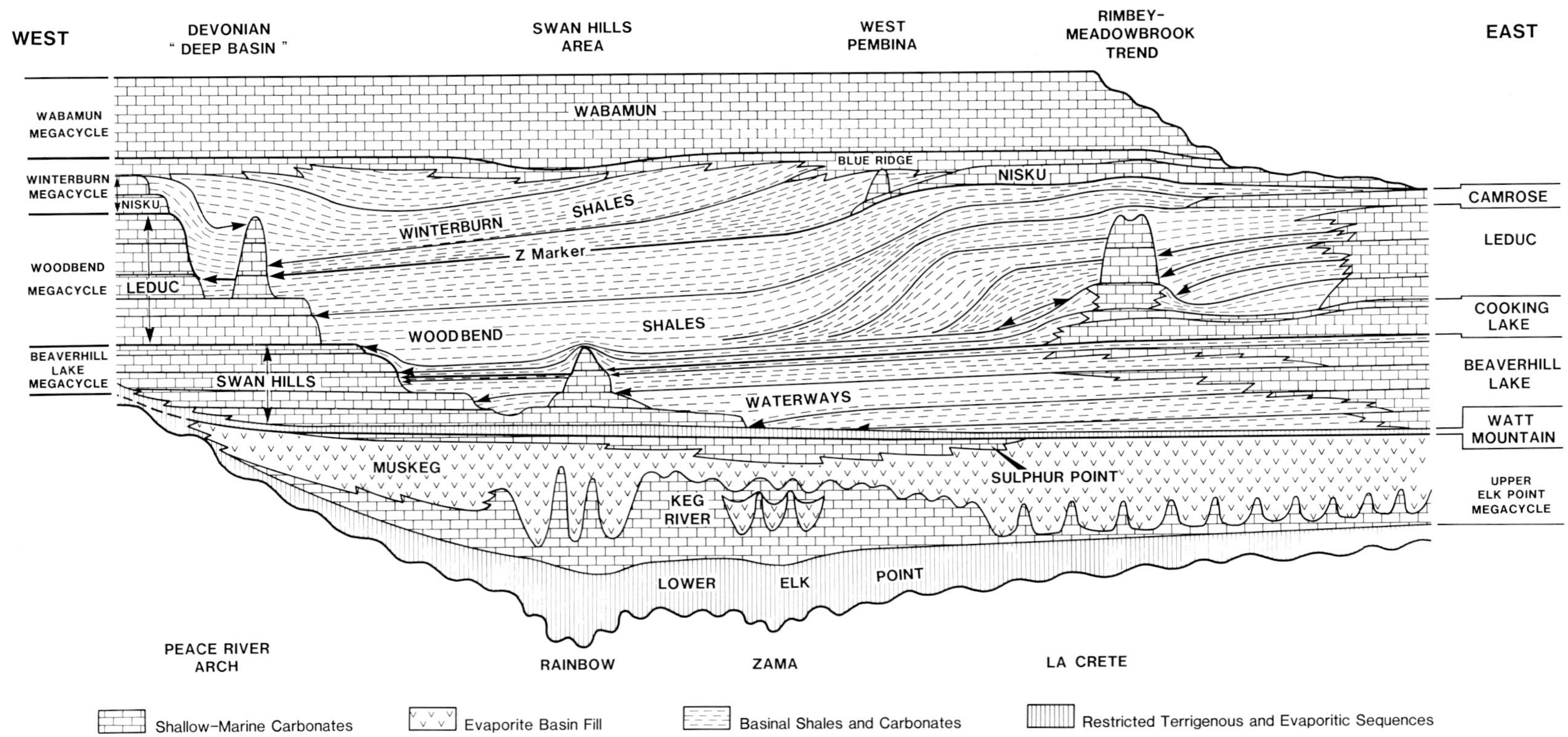


Figure 1. Composite schematic cross section illustrating cyclicity of Devonian sequences and the distribution of their major facies.

Peace River Arch through the Rainbow and Zama sub-basins to the La Crete Basin of northeastern Alberta. The transect representing the section above the base of the Watt Mountain Formation is across the shale basin in central Alberta.

Devonian cycles in the WCSB can be considered in terms of three relative, but not absolute levels that are easily definable. First-order Devonian cycles correspond to the major shifts in facies belts and consist of mega transgressive-regressive sequences that can be correlated across the Alberta Basin. The transgressive part of these cycles corresponds to a cratonward shift of shallow-water facies. The regressive portion of first-order cycles corresponds to a basinward shift of shallow-water facies, and generally implies sediment progradation. These first-order cycles split the Devonian section above the Lower Elk Point succession into five major sequences: the Upper Elk Point, Beaverhill Lake, Woodbend, Winterburn and Wabamun. These divisions approximate the group-level stratigraphy established in the basin. However, they do not necessarily correspond to equal increments of time. The evolution of each of these megasequences is considered in following chapters. Second-order cycles are defined by significant shifts or backsteps of facies belts. Shallow-water sequences bounded by second-order cycle boundaries are in general about 10 to 20 m in thickness. Both first and some second-order cycles are depicted on Figure 1. Third-order cycles are 1 to 4 m thick in shallow-water settings. They usually are not characterized by considerable shifts of facies belts and normally cannot be traced into platform or reef-margin settings. These cycles are not depicted on Figure 1.

We emphasize that this is a working model of Devonian strata in the WCSB. Other cycles can potentially be correlated throughout the Alberta Basin; we simply have not done so. Some of these cycles may correspond, in more of an absolute sense, to an intermediate order between what we define as our first and second-order cycles and are discussed in later sections. We do argue, however, that our first-order cycles correspond to the most significant shifts in facies belts. The nature and origin of all the bounding surfaces or correlative conformities are not well understood. Certainly a number of significant questions remain unanswered.

## SECOND AND THIRD-ORDER CYCLES

Second and third-order cycles are the basic building blocks for understanding the evolution of Devonian strata in the WCSB. Figure 2 depicts the occurrence of these cycles in a shallow-water carbonate, in this case a reef complex.

The identifiable extent of these cycles is not uniform across a shallow-water platform or reef. Certain environmental settings inherently contain significantly more cycles than others. Our studies indicate that the expression of cyclicity is a function of the growth potential in certain environmental settings. In "competent" environments sedimentation keeps pace with all but the most rapid rates of relative sea-level rise. In Figure 2, the competent reef margin keeps pace with small increments of relative

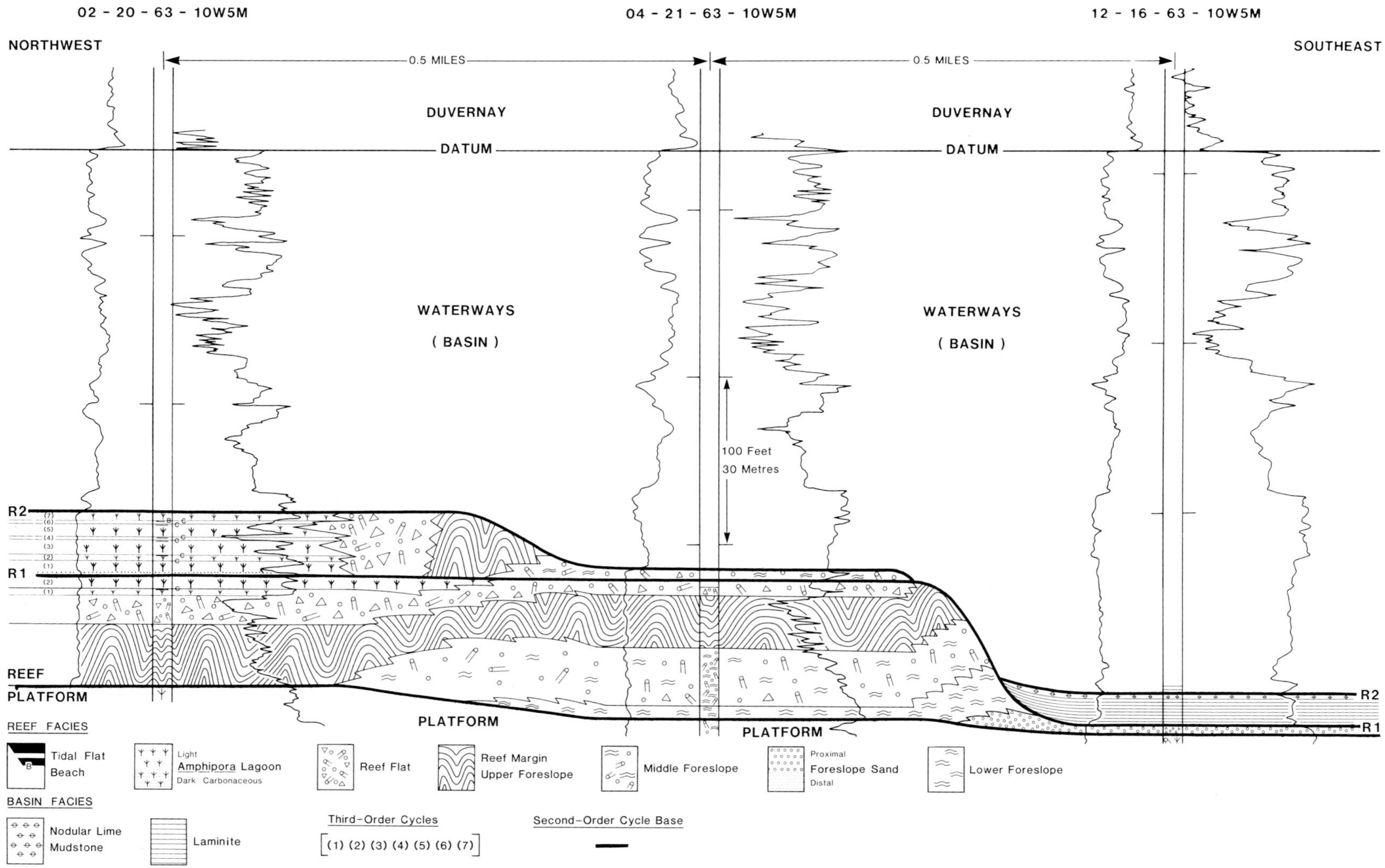


Figure 2. Devonian second and third-order cycles and their facies along a Swan Hills reef margin at Judy Creek in west-central Alberta. Second-order cycles exhibit backstepping of the reef margin and consist of shoaling-upward sequences in foreslope-margin settings. In "less competent" lagoonal settings, second-order cycles breakup into repeated, smaller third-order shoaling-upward cycles. At Judy Creek most of these third-order cycles are hard to correlate laterally.

sea-level rise and shows no lithologic variation corresponding to these changes. Third-order cycles are not expressed in this environment. Conversely, the more "incompetent" lagoon does not keep pace with small increments of relative sea-level rise. Each third-order sequence either becomes submerged, or more deeply submerged, and a new third-order cycle of sedimentation is initiated.

Both second and third-order cycles consist of shoaling-upward sequences. The dominantly asymmetric shoaling-upward aspect is not new and has been described and explained by many authors, for example Wilson (1975), Purser (1972) and James (1979). The lithological successions of certain second and third-order cycles, in this case from a Swan Hills reef complex, are illustrated in Figures 3 and 4, respectively. The shoaling nature of these cycles reflects a combination of sediment progradation and aggradation. The aggradational aspect is a consequence of *in situ* organic growth and is one significant difference between carbonates and siliciclastics.

Almost all cycle boundaries are characterized by the occurrence of either deep-water over shallow-water facies or more basinward over more bankward facies. The boundaries vary from gradational to sharp. Where a deep-water facies overlies a second-order cycle boundary, the contact may be a marine-cemented hardground.

Significant lateral shifts in facies belts occur at some second and even third-order cycle contacts. Facies reflecting an intermediate environment or environments between those typifying the facies below and above the contact are missing. These pronounced shifts in carbonate facies belts indicate cyclicity is externally controlled, or allocyclic, as opposed to be internally controlled, or autocyclic.

Eustatic sea-level rise, subsidence, and decrease in sedimentation rate are three factors that can cause deepenings that initiate new cycles of sedimentation (Jervey, 1988 and Galloway, 1989). Accommodation created from eustatic sea-level rise or due to subsidence are difficult to separate. Throughout this short course manual, we have not made this distinction and use the term sea-level change in a relative sense to denote accommodation created by some combination of eustasy and subsidence.

The potential of initiating new cycles by diminishing the rate of sedimentation can, however, be evaluated. Second-order cycles in carbonate platforms commonly exhibit a backstepping style of growth. Backstepping necessitates essentially continuous sedimentation in a platformward position coincident with drowning of the more basinward section. Provided the sequences are allocyclic, as previously argued, it is difficult (impossible?) to envision causes other than eustasy or subsidence that affect one position and not the other. Certainly, the adverse effects of anoxia, hypersalinity, excessive nutrient input and turbidity would not be confined to sharply bounded, distinct areas on a relatively flat platform. We conclude, therefore, that the initiation of second-order cycles and many of their component third-order cycles correspond to relative sea-level rises, caused by increased rates of either or both eustatic rise or subsidence.

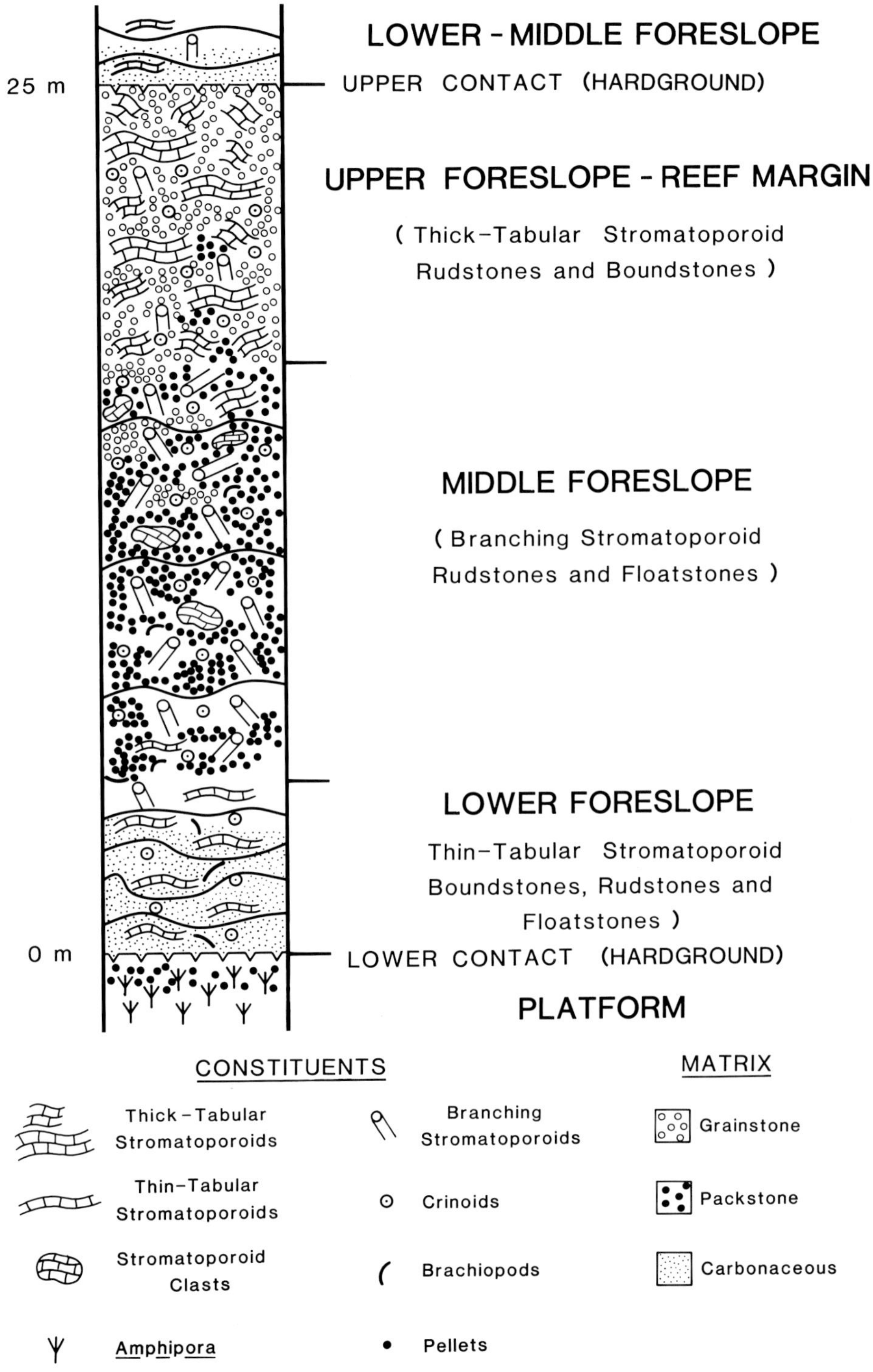


Figure 3. Succession of facies in a Devonian second-order cycle along a Swan Hills reef margin.

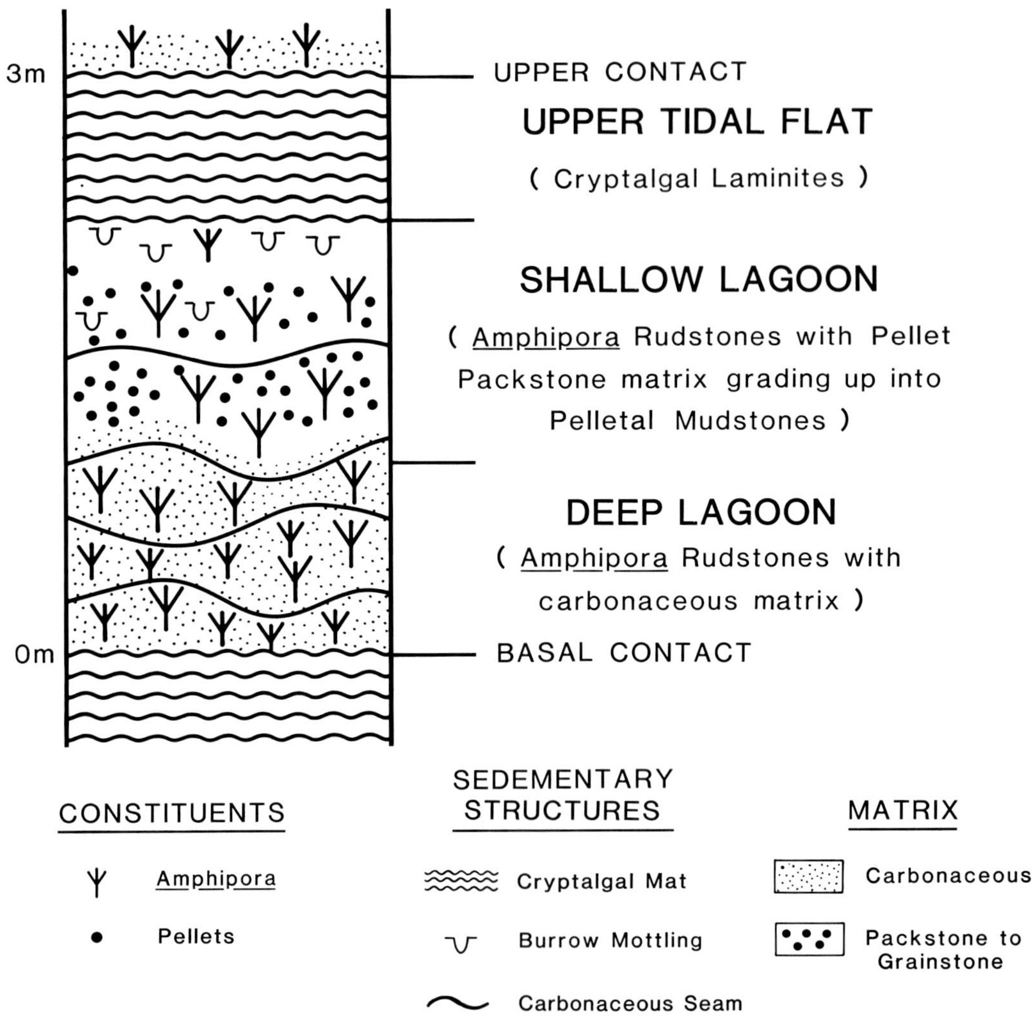


Figure 4. Succession of facies in a "complete" Devonian third-order cycle within the lagoon of a Swan Hills reef complex.

Some third-order cycles may, however, be autocyclic. In isolate reefs, for example, the number of third-order cycles is highly variable between sections. It appears that this variability reflects varying rates of sediment supply. Areas with high rates of sediment supply, such as those closer to the "carbonate factory" along the reef margin, tend to keep pace with a generally rising sea level, and consequently have fewer cycles than those areas with lower sedimentation rates. The initiation of some of these cycles may, therefore, be limited in occurrence to local areas with diminished rates of sediment supply and be independent of any change in relative sea level. As a consequence, these third-order cycles cannot be traced laterally over any significant distance.

## EVOLUTION OF FIRST-ORDER CYCLES

The evolution of a first-order cycle records the stacking patterns of second-order cycles. Four major stacking styles are recognized (Figure 5). Backstepping and retreating both involve a platformward shift in facies belts that differ only in the amount of separation. Backstepping stipulates lateral separation between platform-margin facies of the two cycles and retreating does not. Upstepping corresponds to the vertical stacking of platform-margin facies. Forestepping refers to a basinal shift in facies belts. The growth within individual second-order cycles can be progradational, aggradational or some combination of both.

The first-order Devonian cycles depicted on Figure 1 exhibit a significant degree of variation. The lower and upper megacycles are especially different. Basin-filling evaporites are confined to the Upper Elk Point megacycle and the preceding Lower Elk Point succession. The Wabamun megacycle in the area of the Alberta Basin is a shallow-water carbonate. To the west and northwest, outside of the area of the transect depicted on Figure 1, it ramps downward into deeper water shaley carbonates. The evolution of both of these megasequences is considered in a later section of this manual.

Figure 6 provides a general model that explains the evolution of the Beaverhill Lake, Woodbend and Winterburn megasequences. The diagram most closely approximates the evolution of the Beaverhill Lake megacycle. Each of these megasequences can be split into lower transgressive and upper progradational or regressive phases.

The lower transgressive phase is marked by the occurrence of widespread carbonate platforms and overlying reef complexes. The degree of contemporaneous basin infill is minimal. As a consequence, significant platform to basin relief is created. In the earliest stage, the platform backsteps away from the basin. At some stage in platform evolution, a subtle but distinct accretionary high on the platform is developed. A subsequent backstep shifts platform deposition bankward and results in drowning of most of the basinward portion of the preceding platform. Isolate reefs become localized on the aborted platform highs. Isolate Swan Hills reefs are mainly limited to the west side of the basin; Leduc reefal complexes occur on both sides of the basin. Subsequently, deposition in the later stage of the transgressive phase changes to more of an upstepping style. Some

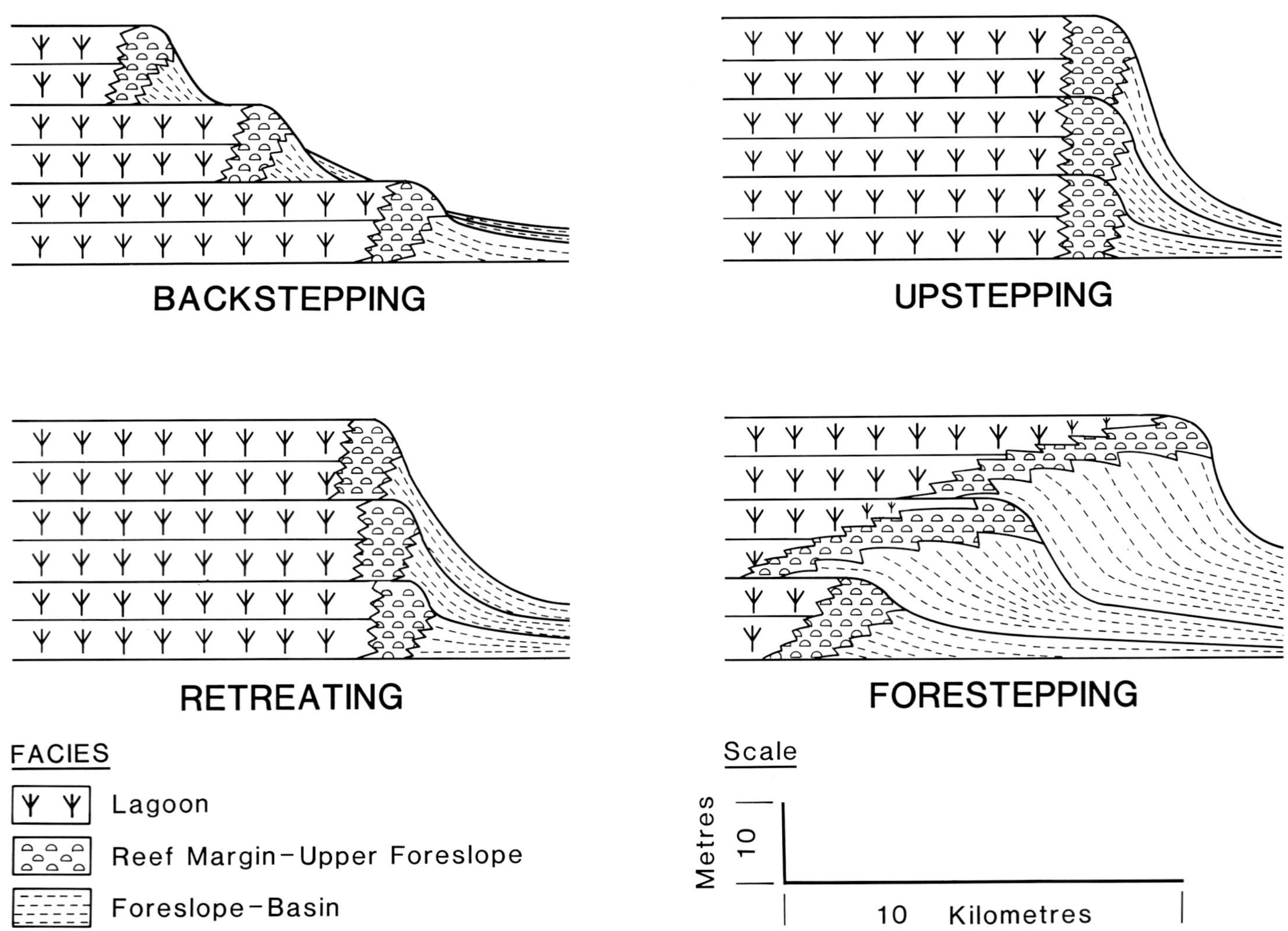


Figure 5. Stacking styles of Devonian platform and reef complexes.

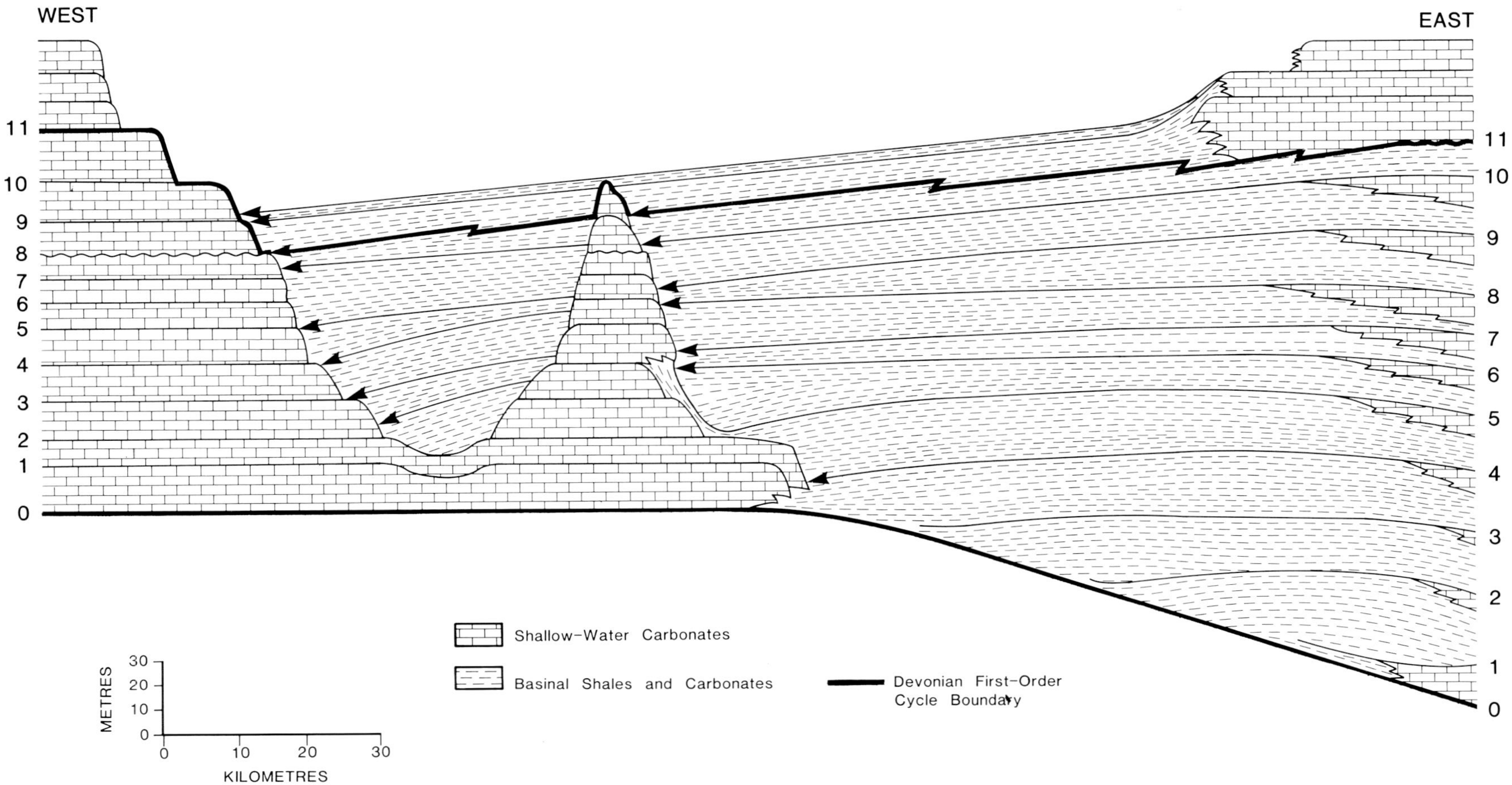


Figure 6. Model of a first-order Devonian cycle. The evolution depicted in this diagram most closely approximates that of the Beaverhill Lake megacycle.

of the isolate reefs are drowned during this phase; others continue to grow up into the upper progradational phase.

The nature and geometry of the depositional elements in the transgressive portion of the Winterburn megasequence are somewhat different from those in the Beaverhill Lake and Woodbend megasequences. The Nisku "shelf" differs from the platform beneath the Swan Hills and Woodbend reef complexes in that it has a ramp configuration. Backstepping in the early phase of this ramp leads only to the formation of small highs on the intermediate portion of the ramp. Coincident with subsequent backstepping of the ramp, isolate reefs are localized on these highs. Because these reefs initially inherit their size from the underlying highs, they attain only a small areal extent, normally less than a section in extent.

The upper regressive or progradational phase is marked by pronounced asymmetric infilling of the basin. Carbonate platforms on the east side of the basin build out, in a piggyback manner, over westerly clinoforming basinal limestones and shales. Carbonate platforms and reefs continue to grow, in generally an upstepping style, on the west side of the basin away from the clinoforming basinal sediment. Many reefs on the east side of the basin are killed by increasing turbid-water conditions. Those on the west side of the basin continue to backstep, become smaller in areal extent and ultimately drown.

These transgressive-regressive megacycles correspond in a general manner to the sequences of Jervey (1988) created by an overall gradual rise and fall of eustatic sea level. In addition, the westerly directed trade winds also influenced depositional patterns of Devonian strata in the WCSB. Basinal limestones and shales asymmetrically clinoform and fill the basin from east to west. The widespread extent of the Devonian first-order cycles places constraints on how they formed. The relative rise of sea level must have been at least regional in scope, whether driven by eustasy or tectonic subsidence.

## UTILITY OF CYCLICITY TO HYDROCARBON EXPLORATION AND EXPLOITATION

The cyclic context of Devonian strata in the WCSB provides a framework useful in all stages of hydrocarbon exploration and exploitation. This section merely outlines these uses. The details and methodology of these uses are discussed in later chapters.

An understanding of the first-order context of these cycles provides a general model that can be used to highgrade and evaluate potential reservoir, source and seal facies. Reservoir facies include platforms fringing the basin and isolate reefs on aborted platforms and ramps. Deep-water, organic-rich laminites of varying composition are potential sources of hydrocarbons and occur in the basal, "condensed" positions of megasequences in basin-centre locations. These strata intertongue on more shelfal areas with carbonate platforms and isolate reefs. Clinoforming shales and limestones in the upper progradational phase of first-order cycles provide excellent seals.

Stacking patterns expressed at the second-order cycle level constitute the exploration play types within the megasequences. The position and configuration of shallow-water, carbonate reservoir facies is especially important to exploration and can be determined through second-order sequence analysis. The localization of isolate reefs, for example, is governed by the second-order cyclic evolution of the underlying carbonate platforms and ramps. The position of many platform-margin reservoirs, such as the Swan Hills Caroline pool, is governed by the backstepping style of second-order cycles.

Understanding the nature of both second and third-order cycles is a key to understanding the distribution and quality of reservoir units in hydrocarbon pools. The exploitation history of many of the large Devonian limestone pools in the WCSB underscores the significance of discerning reservoir continuity. The expression of cyclicity and the continuity of reservoir facies vary according to depositional setting. Along reef or platform margins, second-order cycles contain thick porous and permeable deposits and, as a consequence, have good vertical reservoir continuity. Isolate reef and platform interiors, conversely, are broken into thin, third-order cycles. This finer level cyclicity produces stratified thin reservoirs with poor vertical continuity.

## REFERENCES

BLACKWELDER, E., 1909, The valuation of unconformities: Journal of Geology v. 17, p. 289-299.

GALLOWAY, W.E., 1989, Genetic sequences in basin analysis 1: Architecture and genesis of flooding-surface bounded depositional units: American Association of Petroleum Geologists Bulletin, v. 73, no. 2, p. 125-142.

HANDFORD, C.R., AND LOUCKS, R.G., 1990, Dynamic response of carbonate systems tracts to relative sea-level changes and the development of carbonate depositional systems in platforms and ramps: American Association of Petroleum Geologists Bulletin, v. 74, no. 5, p. 669.

JAMES, N.P., 1979, Shallowing-upward sequences in carbonates, *in* Walker, R.G., ed., Facies Models: Geoscience Canada, Reprint Series 1, p. 121-132.

JERVEY, M.T., 1988, Quantitative geological modelling of siliciclastic rock sequences and their seismic expression, *in* Wilgus, C.K., Hastings, B.S., Kendall, C.G. St.C., Posamentier, H.W., Ross, C.A., and Van Wagoner, J.C., eds., Sea-Level Changes: An Integrated Approach: Society of Economic Paleontologists and Mineralogists, Special Publication no. 42, p. 47-70.

KINSMAN, D.J.J., 1969, Modes of formation, sedimentary associations and diagnostic features of shallow-water and supratidal evaporites: American Association of Petroleum Geologists Bulletin, v. 53, p. 830-840.

POSAMENTIER, H.W., JERVEY, M.T., AND VAIL, P.R., 1988, Eustatic controls on clastic deposition 1 - conceptual framework, *in* Wilgus, C.K., Hastings, B.S., Kendall, C.G. St.C., Posamentier, H.W., Ross, C.A., and Van Wagoner, J.C., eds., Sea-Level Changes: An Integrated Approach: Society of Economic Paleontologists and Mineralogists, Special publication no. 42, p. 109-124.

POSAMENTIER, H.W., AND VAIL, P.R., 1988, Eustatic controls on clastic deposition 2 - sequence and system tract models, *in* Wilgus, C.K., Hastings, B.S., Kendall, C.G. St.C., Posamentier, H.W., Ross, C.A., and Van Wagoner, J.C., eds., Sea-Level Changes: An Integrated Approach: Society of Economic Paleontologists and Mineralogists, Special Publication no. 42, p. 125-154.

PURSER, B.H., 1972, Subdivision et interpretation des sequences carbonates: Mem. B.R.G., M. 77, p. 679-698.

RICKETTS, B.R., ed., 1989, Western Canada Sedimentary Basin - A Case History: Canadian Society of Petroleum Geologists, Special Publication no. 30, 320 p.

ROEHL, P.O., 1967, Stony Mountain (Ordovician) and Interlake (Silurian) facies analogues of recent low-energy marine and subaerial carbonates, Bahamas: American Association of American Geologists Bulletin, v.51, p. 1979-2032.

SARG, J.F., 1988, Carbonate Sequence stratigraphy, *in* Wilgus, C.K.., Hastings, B.S., Kendall, C.G. St.C., Posamentier, H.W., Ross, C.A., and Van Wagoner, J.C., eds., Sea-Level Changes: An Integrated Approach: Society of Economic Paleontologists and Mineralogists, Special Publication no. 42, p. 155-182.

SCHLAGER, W., 1989, Drowning unconformities on carbonate platforms, *in* Crevello, P.D., Wilson, J.C., Sarg, J.F., and Read, J.F., eds., Controls on Carbonate Platform and Basin Development: Society of Economic Paleontologists and Mineralogists, Special Publication no. 44, p. 15-26.

SHINN, E.A., LLOYD, R.M., AND GINSBURG, R.N., 1969, Anatomy of a modern tidal flat, Andros Island, Bahamas: Journal of Sedimentary and Petroleum Geology, v.39, p. 1202-1228.

STOAKES, F.A., AND WENDTE, J.C., 1987, The Woodbend Group, *in* Burrowes, O.G., and Krausse, F.F., eds., Devonian Lithofacies and Reservoir Styles in Alberta: 13th Canadian Society of Petroleum Geologists Core Conference and Second International Symposium on the Devonian System, p. 153-170.

WENDTE, J.C., WONG, P.K., MUIR, I.D., WEISSENBERGER, J., GILHOOLY, M.G, AND STOAKES, F.A., 1990, Devonian carbonate sequence evolution of Western Canada: American Association of Petroleum Geologists Bulletin, v. 74, no. 5, p. 788.

VAIL, P.R., MITCHUM, R.M., TODD, R.G., WIDMIER, J.M., THOMPSON, S., SANGREE, J.B., BUBB, J.N., AND HATELID, W.G., 1977, Seismic stratigraphy and global changes of sea level, *in* Clayton, C.E., ed., Seismic stratigraphy - Applications to hydrocarbon exploration: American Association of Petroleum Geologists, Memoir 26, p. 49-212.

WILSON, J.L., 1967a, Cyclic and reciprocal sedimentation in Virgilian strata of southern New Mexico: Geological Society of America Bulletin, v. 78, p. 805-818.

WILSON, J.L., 1967b, Carbonate-evaporite cycles in lower Duperow Formation of Williston Basin: Bulletin of Canadian Petroleum Geology, v. 15, p. 230-312.

WILSON, J.L., 1975, Carbonate Facies in Geologic History: New York, Springer-Verlag, 471 p.

CHAPTER 3

# PLATFORM EVOLUTION AND ITS CONTROL ON REEF INCEPTION AND LOCALIZATION

J.C. WENDTE

## INTRODUCTION

Many ancient biohermal carbonate reefs occur on areally extensive, shallow marine carbonate platforms. Because these reefs are imposing structures in outcrop and contain significant hydrocarbon reserves in the subsurface, they have been studied in detail. These studies have concentrated on their facies, paleontology, reservoir attributes, and more recently on reef evolution. Only a few studies have focused on their origin. Furthermore, the internal anatomy of carbonate platforms, from which these reefs evolved, has received only minor attention in the WCSB. As a consequence, models explaining the temporal and spatial distribution of biohermal reefs are very speculative.

This paper addresses the question of the origin of large Devonian platform-reef complexes in the WCSB. Why do these reefs form during specific episodes and in certain locations?

Large areally extensive biohermal reefs occur in two main stratigraphic intervals in the WCSB: the late Givetian Swan Hills reef complexes of west-central Alberta and the early to middle Frasnian reefs of both east-central and west-central Alberta. These reefs are distinctly different from either earlier Keg River reefs or later Nisku reefs in that they attained much greater areal extent, commonly in excess of 10 $km^2$ and up to 600 $km^2$, and nucleated on shallow-water carbonate platforms. Keg River and Nisku reefs, on the other hand, are normally less than 1 $km^2$ in size and formed on underlying carbonate ramps.

This section focuses on the origin of Leduc reefs in east-central Alberta from the perspective of understanding the evolution of the underlying carbonate platform. In east-central Alberta, Leduc reefs are underlain by platformal carbonates of the Cooking Lake Formation (Figure 1). The results reported here are based on a detailed subsurface investigation of the Cooking Lake platform beneath and away from the Leduc reef complex at Redwater (Figure 2). The conclusions from this study are considered applicable not only to the origin of Leduc reef complexes, but to understanding the origin of Swan Hills reefs and other platform reefs of varying ages and occurrences.

## GEOLOGICAL SETTING

The stratigraphic sequence associated with Leduc reefs is most easily explicable in terms of three general facies illustrated in Figure 3: (1) the underlying shallow-marine carbonate platform, (2) the carbonate buildup itself, and (3) the laterally equivalent and overlying basinal fill.

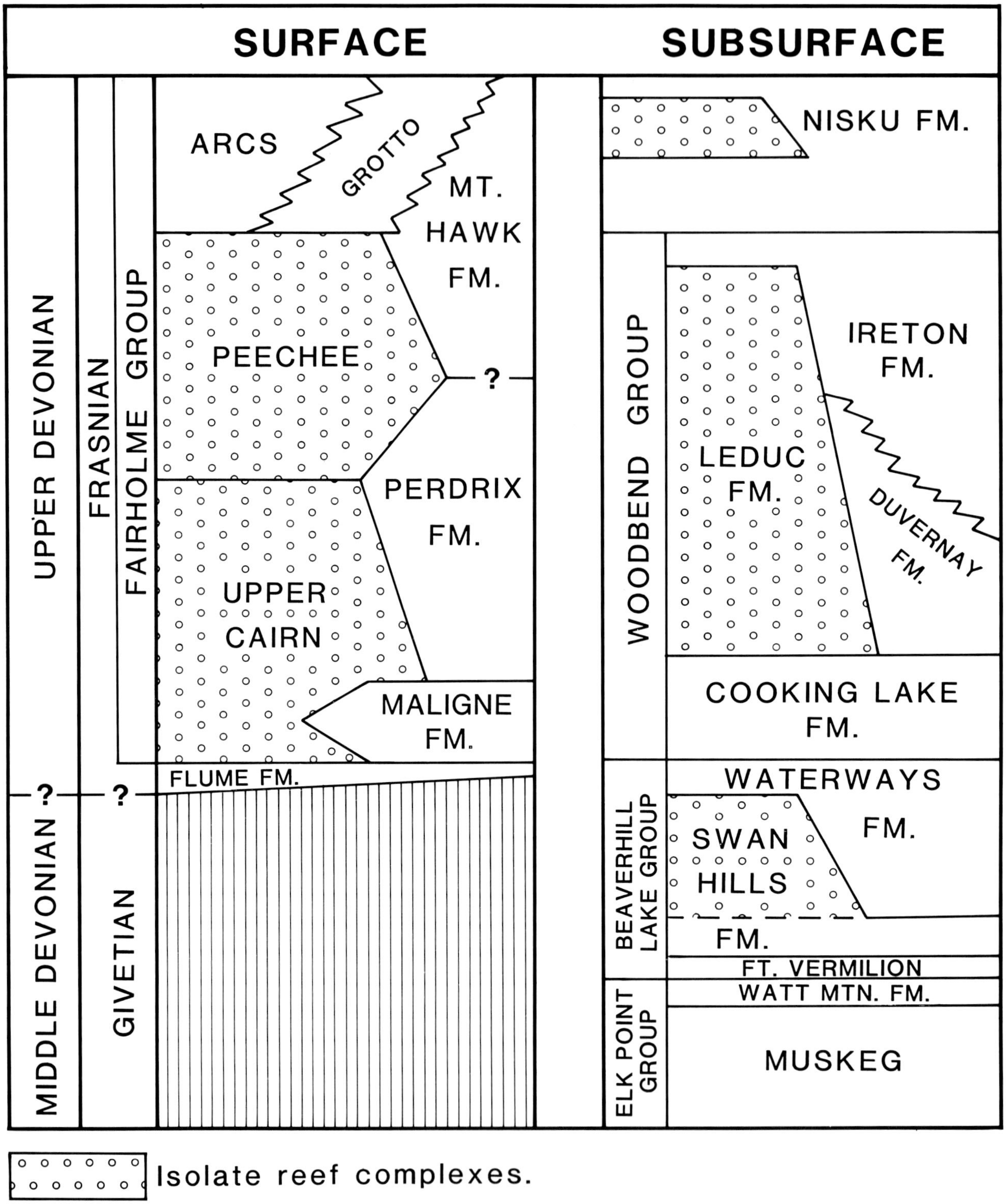


Figure 1. Correlation of Givetian and Frasnian strata between the subsurface of central Alberta and the Cordilleran of western Alberta.

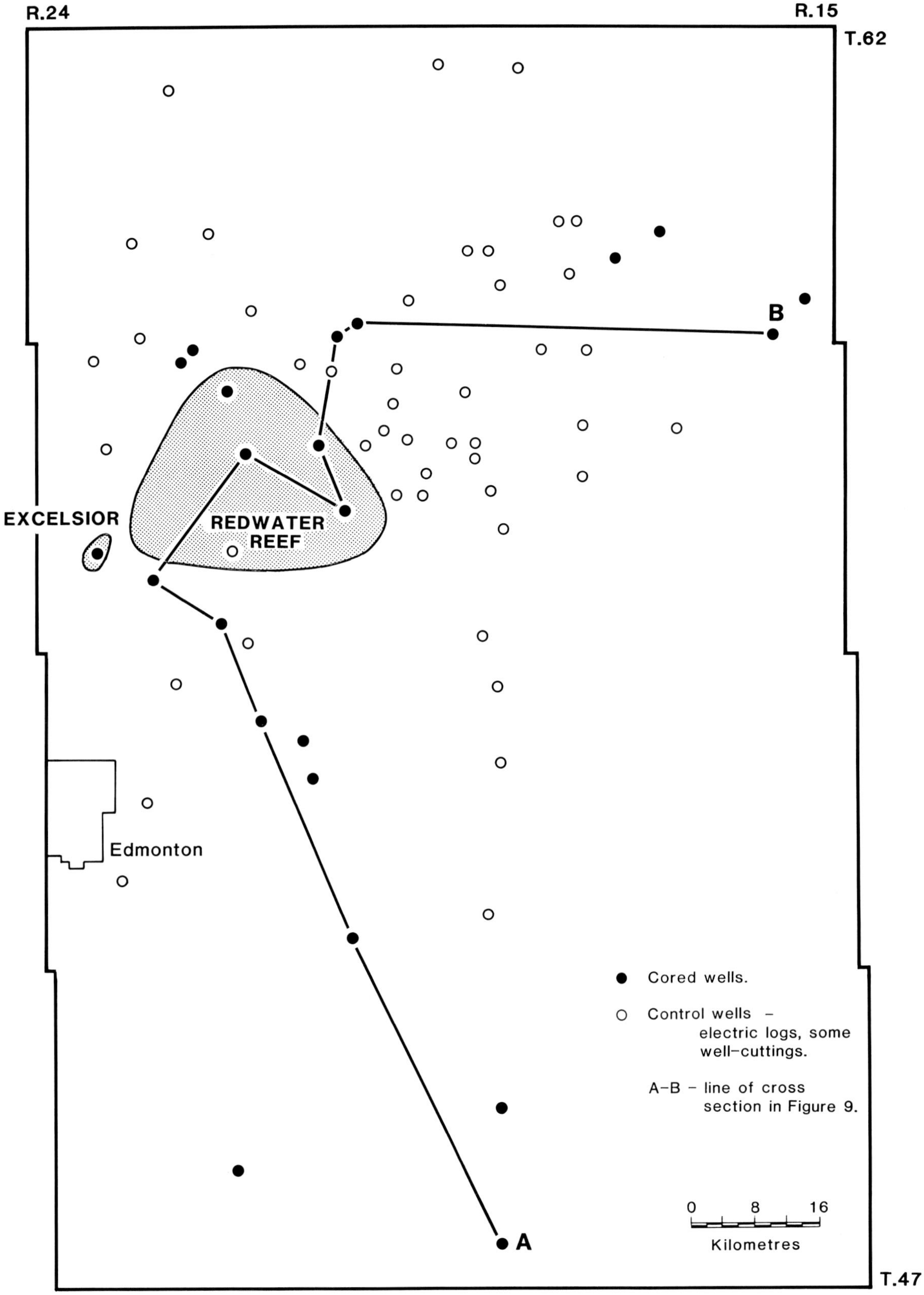


Figure 2. Area of study showing well control.

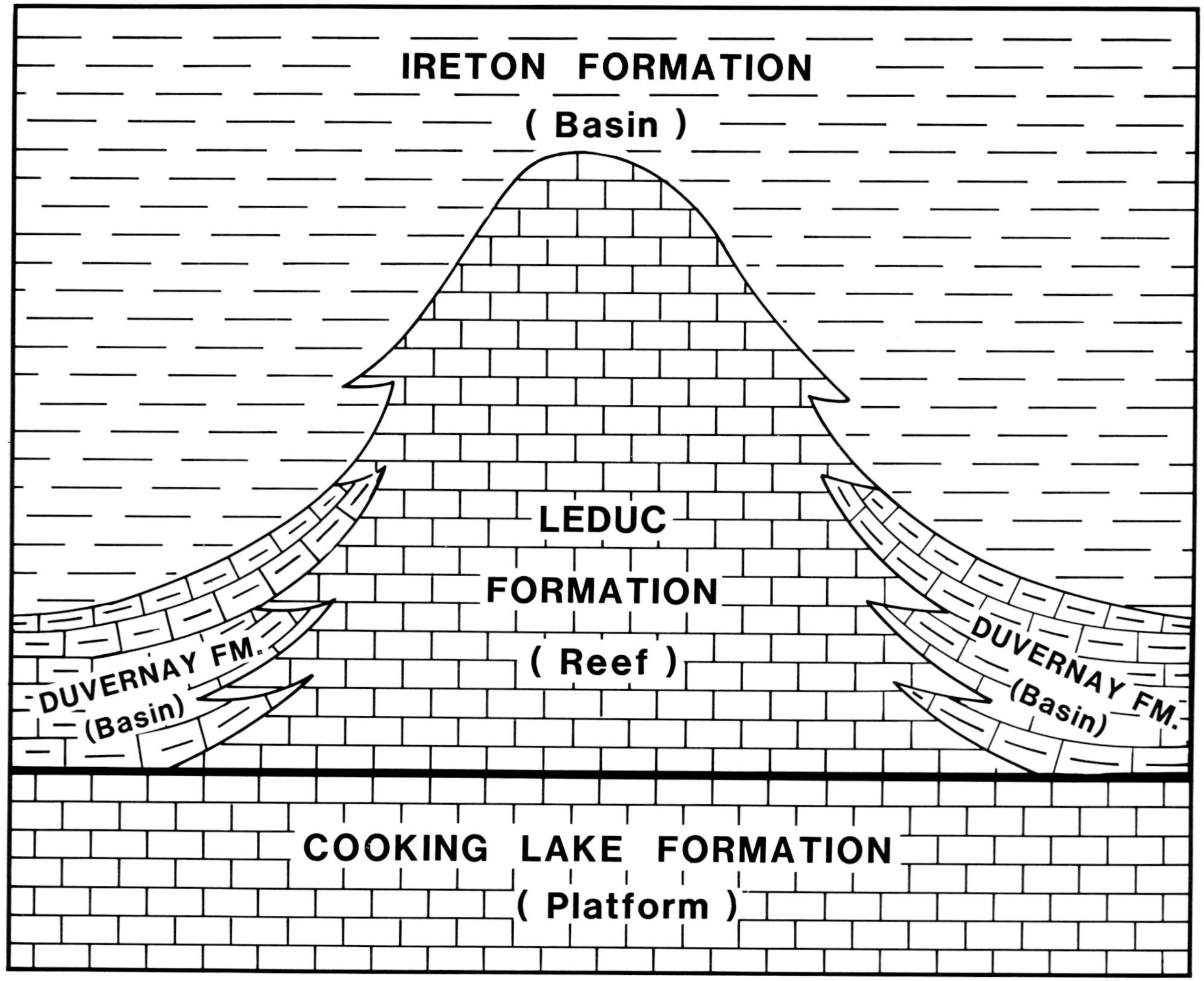


Figure 3. Schematic representation of a Leduc reef complex and associated units and facies.

The distribution of Cooking Lake and approximate time-correlative Fairholme carbonate platforms in Alberta is shown in Figure 4. The Cooking Lake Formation is a widespread sheet-like unit up to 100 m thick in east-central Alberta. These platform carbonates extend from the eastern margin of a deep-water basin in central Alberta into eastern and southern Alberta where they grade into more restricted limestones, dolomites, and anhydrites of the Duperow Formation. Other Woodbend and Fairholme carbonate platforms occur in the subsurface in west-central Alberta and around the Peace River paleo-landmass, and in outcrops in the Rocky Mountain thrust belt.

Cooking Lake carbonates are predominantly limestones, with variable amounts of dolomite in the lower half of the Formation. Massive dolomites occur only along a belt up to 29 km wide along the western margin of the platform (see Andrichuk, 1958b), outside all but the westernmost part of the study area.

Leduc reef complexes in east-central and west-central Alberta contain major oil and gas reserves. In east-central Alberta these reefs occur in sub-parallel belts along the western margin of the Cooking Lake platform and along the Bashaw-Duhamel trend. A barrier reef also fringes the margin of the southern Alberta shelf further to the southeast (Figure 4).

The Leduc reef at Redwater occurs approximately 24 km east of the western margin of the Cooking Lake platform, in an apparent platform-interior position (Figure 4). This reef has a triangular shape, a thickness of approximately 275 m, and a total areal extent of close to 600 $km^2$. The upper, northeast perimeter of this complex contains a large oil field with ultimate recoverable reserves of approximately 135 x $10^6 m^3$ (857 million barrels). Basinal limestones and shales of the Duvernay and Ireton formations intertongue, downlap and cover the reef complex.

## PREVIOUS STUDIES

Although many detailed studies on various aspects of Leduc and Fairholme reef complexes have been published (Andrichuk, 1958a; Klovan, 1964; Mountjoy, 1967; Dolphin and Klovan, 1970; Walls and others, 1979; and others), only two published studies have been undertaken where the primary goal was to determine and interpret the depositional facies of an underlying Woodbend platform.

Andrichuk (1958b) analyzed cores, well-cuttings, and mechanical logs from the Cooking Lake Formation in a regional study of about 21,000 $km^2$ in the Edmonton area. He concluded that these platform carbonates were deposited on or between shoals. He found Leduc reef complexes selectively developed on the shoals.

Cook (1972) studied widespread platform carbonates of the Flume Formation beneath and away from the Fairholme Miette buildup, in the Rocky Mountains near Jasper, Alberta. He concluded that the Miette buildup formed on a platform shoal, inherited from a pre-Devonian erosional high. He postulated that this shoal was maintained throughout platform evolution by deposition, and possibly by structural deformation.

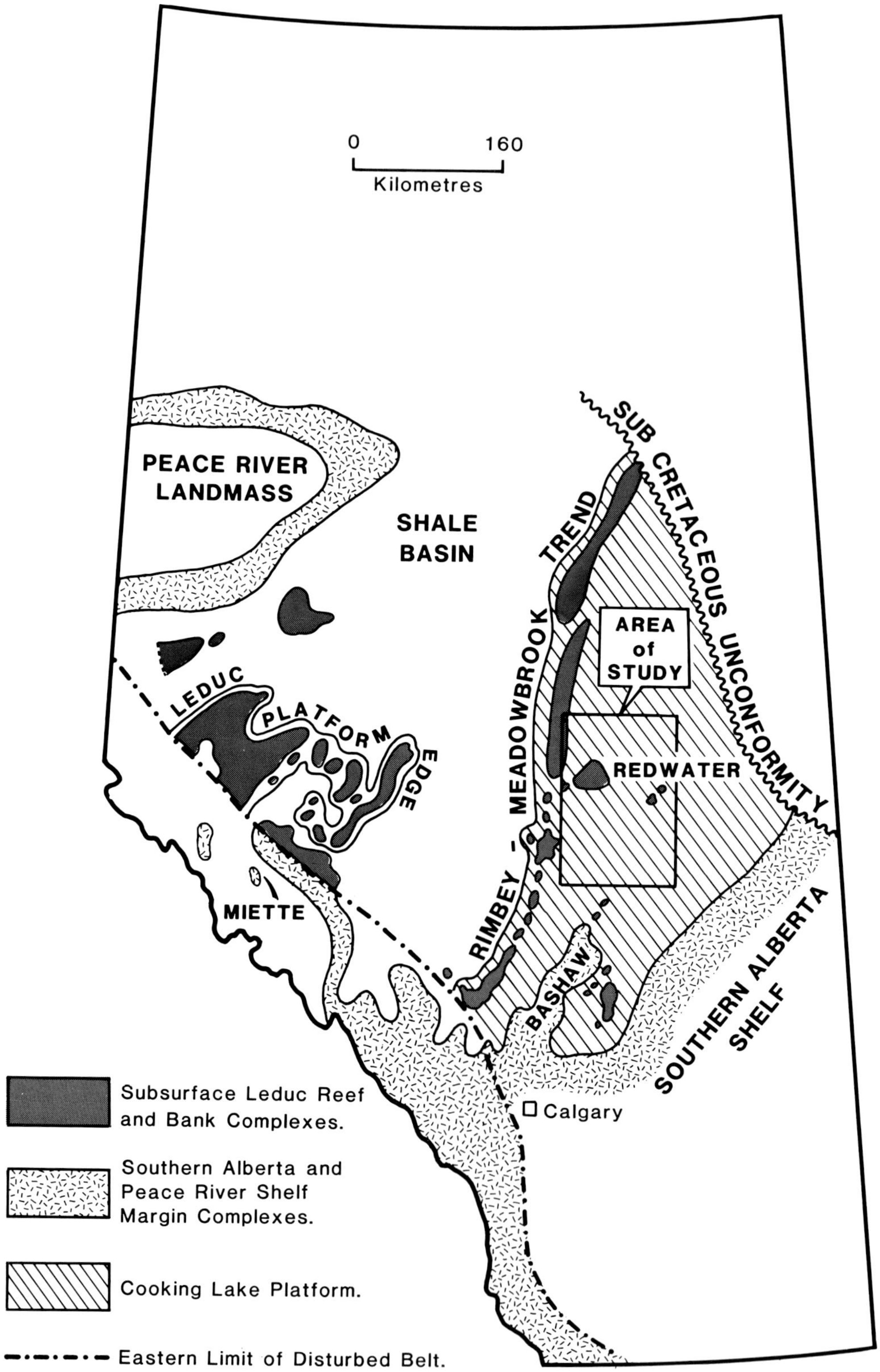


Figure 4. General distribution of Fairholme-Woodbend carbonate complexes in Alberta.

## STRATIGRAPHY

The Cooking Lake Formation is predominantly a clean carbonate that overlies argillaceous carbonates of the Waterways Formation. The Cooking Lake is either overlain by off-reef limestones of the Duvernay Formation or by reefal carbonates of the Leduc Formation. Figure 5 shows a 3-well cross section illustrating the electrical log signature of the Cooking Lake Formation in the study area. This cross section is intended to demonstrate the stratigraphic relationships between the Cooking Lake Formation and the adjacent Waterways, Duvernay and Leduc formations. The logs on either end of the section correspond to off-reef wells. 4-14-51-21W4 is south of Redwater and is the type section of the Cooking Lake Formation. 14-5-54-15W4 is east of Redwater. The log in the centre corresponds to a section (12-28-57-21W4) beneath the Redwater reef complex.

The lower boundary of the Cooking Lake Formation shows as one of the most widespread electrical-log markers in east-central Alberta. It corresponds to a change from argillaceous lime mudstones of the underlying Waterways Formation to relatively clean limestones above. In the study area this change varies from gradational beneath the Redwater complex to abrupt east of the reef complex. Where cored in the area east of Redwater, the contact is a sharp lithified surface (Plate 1).

The transitional nature of the base of the Cooking Lake Formation over much of the study area makes it unsuitable for use as a stratigraphic datum. Instead, an overlying small but widely correlatable peak on the SP log was used as a datum. This log signature corresponds to the change from deeper water, open-marine carbonates below to shallower water, platformal carbonates above. In the context of this study, this level is defined as the base of the Cooking Lake <u>platform</u>. Carbonates below this marker are mainly crinoidal lime mudstones and wackestones (Plate 2A).

The upper contact of the Cooking Lake Formation is always abrupt in the study area. In off-reef positions, this surface corresponds to a change from clean limestones to argillaceous limestones of the Duvernay Formation and is easily recognizable on electrical logs. At Redwater, the upper contact of the Cooking Lake Formation is depicted on SP logs by a shift toward the "shale base line" because of the tight nature of the basal Leduc limestones. Deposits immediately below this surface show evidence of syndepositional lithification.

Isopach maps of the Cooking Lake <u>platform</u> and that of the open-marine carbonates of the basal Cooking Lake Formation (pre-platform) are illustrated in Figures 6 and 7, respectively. The platform isopach map shows a "thick" underneath and around the perimeter of the Redwater reef complex and a pronounced "thin" to the immediate east and north. Because the base of the Cooking Lake platform is essentially a depositionally flat surface, thickness variations in this unit correspond to depositional topography at the top of the platform. Platform "thicks" therefore correspond to highs on the platform. The high at Redwater is referred to as the Redwater Shoal. Another Cooking Lake platform "thick" occurs to the immediate northeast of the pronounced "thin". This high is

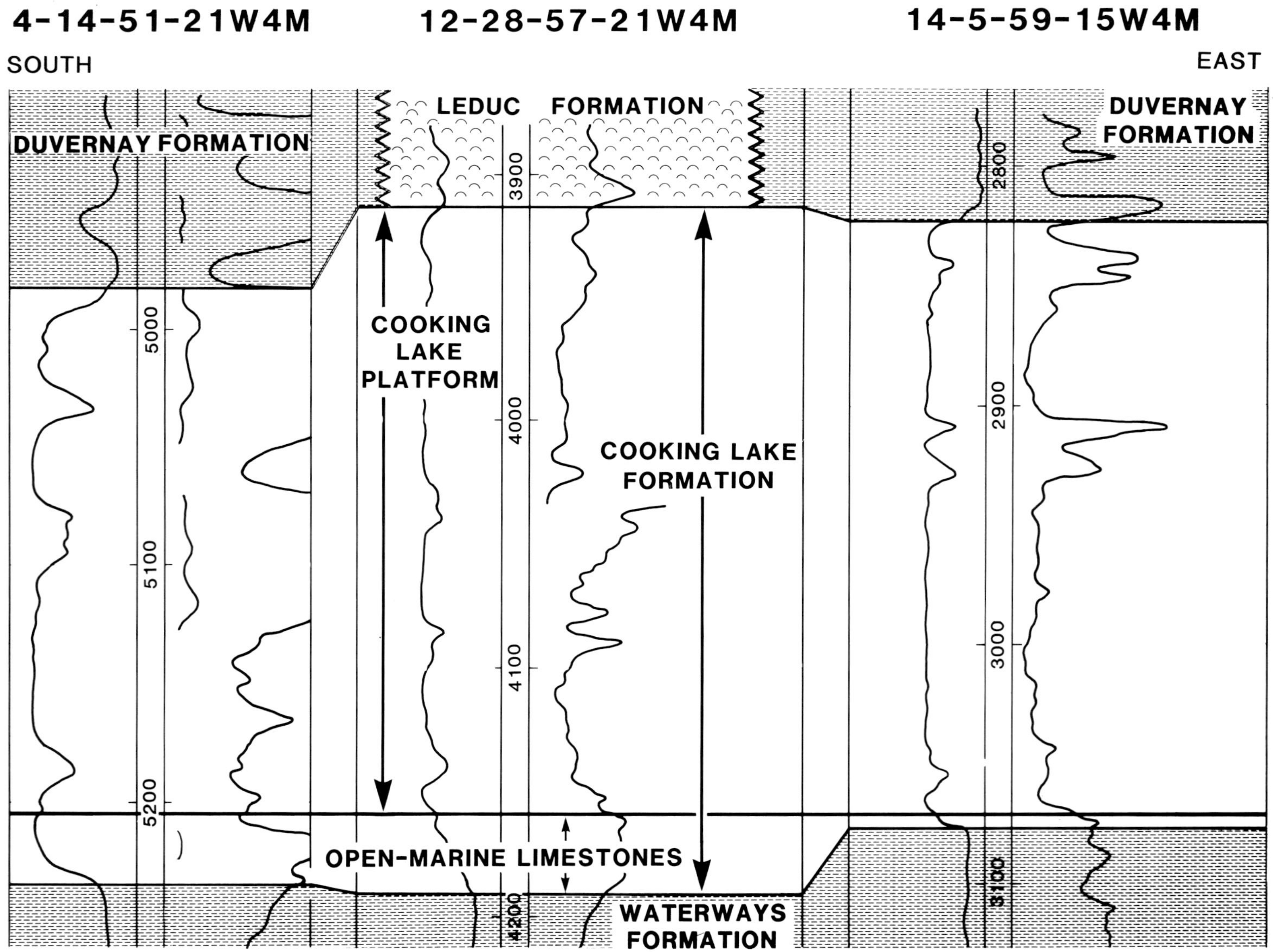


Figure 5. Cross section showing the electrical log signature of the base and top of both the Cooking Lake Formation and platform.

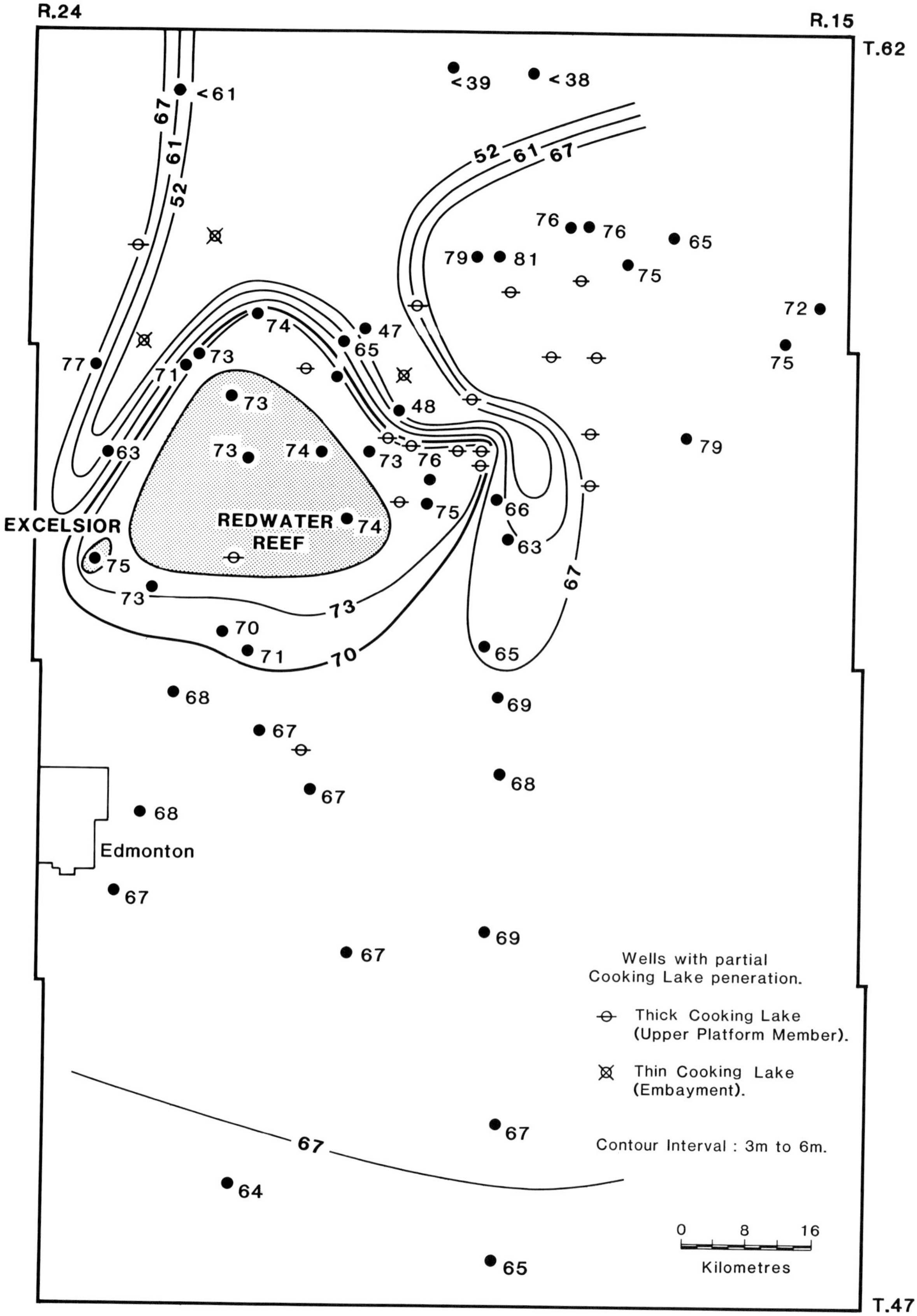


Figure 6. Isopach map of the Cooking Lake platform.

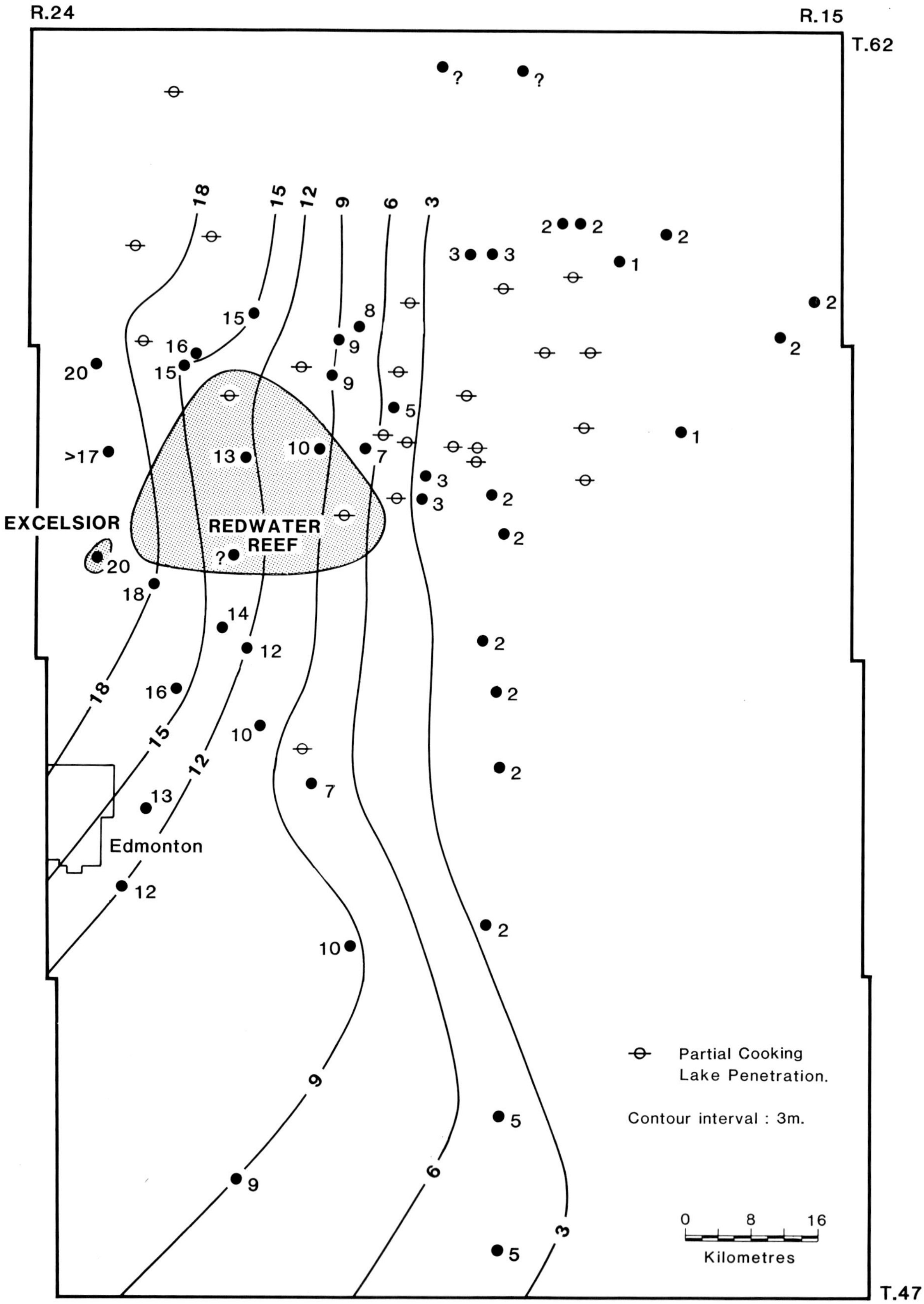


Figure 7. Isopach map of the open-marine facies of the basal Cooking Lake Formation.

referred to as the Smoky Lake Shoal.

The isopach of the basal open-marine carbonate demonstrates a westward thickening (Figure 7). This progressive thickening is interpreted to reflect depositional relief along the top of the underlying Waterways Formation. The pattern of thickening corresponds to a westward sloping ramp.

## PLATFORM FACIES

Platform limestones exhibit a wide variety of organic and inorganic constituents, depositional textures, primary sedimentary structures and colour. Genetically related rock types, defined according to these attributes, are grouped into broader depositional facies.

Eleven facies are recognized. These are named according to Embry and Klovan's modification (1971) of Dunham's (1962) carbonate rock classification. Environmental interpretations of these facies are based on the depositional attributes, mentioned above, comparison with recent analogues, and on vertical and lateral facies associations.

Paleo-water depths of the facies were estimated according to their relative position on a cross section datumed on a marker that approximates a time-synchronous, depositionally flat surface. Thus, the vertical relief measured along a cycle boundary between a given facies and a reference limestone deposited at or very near sea level, such as a tidal flat laminite, approximates the original water depth of the given facies.

Emphasis in interpreting these facies was also placed on the growth forms of stromatoporoids, the dominant organism and constituent in some of the platform limestones. The growth forms of these organisms, like those of present day hermatypic corals, are a response to such controls as nutrient supply, water energy, water turbidity, sediment supply and water depth, and hence reflect the environment of deposition. As the growth forms of stromatoporoids are of such paramount importance, the terminology used in this paper is reviewed here:

Tabular - flat, plate-like forms varying from a few mm to several cm thick.

Bulbous - small head-shaped forms with diameters less than 8 cm.

Hemispherical - large head-shaped forms with diameters greater than 8 cm.

Cylindrical - branching dendroid forms excluding *Amphipora* and *Euryamphipora*; most commonly *Stachyodes*.

*Amphipora* - a distinctive finger-shaped stromatoporoid, usually 2 to 4 mm in diameter and with a central canal and a peripheral ring of vesicles in well-preserved forms.

*Euryamphipora* - forms with a similar cell structure to *Amphipora* but more fan shaped and lacking a central canal.

Irregular - irregular forms not assigned to other growth habits, commonly adapted to an encrusting habitat.

The major facies are summarized in Figure 8 and are discussed below.

| FACIES | ROCK TYPES | IMPORTANT CONSTITUENTS | SEDIMENTARY STRUCTURES | ENVIRONMENT |
|---|---|---|---|---|
| Stromatolitic and Planar-Fenestral Limestones | Light-coloured stromatolitic lime mudstones-wackestones and planar-fenestral lime grainstones-packstones. Common flat-pebble breccia and Amphipora-pellet lime grainstone-packstone interbeds. | Pellets, flat-pebble clasts, Amphipora. | Stromatolitic laminae, planar fenestrae. | Upper intertidal to supratidal. |
| Pelletal Lime Mudstones | Light-coloured pelletal lime mudstones-wackestones with thin intraclast and pelletal lime grainstone-packstone interbeds. | Mashed pellets, hardened pellets, intraclasts. | Graded and cross bedding, scour surfaces. | Intertidal to adjacent subtidal. |
| Nodular Lime Mudstones | Light-coloured nodular lime mudstones and wackestones. | Ostracodes, gastropods. | Extensive burrowing. | Subtidal, upslope ramp. |
| Shallow-Shoal Lime Grainstones | Light-coloured intraclast-lump-pellet lime grainstones and packstones. | Abundant intraclasts; common lumps, cemented-grain clasts and pellets; some algal-coated and superficial oolitic-coated grains. Common Parathuramminid foraminifera, calcispheres and Amphipora. | Common horizontal and low-angle laminations, minor ovoid fenestrae. | Shallow grain shoal. |
| Deep-Shoal Lime Grainstones and Packstones | Light-coloured intraclast and pellet lime grainstones and packstones. | Abundant intraclasts and pellets, common calcispheres and Parathuramminid foraminifera. | Some burrowing and wispy carbonaceous partings. | Intermediate-depth downslope ramp. |
| Amphipora Rudstones and Floatstones | Light-to dark-coloured Amphipora lime rudstones and floatstones, minor Amphipora lime wackestones. | Abundant Amphipora; common calcispheres and Parathuramminid foraminifera; some bulbous and cylindrical stromatoporoids. | Some burrowing. | Lagoonal environment. |
| Bulbous-Hemispherical-Irregular Encrusting Stromatoporoid Rudstones and Floatstones | Light-to medium-coloured bulbous, hemispherical and irregular-encrusting stromatoporoid lime rudstones and floatstones; minor boundstones. | Abundant bulbous, hemispherical and irregular-encrusting stromatoporoids; common Amphipora; some cylindrical and tabular stromatoporoids. | Some burrowing. | Well-circulated lagoon to platform-margin environment. |
| Cylindrical Stromatoporoid Rudstones and Floatstones | Light-to medium-coloured cylindrical stromatoporoid lime rudstones and floatstones. | Abundant cylindrical stromatoporoids; common tabular, hemispherical and irregular-encrusting stromatoporoids. Common Amphipora and massive, stick and cup corals. | Some burrowing. | Well-circulated lagoon to ramp-margin environment. |
| Tabular Stromatoporoid Boundstones and Rudstones | Medium-to dark-coloured tabular stromatoporoid lime boundstones and rudstones. | Abundant tabular stromatoporoids, common hemispherical, irregular encrusting and cylindrical stromatoporoids. Common massive, stick and cup corals; crinoids and brachiopods. | Abundant binding, some burrowing. | Platform-Margin and foreslope. |
| Skeletal Lime Wackestones | Medium-to dark-coloured brachiopod-crinoidal lime wackestones. | Brachiopods, crinoids, some cricoconarids. | Extensive burrowing. | Open-marine, relatively deep-water ramp. |
| Argillaceous Lime Wackestones | Dark-coloured argillaceous lime wackestones and mudstones. | Brachiopods, crinoids, some corals and cricoconarids. | Extensive burrowing. | Deep-water ramp. |

Figure 8. Summary of platform facies.

### *Stromatolitic and Planar-Fenestral Limestones*

These two rock types are grouped together because both are interpreted as having formed just below or above normal high tide. Deposition of these limestones is inferred to have occurred during storms or periods of unusually high tides.

Stromatolitic limestones are lime mudstones and wackestones which consist of alternating thin bituminous laminae (less than 0.2 mm) and thicker, lighter coloured carbonate laminae (0.5 to 5 mm) (Plate 2B). The laminations vary from flat to highly irregular. On a microscale, these laminae contain small unconformities, crinkles and small domes with relief up to 1.5 cm. The nature of the bituminous laminae suggest that many of the laminated carbonates may be the remains of algal films with carbonate deposited on a sediment trapping surface. Similar limestones are forming today in upper intertidal and lower supratidal environments.

Thin beds of flat-pebble conglomerates and breccias commonly overlie the stromatolitic limestones (Plate 2C). These pebbles are generally elongate and laminated, similar lithologically to the underlying sediment. These deposits are attributed to the reworking of desiccated supratidal flats by storms or higher than normal flood tides.

Planar-fenestral limestones are mainly thinly laminated (0.3 to 3 mm), and consist of light brown, lime grainstones and packstones with elongate voids or spar-filled voids too large to be depositional (Tebbutt and others, 1965) (Plate 2D). The planar-fenestrae are thought to form either by the decay of algal films or by shrinkage of sediments due to alternate periods of wetting and drying (Shinn, 1968). In Recent, normal salinity environments, they occur most commonly in the upper parts of tidal flats (Shinn, 1968). Thicker beds (up to 4 cm thick) of *Amphipora*-pellet grainstones locally occur interstratified with the finer laminated sediment and are interpreted to be storm deposits.

### *Pelletal Lime Mudstones*

This facies consists of light brown lime mudstones and wackestones with some thin (few cm) interbeds of intraclast and pellet lime grainstone and packstone (Plate 3). These limestones are interpreted to have formed slightly above or below normal low tide.

The lime mudstones and wackestones are sparsely fossiliferous. Although ostracodes predominate, they only make up between one and two percent of the rock. Calcispheres, Parathuramminid foraminifera, *Amphipora*, gastropods, brachiopods and crinoids are less abundant.

The micrite in these limestones is typically pelleted. The pellets are commonly mashed together, obscuring, if not obliterating, original grain contacts (Plate 3B). The pellets were initially soft and, unless partially cemented, compacted under only slight burial. These pelletal lime mudstones and wackestones reflect the normal, low-energy conditions that prevailed on intertidal flats and on the adjacent subtidal sea floor. Similar sediments are presently forming in analogous environments on the Great Bahama bank.

Intraclast and pelletal lime grainstones and packstones typically overlie erosional surfaces. These sediments are commonly graded, from basal grainstones to overlying packstones (Plate 3C). The grading is considered to reflect deposition following high-energy storm events. Other interstratified grainstones are cross-laminated (Plate 3D). Laporte (1967) and Cook (1972) have interpreted similar ancient alternating grain- and mud-supported carbonates as intertidal deposits. The nature of these limestones and their interbedding with stromatolitic lime mudstones and wackestones supports a similar interpretation here.

### *Nodular Lime Mudstones*

This facies consists of light to medium brown lime mudstones and minor wackestones, with either an extensively burrowed mottled, nodular, pinch and swell stratification (Plates 4A and 4B), or less commonly a more irregular nodular fabric (Plate 4C).

The fauna is rather sparse and of low diversity, with gastropods and ostracodes the most common fossils. The more irregular nodular limestones contain a somewhat more diverse and abundant fauna, with some brachiopods and a few crinoids in addition to the gastropods and ostracodes.

The occurrence of these nodular lime mudstones beneath and interstratified with interpreted supratidal and intertidal limestones indicates a low-energy, subtidal environment of deposition. Additionally, the low faunal diversity and the general absence of open-marine organisms such as crinoids suggests restrictive, perhaps slightly above normal marine salinity conditions.

### *Shallow-Shoal Lime Grainstones*

These limestones are light brown, non-skeletal lime grainstones, and less commonly lime packstones with only minor dark carbonaceous partings (Plates 4D and 5). The fauna is generally sparse and non-diverse, consisting mainly of calcispheres, Parathuramminid foraminifera, *Amphipora*, some small spirifer brachiopods and a few bulbous stromatoporoids. Deposition of these limestones occurred on extensive, shallow grain flats interpreted as varying from mainly muddy, stable grain flats to some high-energy, mobile grain shoals and beaches.

A variety of non-skeletal grains are present in these lime grainstones and packstones. The most common are irregular shaped intraclasts varying in composition from entirely micrite to a combination of micrite, calcispheres and Parathuramminid foraminifera (Plate 4D). Rounded intraclasts of similar composition (Plate 5A), pellets, superficial and very small oolites (Plate 5B), oncolites, lumps (accretional composite-aggregate grains similar to Bahamian grapestones) (Plate 5C), and cemented-grain clasts (Plate 5D) occur less commonly.

The irregular-shaped intraclast lime grainstones and subordinate lime packstones contain some pellets and lumps and are interpreted as having been deposited in a shallow shelf-lagoon, similar to the grapestone and hardened pellet flats on the Great Bahama bank. These deposits reflect alternating prevailing low to intermediate and episodic high-energy conditions. During the normal prevailing low and intermediate energy conditions, lime mud, calcispheres, and Parathuramminid foraminifera were deposited. Pellets were formed by excretion from burrowing and browsing organisms and hardened by internal cementation. In some places hardened pellets, calcispheres, and Parathuramminid foraminifera were aggregated together to form lumps. Scour during intermittent storms broke the sediment into irregular-shaped intraclasts, and deposited them with the pellets and lumps. Subsequent return to low or intermediate energy conditions prevented prolonged reworking and preserved the irregular form of the intraclasts.

Lime grainstones consisting of rounded intraclasts, superficial oolites, oncolites, lumps, pellets, and cemented-grain clasts occur less commonly and are thought to have been deposited on scattered high-energy shoals and beaches. Deposition in these very shallow to emergent settings is reflected by the presence of low angle (beach?) laminations and ovoid fenestrae (Plate 6A). In these environments grains were reworked and rounded by waves and/or currents. Agitation also resulted in the formation of algal and oolitic coats around these grains. Intermittent stabilization of parts of these shoals resulted in the formation of cemented-grain crusts or hardgrounds, analogous to that in the Schooners Cay oolite belt in the Bahamas (Dravis, 1979). Subsequent erosion of these crusts led to the formation of cemented-grain clasts as shown in Plate 5D.

### *Deep-Shoal Lime Grainstones and Packstones*

These limestones are mainly light to medium brown, pellet and intraclast lime grainstones and packstones. They commonly have a nodular structure (Plate 6B), carbonaceous partings and admixtures of lime mudstone and wackestone.

These limestones are generally similar to the shallow-shoal lime grainstones. The fauna is sparse and non-diverse, containing calcispheres, Parathuramminid foraminifera and, in places, brachiopods. Grains are mainly pellets and irregular-shaped intraclasts, consisting of micrite or combinations of micrite, calcispheres, and Parathuramminid foraminifera.

Indicators of extremely shallow or agitated conditions, however, are absent. Superficial and small oolites as well as cemented-grain clasts and lumps are not present. Only a subordinate number of intraclasts are well rounded, indicating a lack of prolonged sediment reworking. Low angle laminations and ovoid fenestrae, indicative of shallow and emergent shoal conditions, do not occur. The absence of these shallow shoal indicators, the presence of carbonaceous partings and the lateral occurrence of these limestones between shallow-shoal lime grainstones and deeper water micritic limestones all indicate a relatively deep-water downslope ramp environment. Deposition of these limestones probably occurred in water depths of 5 to 15 m.

### *Amphipora Rudstones and Floatstones*

The limestones in this facies are light to medium brown, *Amphipora* (or less commonly *Euryamphipora*) lime rudstones, floatstones and minor wackestones (Plate 6C). The matrix is normally micritic, varying from lime mudstone to packstone. Spar-filled and intraclast-pellet lime grainstone matrices occur less commonly.

*Amphipora* limestones are characterized by low faunal diversity but large numbers of individual forms. *Amphipora* coenostea are the dominant constituents. In places these coenostea are algally coated. Calcispheres and Parathuramminid foraminifera occur abundantly in the fine matrix. Bulbous and cylindrical stromatoporoids are commonly present as minor constituents. A few brachiopods, gastropods, cup corals, crinoids and solenoporid and dasycladacean algae also may be present locally.

These limestones are commonly associated with supratidal limestones. This occurrence, the low diversity fauna, and the common micritic matrix indicate a restricted, low energy, shallow lagoon environment. *Amphipora* limestones with lime grainstone matrices formed in slightly higher energy environments where lime mud was winnowed.

### *Bulbous, Hemispherical and Irregular-Encrusting Stromatoporoid Rudstones and Floatstones*

These limestones are mainly bulbous, hemispherical and irregular-encrusting stromatoporoid lime rudstones and floatstones (Plates 6D and 7A). The matrix varies from a skeletal lime wackestone to grainstone.

The fauna is rather abundant and diverse. Bulbous, hemispherical and irregular encrusting stromatoporoids are the dominant constituents. Cylindrical and tabular stromatoporoids, *Amphipora*, and encrusting corals occur in subordinate amounts. Solenoporid algae, brachiopods, gastropods, crinoids, ostracodes, calcispheres, and Parathuramminid foraminifera are minor constituents.

Deposition of these limestones occurred in well-circulated settings both along and behind the platform margin. Limestones found along the platform margin are predominantly abraded, hemispherical, bulbous and irregular-encrusting stromatoporoid lime rudstones with a skeletal grainstone to packstone matrix (Plate 7A). These characteristics indicate deposition in a wave-swept environment.

Limestones occurring in the platform interior are mainly bulbous stromatoporoid lime rudstones (Plate 6D) and floatstones with a more micritic lime packstone and wackestone matrix, indicative of a lower energy environment. Some of these limestones may have been deposited along well-circulated portions of ramps.

### *Cylindrical Stromatoporoid Rudstones and Floatstones*

This facies consists of light to medium brown cylindrical stromatoporoid lime rudstones and subordinate floatstones with a skeletal wackestone to grainstone matrix (Plate 7C).

The fauna is abundant and diverse. Cylindrical stromatoporoids are the dominant organism. Tabular, hemispherical, bulbous and irregular stromatoporoids, *Amphipora*, and massive, stick and cup corals are common subordinate constituents. Crinoids, brachiopods, solenoporid algae, gastropods, and ostracodes occur in minor amounts.

Cylindrical stromatoporoid lime rudstones and floatstones occur associated with deeper water, more open-marine limestones at the base of the platform, and with other stromatoporoid bearing limestones. The cylindrical stromatoporoid rudstones and floatstones at the base of the Cooking Lake platform are overlain by relatively restricted, nodular lime wackestones. This occurrence and the diverse, open-marine fauna of these limestones indicate a relatively shallow, well-circulated "ramp-margin" setting.

### *Tabular Stromatoporoid Boundstones and Rudstones*

These limestones vary from medium to dark brown, tabular stromatoporoid lime boundstones to rudstones with a skeletal wackestone to packstone matrix.

The fauna is the most diverse and abundant of any platform facies. Tabular stromatoporoids are the most abundant organism and commonly occur in growth position encrusting various types of stromatoporoids and massive corals. Massive, stick and cup corals, hemispherical, irregular and cylindrical stromatoporoids, crinoids and brachiopods are common secondary constituents. *Amphipora*, solenoporid algae, calcispheres, Parathuramminid foraminifera and ostracodes occur in minor amounts. Deposition of these limestones occurred both along the platform margin and foreslope.

Platform-margin limestones are typified by relatively thick, *in situ* tabular stromatoporoids (Plates 7B and 7D) and associated stromatoporoid rubble (Plate 7A). These sediments formed in high-energy, wave swept, turbulent settings along a distinct slope break. The thick-tabular stromatoporoids were adapted to an encrusting habitat, generally comparable to that of the coralline red algae *Lithothamnion* in Recent reefs.

Platform-foreslope limestones are characterized by fragmented tabular stromatoporoids set in a dark, carbonaceous micritic matrix (Plate 8A). These limestones grade up into brecciated hemispherical and irregular-encrusting stromatoporoid rudstones that formed along the platform margin. The tabular stromatoporoid rudstones represent allochthonous reef debris deposited on a gently sloping foreslope. Stratigraphic evidence indicates deposition in depths down to only 12 m.

### *Skeletal Lime Wackestones*

These limestones are mainly medium brown, extensively bioturbated, skeletal lime wackestones, mudstones and minor packstones (Plates 8B and 8C). Most have a nodular fabric, with a dark carbonaceous matrix between medium brown nodules. A few contain thin zones, up to 2 cm thick, of subparallel bituminous partings alternating with more burrowed limestones.

The fauna is distinctively open marine. Disarticulated brachiopods and crinoids are the most abundant fossils. High spired gastropods are common in some limestones. Bivalves, stromatoporoids, corals, ostracodes and cricoconarids occur in minor amounts. Algal-coated grains are present in some limestones.

These limestones occur associated with deep-shoal lime grainstones and packstones on positions on the platform away from tidal flats. This occurrence, along with the micritic nature, the characteristic carbonaceous partings and the generally open-marine fauna indicate a relatively deep, open-marine setting. Stratigraphic evidence suggests deposition along a carbonate ramp in depths generally greater than 10 m.

### *Argillaceous Lime Wackestones*

These limestones are similar to skeletal lime wackestones and mudstones except that they contain a relatively high concentration of argillaceous matter (Plate 8D). The clay occurs in sufficient amounts to cause a deflection toward the shale base line on the spontaneous potential log or to the right on the gamma-ray log.

Disarticulated brachiopods and crinoids are the most common constituents. Irregular stromatoporoids, stick and massive corals, gastropods, ostracodes and cricoconarids occur in minor amounts.

These limestones occur at the base of depositional cycles, underlie skeletal lime wackestones, and are present on parts of the platform farthest from tidal-flat sediments. Deposition of these limestones is ascribed to a relatively deep, oxygenated environment along a carbonate ramp. Prevailing slow rates of carbonate sedimentation in these settings permitted a significant amount of terrigenous clay to accumulate. Deposition generally occurred in water depths greater than 10 m.

## DISTRIBUTION OF PLATFORM FACIES

Figure 9 shows the distribution of Cooking Lake and time-equivalent Duvernay facies along a cross section from 88 km south of Redwater, to and under the reef complex, and ultimately to 56 km east of Redwater (see Figure 2).

This cross section shows that the lateral distribution of facies does not remain uniform throughout platform evolution. The type and distribution of facies change vertically as well as laterally. Shifts in facies belts correspond to cycles of sedimentation. These cycles are approximately 10 m thick and correspond to the relative second-order Devonian cycles defined in the previous chapter. Deposition within these cycles is generally upward shoaling, reflecting either the aggradation or progradation of shallower water facies over more basinward and deeper water sediments. Cycle tops are thinly transitional to sharp and are considered to be time-synchronous surfaces. Marked shifts in facies belts or the appearance of new facies occur at two horizons. These divisions allow the Cooking Lake Formation to be split into informal Lower, Middle and Upper Members.

In certain settings, second-order Devonian cycles in the

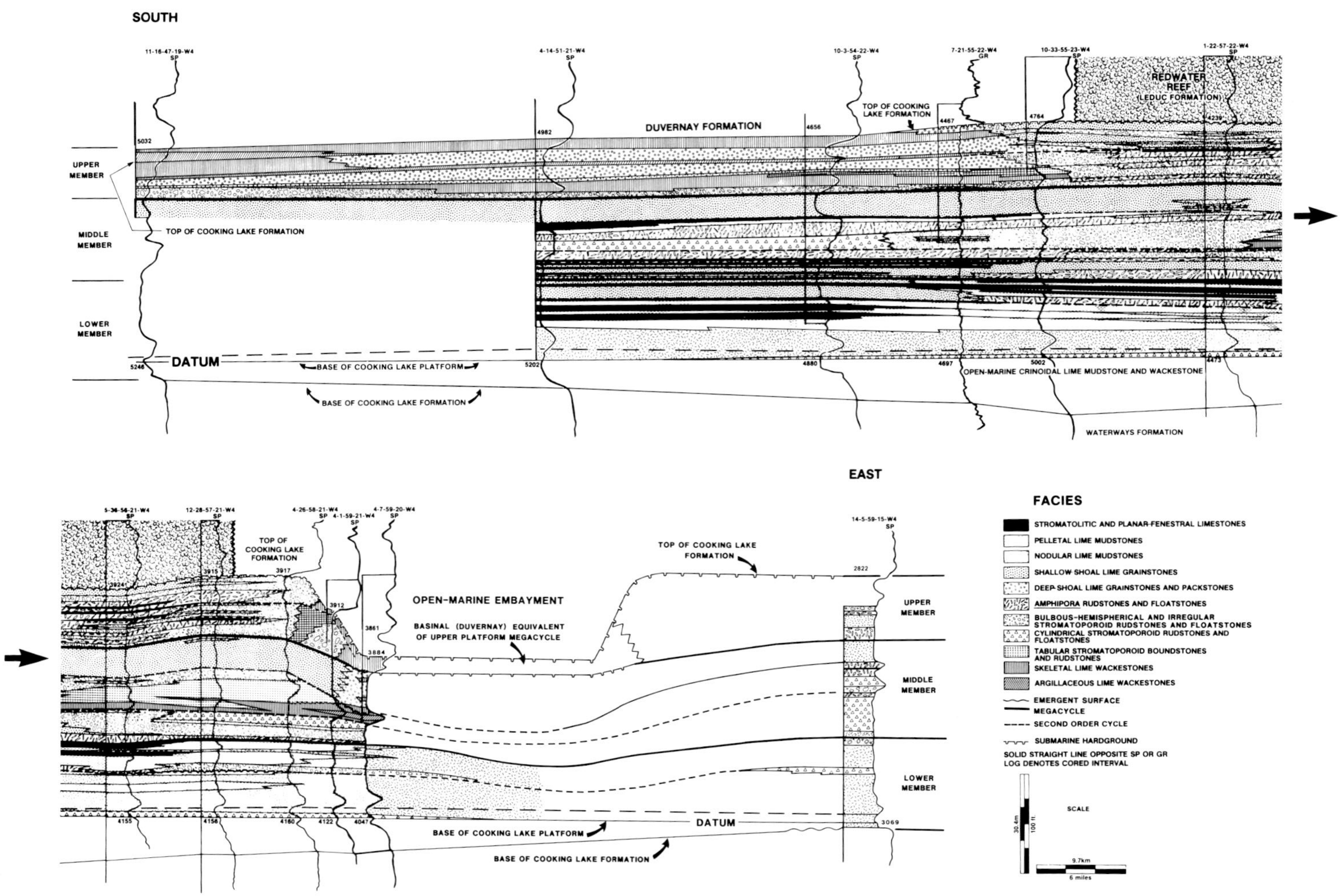


Figure 9. Sequence-facies cross section of the Cooking Lake Formation beneath and away from the Redwater reef complex.

Cooking Lake Formation consist of repeated, thinner upward-shoaling sequences. These sequences are only a few m thick and correspond to relative third-order Devonian cycles. These third-order cycles are best expressed in "incompetent" settings especially sensitive to relative sea-level rises. These cycles are most evident in peritidal and downslope-ramp environments.

The division of the Cooking Lake Formation into Lower, Middle and Upper Members provides a framework useful in summarizing the distribution of facies.

### *Lower Member*

In the Redwater area, this member comprises a shoaling-upward succession of facies conformable with underlying argillaceous lime mudstones and wackestones of the Waterways Formation. In the area east of Redwater, the contact between the Waterways and Cooking Lake formations is apparently unconformable.

The Lower Member attains a thickness up to 40 m. In the area beneath and to the south of the Redwater reef, basal open-marine wackestones grade up into peritidal deposits in the upper part of the succession. To the immediate north and east of Redwater, the deposits are mainly nodular lime mudstones and wackestones, indicative of continuous subtidal conditions.

### *Middle Member*

This member comprises three second-order Devonian cycles. Collectively these sequences vary in thickness from as much as 30 m in the Redwater and Smoky Lake areas to as little as 20 m in the area immediately east and north of Redwater. This latter area corresponds to the region of the Cooking Lake isopach "thin" (<52 m) on Figure 6. This area is interpreted as a shallow-marine embayment during the formation of the Middle Member.

The lower cycle consists of mainly bulbous and cylindrical stromatoporoid rudstones along the western margin of the shallow-marine embayment east of Redwater. These units are interpreted to represent the *in situ* growth of stromatoporoids along the gentle, ramp-like, break-in-slope. These deposits grade laterally into more restricted subtidal and peritidal deposits to the south of Redwater.

The middle cycle corresponds to an expansion of the marine embayment. Relatively deep-water skeletal limestones that occur at the base of this cycle extend onto the Redwater Shoal from the shallow marine embayment to the east. Stromatoporoid boundstones and rudstones formed along the margin and foreslope of this extended embayment. These deposits initially aggraded upwards and then prograded back, toward the marine embayment, over the deeper water, micritic sediments. In the area behind the platform margin, *Amphipora* and other stromatoporoid bearing limestones accumulated in a shallow, more protected lagoon. Thin tidal-flat laminites cap these deposits.

The upper cycle consists of shallow-water lime grainstones that formed on either side of the embayment. The scattered occurrence of low-angle beach laminae and ovoid fenestrae indicate the presence of small, emergent shoals in this grainstone unit. The shallow-water lime grainstones grade into deeper water, peloidal lime packstones and grainstones that fringe the western edge of the marine embayment.

### *Upper Member*

Limestones of the Upper Member of the Cooking Lake Formation are up to 20 m thick and are restricted in occurrence to the platform on either side of the marine embayment. Time-equivalent strata within the embayment are deeper water deposits of the basal part of the Duvernay Formation. Along the western margin of the embayment (4-7-59-20W4), these deposits vary from bioturbated lime mudstones and wackestones to laminated limestones with washed-in *Amphipora*. The nature of these deposits indicates deeper water conditions in the embayment during the formation of the Upper Member than during the Middle Member of the Cooking Lake Formation.

Platform facies within this member vary laterally. Stromatoporoid boundstones and rudstones occur along the margin of the open-marine embayment. Behind the barrier reef and beneath the Redwater reef complex, the platform consists of a succession of facies generally grading from stromatoporoid-bearing limestones up into peritidal deposits and shallow-water lime grainstones. Facies at the top of the platform indicate deposition near sea level. These shallow-water facies grade, in a ramp-like manner, into deeper water sediments south of Redwater. A slight break in slope a few km south of the Redwater reef marks the occurrence of poorly developed stromatoporoid buildups. Deeper water facies further south include argillaceous and non-argillaceous lime wackestones and deep-shoal lime packstones and grainstones. Units of these deposits are repeated in a cyclic manner, both at a second and a third-order cycle level. The overall vertical succession of facies in the Upper Member in this area indicates progressively deeper water conditions.

The isopach map of the Cooking Lake platform in Figure 6 depicts the configuration of the top of the platform. An open-marine embayment extended into the study area from the north. Thickness variations indicate water depths in the southerly portion of the embayment of at least 25 m. The large Redwater and Smoky Lake Shoals formed on either side of the embayment. These shoals grew up to or close to sea level. From the Redwater Shoal, the sea floor gradually sloped down to the south. In the area east of Edmonton, water depths were approximately 10 m.

## UPPER CONTACT

The top of the Cooking Lake platform beneath and away from the Redwater reef complex and that of the time-synchronous Duvernay unit in the open-marine embayment is abrupt. In many, but not all sections, the sediment below this surface shows evidence of lithification prior to the deposition of the overlying sediment. This interformational break, or discontinuity surface, records a break in sedimentation between the platform and reef or stratigraphic-equivalent off-reef sediment.

The discontinuity surface is widespread and overlies many different platformal facies. Deposits overlying the surface vary from argillaceous skeletal lime wackestones in off-reef positions to skeletal lime wackestones and floatstones containing fragmented tabular stromatoporoids in reef sections. These latter lime wackestones and floatstones grade up within a metre or two into *in situ* tabular stromatoporoid boundstones.

The visual nature of this surface in core is illustrated in Plate 9. Plate 9A shows the contact beneath the Redwater reef. Plates 9B, 9C and 9D illustrate the contact in off-reef sections. Thin-section and peel photomicrographs of this surface are shown in Plates 10A, 10B and 10C.

The macroscopic configuration of this surface is planar to hummocky and, on a microscale, smooth to jagged. Vertical burrows extend down from some of these surfaces (Plates 9A and 9B). The samples in Plates 9A, 9B and 9C all exhibit evidence of syndepositional lithification, such as razor sharp surfaces, truncation of underlying grains, scalloped surfaces and pyrite mineralization. The truncation of cements as well as grains is visible on the photomicrograph in Plate 10A. A possible boring is illustrated in Plate 10B. The contact shown in Plate 9D, however, is fuzzy and shows no evidence of syndepositional lithification.

Two environments of lithification, marine and subaerial, are possible. The following macro and micro-scale evidence indicate marine lithification along the top surface of the platform. Firstly, the pattern of varying degrees of lithification is most easily explained by marine processes. Marine lithification is favoured in Holocene settings by better water circulation, coarser sediment and less lime mud or argillaceous matter. Variations result in patchy cementation. Surficial subaerial lithification, on the other hand, is triggered by evaporation of capillary water. Given a relative sea-level drop down to a specified level, surficial lithification should be continuous in all emergent carbonate sediment. As an example, witness the ubiquitous subaerial lithification of the youngest Pleistocene deposits now submerged on the Great Bahama Bank. Secondly, the depth of lithification below the surface is only a couple of centimetres. This is most evident along the walls of vertical burrows extending down from the surface (Plates 9A and 9B). This thin zone of lithification corresponds to that of many documented Holocene and ancient marine-cemented hardgrounds. Thirdly, where cements are present, the crystals are bladed, similar to high-magnesium calcite cements associated with some Holocene hardgrounds, and in places are covered by internal marine sediment (Plate 10C). In this sample, the first-generation cements have grown competitively and hence synchronously with the internal marine sediment.

Stratigraphic and petrographic evidence indicate syndepositional marine lithification of the top of the Cooking Lake platform. The widespread, though not continuous, nature of this hardground, its coincidence with a major cycle contact and the occurrence of overlying deeper water sediments all indicate lithification took place during the "drowning" of the Cooking Lake platform. Plate 10D shows an analogous hardground capping a second-order cycle in the Middle Member. Similar "drowning hardgrounds" were initially recognized by Purser (1969).

## PLATFORM EVOLUTION AND REEF INCEPTION

Division of the Cooking Lake Formation into major cycles of sedimentation permit the sequential reconstruction of the platform, as depicted in Figure 10. The major cycles are grouped into larger sets of cycles corresponding to the Lower, Middle and Upper Members of the Cooking Lake Formation for the purpose of the following discussion.

### *Lower Member*

Deposition of the Cooking Lake Formation began during a very gradual lowering of relative sea level. During this lowering, crinoidal mudstones and wackestones prograded basinward over more argillaceous sediments of the Waterways Formation, establishing an open-marine carbonate setting in the Redwater area. At the same time, the upslope portion of the Waterways ramp to the east became emergent and cemented. Progradation of the open-marine carbonates continued even as sea level rose gradually and marine waters covered the previously emergent portion of the Waterways ramp.

Ensuing progradation led to further shallowing. In the area around Redwater and to the south, peritidal conditions became established. In the area to the immediate north and east of Redwater, conditions remained submergent. At the end of this stage of growth, the profile of the platform was one of a gently sloping ramp from the Redwater Shoal to the immediate north and east.

### *Middle Member*

A rapid rise in relative sea level caused marine inundation of the widespread tidal flats around Redwater. During the initial open-marine incursion, stromatoporoidal limestones formed along a very subtle break-in-slope immediately north and east of Redwater. The deposition of these limestones marks the initiation of a shallow open-marine embayment that extended into to the Redwater area from the north.

A subsequent, possibly even more rapid increase in the rate of sea-level rise led to inundation of open-marine waters onto the Redwater Shoal. A stromatoporoid-constructed, barrier reef formed along the periphery of this extended embayment. The shallow-water, reef fringed platform gradually built upwards and then prograded back, toward the marine embayment, over open-marine lime wackestones.

Another rapid rise in relative sea level again led to open-marine inundation of the Redwater Shoal. Shallow-shoal lime grainstones were deposited over widespread areas around and to the south of Redwater, while deeper water pelloidal lime packstones and grainstones were deposited along the fringe of the shallow open-marine embayment to the immediate north and east of Redwater.

At the end of this stage of growth, the profile of the Cooking

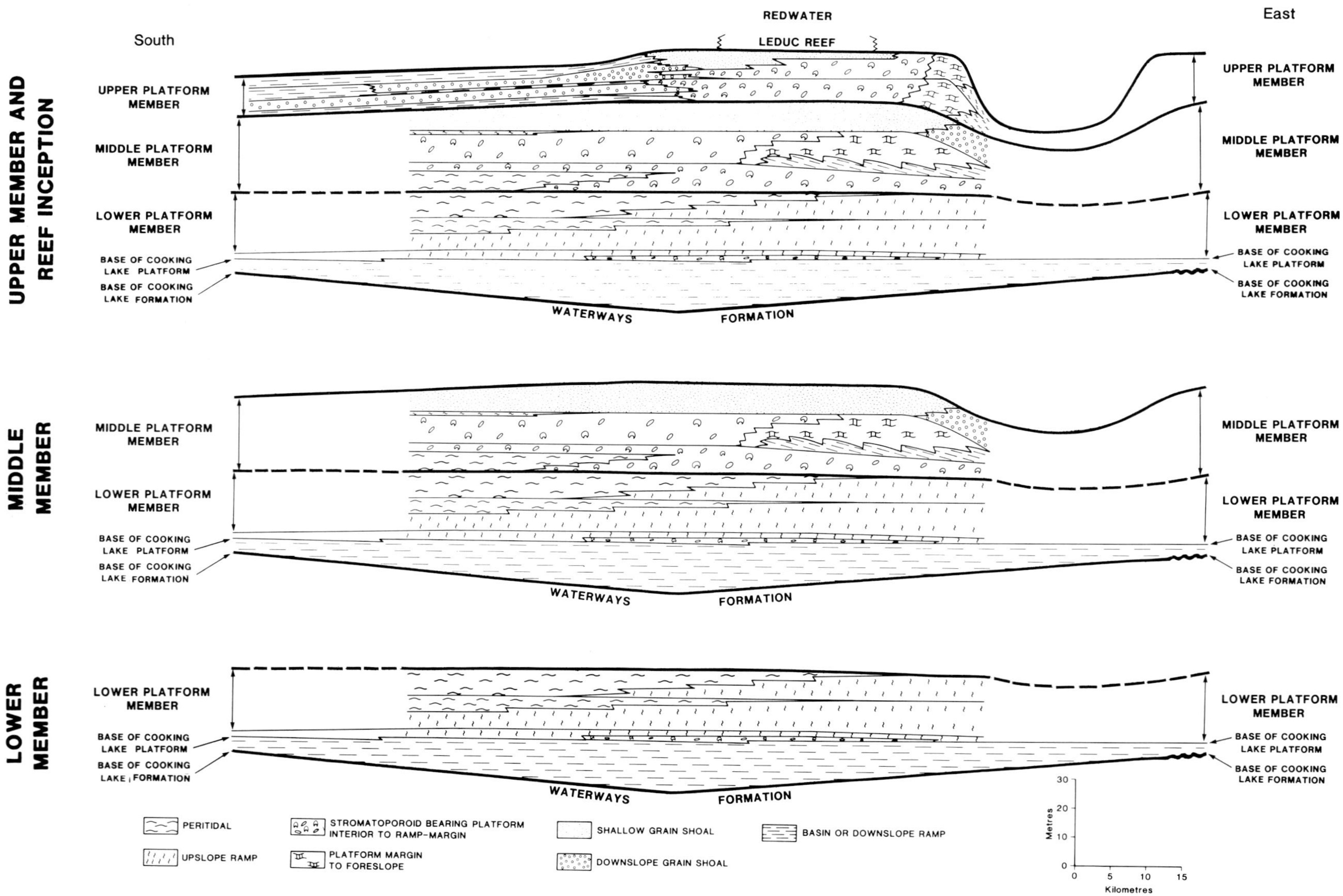


Figure 10. Sequential evolution of the Cooking Lake Formation and its relationship to Leduc reef inception and localization.

Lake platform was one of a broad platform, incised into by a shallow marine embayment. Slopes along the margin of this embayment were very gentle, less than one degree. Water depth along the western periphery of this embayment was approximately 12 m.

### *Upper Member*

Relative sea level again rose quickly and more open-marine waters covered the Redwater Shoal. During this rise, stromatoporoid boundstones formed along the margin of the marine embayment, creating an abrupt platform margin with foreslopes of a few degrees. Deeper water basinal limestones of the Duvernay Formation were deposited in the embayment.

Suitable water circulation for habitation and growth of other stromatoporoids became confined to an area up to 42 km wide behind the platform margin. Stromatoporoid deposits and overlying shoal-water lime grainstones in this area were able to grow up to or very close to sea level. The area to the south, where stromatoporoids could not sustain growth, became more deeply submerged as a result of the slower sedimentation rates. Continued small increments of relative sea-level rise led to even deeper water conditions and the deposition of lime mudstones and wackestones in the southern part of the study area. As a consequence, shallow-water platform sedimentation became confined to an area extending from a position approximately 11 km south of the Redwater complex, north and east to the marine embayment. Thus, as a result of selective carbonate accretion in the area behind the margin of an open-marine embayment, a platform high was formed.

At the end of this stage of growth, the profile of the Cooking Lake was one of a platform incised into by a relatively deep open-marine embayment with water depths of at least 25 m. To the south and west of this embayment, carbonates built a shoal up to or near sea level. South of the Redwater Shoal, the sea floor gradually sloped in a ramp-like manner down to water depths of approximately 10 m.

### *Platform Termination and Reef Inception*

Another rapid rise in sea level caused drowning of the Cooking Lake platform. Very low rates of carbonate supply and sediment stability during this deepening phase allowed widespread cementation and hardground formation at the top of the platform. Eventually, stromatoporoids selectively recolonized the highest portion of the underlying Cooking Lake platform. These stromatoporoids formed the initial deposit of the reef complex at Redwater. Lime mud covered lower lying areas of the platform in off-reef settings.

The area of the reef at Redwater corresponds to a slightly reduced portion of the underlying Cooking Lake high (see Figure 6). Thus, even though there was a break in deposition between that of the platform and the reef, the localization and shape of the reef was largely inherited from that of the underlying platform.

## SUMMARY

This study was undertaken in an attempt to answer two general questions. Firstly, why do large reef complexes grow where they do? Secondly, why do large reef complexes grow when they do? This investigation of the Cooking Lake Formation in east-central Alberta indicates that an improved understanding of the origin and localization of large platform reefs can be attained by intensive examination of platform rocks beneath and away from these buildups.

The results of this study indicate that platform reefs form when relative sea-level rises coincide with the development of localized accretionary stages of platform growth. Significant sea-level rises controlling platform evolution and reef inception appear to be in increments of approximately 10 m. Smaller increments of relative sea-level rise are, however, important in controlling the configuration of the platform in less competent, deeper water environments. These smaller increments of sea-level rise helped create the necessary relief on the south or "backside" of the Redwater Shoal needed to localize later reef growth.

The formation of the discrete Redwater Shoal is the final stage in the development of the Cooking Lake platform. Platform evolution reflects both sea-level and biogenic depositional controls. During sea-level rises marking the onset of the initial two cycles of the Middle Member and the initial cycle of the Upper Member, stromatoporoids selectively colonized and accentuated a pre-existing topographic high. As a result of this platform growth, a very subtle and broad lower Cooking Lake topographic high became progressively more pronounced and localized during platform evolution. By the end of Cooking Lake deposition, a discrete triangular-shaped topographic high up to 42 km across existed on the platform adjacent to a marine embayment that was incised into the platform.

A relative sea-level rise coincident with this platform high initiated reef growth at Redwater. Surrounding low-lying areas of the platform became more deeply submerged and the site for deposition of basinal limestones. Preceding sea-level rises, of the same magnitude but not coincident with localized buildup stages of platform growth, did not localize shallow-water sediments but merely initiated new cycles of widespread platform sedimentation.

PLATE 1 :
Waterways - Cooking Lake contact, core slab 14-5-59-15W4M, 3074 feet

Sharp contact between the Waterways and Cooking Lake formations.
The Waterways is a medium gray argillaceous lime mudstone.
The basal Cooking Lake is a lime packstone containing algally coated intraclasts, abundant Charophytes,common brachiopod fragments ,some crinoid columnals and a few Solenoporid algae.
The intraclasts are rounded micritic grains with some skeletal particles and were derived from the underlying Waterways.
Some intraclasts contain truncated skeletal fragments along their margins,indicating lithification prior to erosion.

PLATE 2

A. Nodular crinoidal wackestone, basal part of Cooking Lake Formation. Core slab. Bar length equals 3 cm. 12-28-57-21W4, 4172 feet.

B. Crinkly stromatolitic laminated limestone with alternating bituminous and sediment laminae. Core slab. Bar equals 3 cm. 1-22-57-22W4, 4399 feet.

C. Flat-pebble breccia overlying stromatolitic laminated limestone. Core slab. Bar length equals 3 cm. 5-36-56-21W4, 4089 feet.

D. Planar fenestral limestone with alternating laminae of lime mudstone and non-skeletal grainstone and packstone. Core slab. Bar equals 3 cm. 5-36-56-21W4, 3957 feet.

a
b
c
d

PLATE 3

A. Extensively burrow-mottled pelletal lime mudstone. Core slab. Bar length equals 3 cm. 1-22-57-22W4, 4093 feet.

B. Pelletal lime mudstone. Note mashed pellets, where original contacts are obscured, and in some cases obliterated. White material is mainly sparry calcite. Thin section, plane light. Bar length equals 740 microns. 12-28-57-21W4, 4095 feet.

C. Graded intraclast grainstone-packstone overlying pelletal lime mudstone. Core slab. Bar length equals 3 cm. 12-28-57-21W4, 4122 feet.

D. Interstratified intraclast grainstones and pelletal lime mudstones. Note cross-lamination in grainstones. Core slab. Bar length equals 3 cm. 12-28-57-21W4, 4126 feet.

a
b
A
A
B
B
B
c
d

PLATE 4

A. Nodular ostracod-bearing mudstone. Note compacted burrows between nodules (top arrow), and uncompacted burrows within nodules (lower two arrows). Carbonaceous matter has been concentrated through pressure solution in areas between nodules. Core slab. Bar length equals 3 cm. 1-22-57-22W4, 4463 feet.

B. Nodular gastropod-bearing wackestone. Note pinch and swell stratification. Carbonaceous matter is concentrated between elongate nodules. Core slab. Bar length equals 3 cm. 12-28-57-21W4, 4134 feet.

C. Irregular nodular lime wackestone. Core slab. Bar length equals 3 cm. 14-5-59-15W4, 2976 feet.

D. Intraclast grainstone. Note highly irregular shape of micritic intraclasts. Some skeletal grains are also present. Thin section, plane light. Bar length equals 2.66 mm. 5-36-56-21W4, 4006 feet.

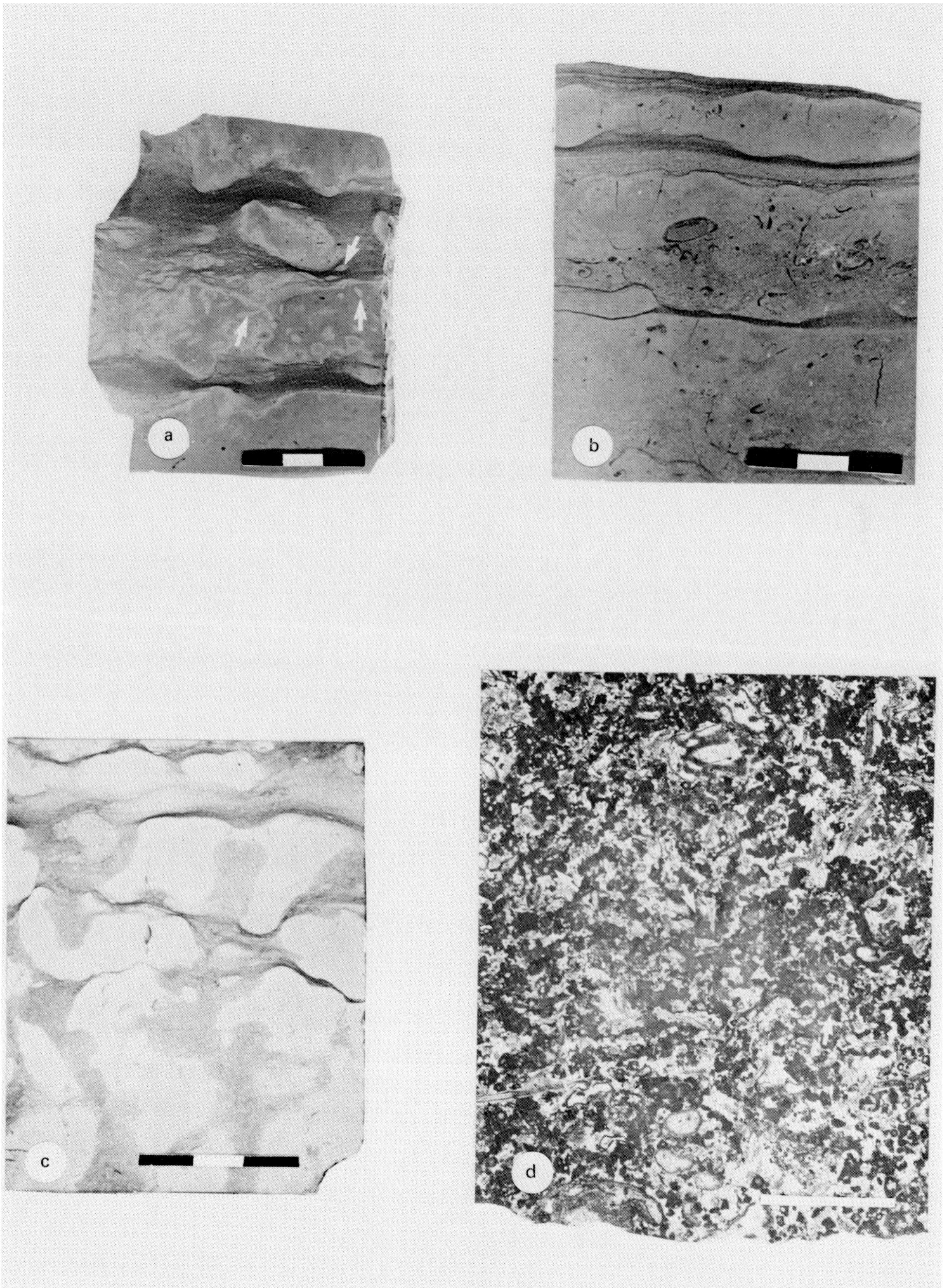
a
b
c
d

PLATE 5

SHALLOW-SHOAL LIME GRAINSTONES

A. Intraclast grainstone. Intraclasts consist of micrite with some calcispheres and Parathuramminid foraminifera. Note rounded shape of intraclasts, as pointed to by arrows. Marine isopachous cements coat grains. 1-22-57-22W4, 4307 feet.

B. Shallow lime grainstone. Grains include cemented-grain clasts (A), small oolites (B), superficial oolites (B), coated lumps (C) and pellets. Individual particles within clasts are coated by marine isopachous fibrous and micrite cements. Thin section, plane light. Bar length equals 1.85 mm. 13-16-58-22W4, 4062 feet.

C. Articulated spirifer brachiopod, with preserved brachidia, in upright presumably growth position in lime grainstone. Note lumps (A and B) and calcispheres (C). Thin section, plane light. Bar length equals 1.53 mm. 12-28-57-21W4, 3991 feet.

D. Lime Grainstone with long and linear cemented-grain clasts (arrows). Core slab. Bar length equals 3 cm. 13-16-58-22W4, 4067 feet.

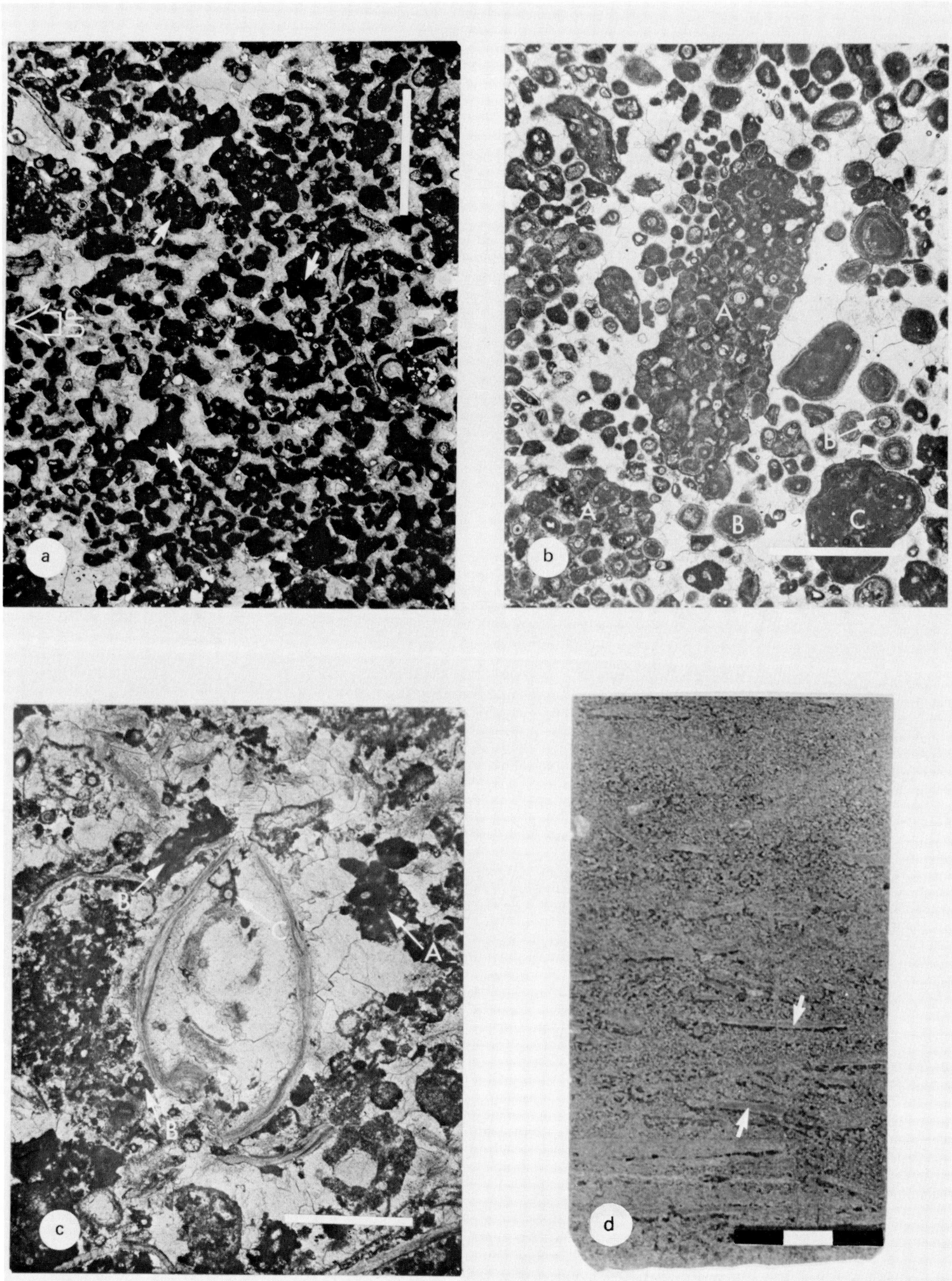
a
A
B
C
b
B
C
A
B
c
d

PLATE 6

A. Grainstone with ovoid fenestrae in bottom half. Arrows point to spar-filled fenestrae. Note hummocky upper contact of fenestral grainstone. Core slab. Bar length equals 3 cm. 12-28-57-21W4, 3986 feet.

B. Nodular peloidal lime packstone with carbonaceous partings. Core slab. Bar length equals 3 cm. 10-3-54-22W4, 4672 feet.

C. *Amphipora* rudstone. Core slab. Bar length equals 3 cm. 5-36-56-21W4, 3940 feet.

D. Bulbous stromatoporoid rudstone. Note borings in bulbous stromatoporoid at top of slab. Core slab. Bar length equals 3 cm. 5-36-56-21W4. 4052 feet.

a
b
c
d

PLATE 7

A. Abraded irregular stromatoporoid rudstone. Irregular stromatoporoids (A and B) are set in a lime grainstone matrix. Core slab. Bar length equals 3 cm. 6-36-58-23W4, 4240 feet.

B. Tabular stromatoporoid boundstone. Core slab. Bar length equals 3 cm. 6-36-58-23W4, 4259 feet.

C. Cylindrical stromatoporoid rudstone. Core slab. Bar length equals 3 cm. 14-5-59-15W4, 2946 feet.

D. Tabular stromatoporoid (B) encrusting hemispherical stromatoporoid (A) and rugose cup coral (C), tabular stromatoporoid limestone facies. Core slab. Bar equals 3 cm. 6-36-58-23W4, 4242 feet.

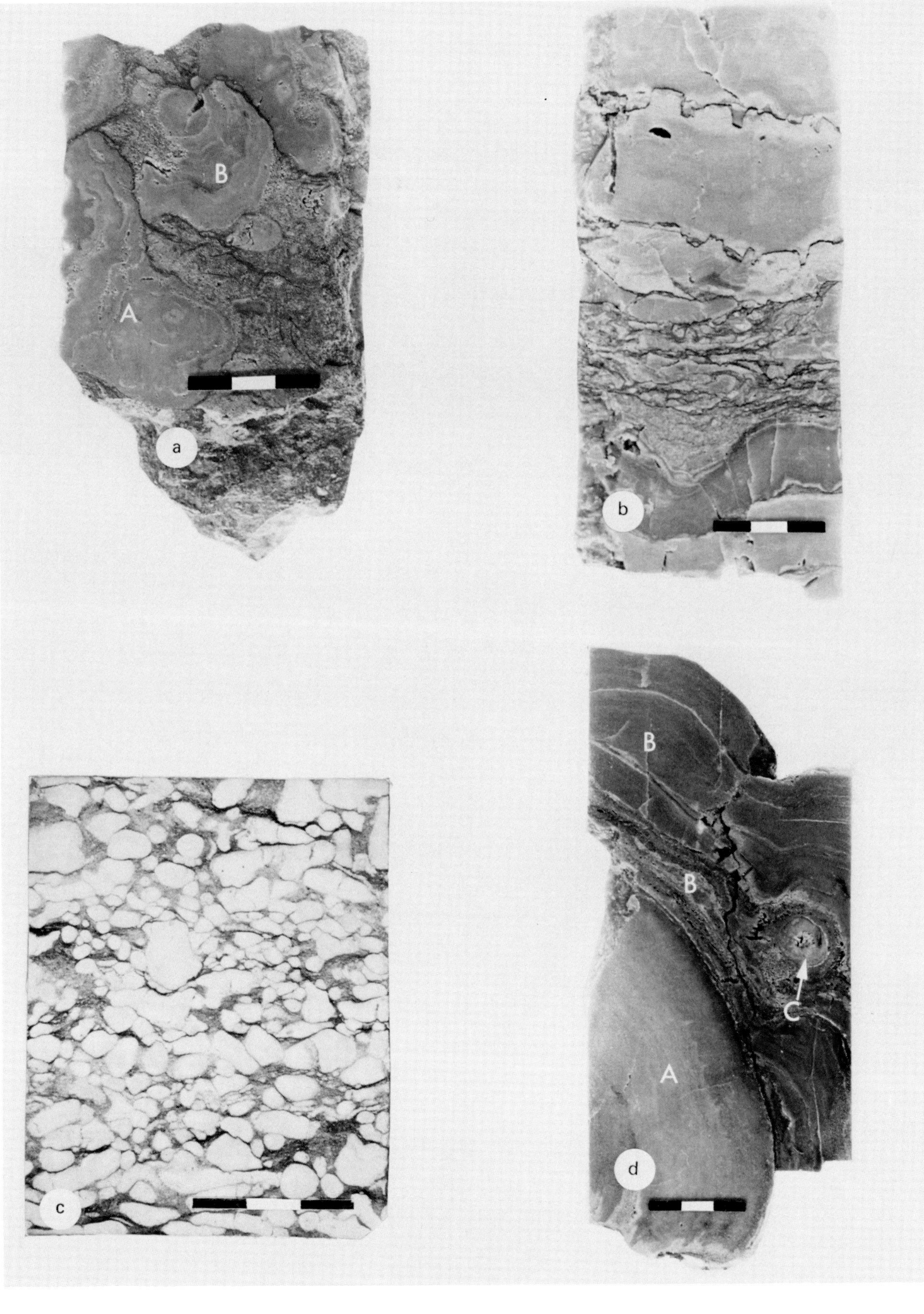
a
A
B
b
c
d
A
B
B
C

PLATE 8

A. Tabular stromatoporoid rudstone. Tabular stromatoporoids (A) and massive colonial corals (B) are fragmented. Note dark carbonaceous matrix. Core slab. Bar equals 3 cm. 12-28-57-21W4, 4024 feet.

B. Nodular brachiopod wackestone. Note concentrations of wispy stylolites and carbonaceous matter between nodules (arrows). Core slab. Bar equals 3 cm. 12-28-57-2W4, 4038 feet.

C. Nodular brachiopod-crinoid wackestone. Arrows point to nodules. Core slab. Bar length equals 3 cm. 12-28-57-21W4, 4042 feet.

D. Argillaceous skeletal wackestone. Note abundant brachiopod fragments. Light colour due to etching, emphasizing argillaceous content. Core slab. Bar equals 3 cm. 15-14-48-23W4, 5760 feet.

A
A
B
a
b
c
d

PLATE 9

DISCONTINUITY SURFACE AT TOP OF COOKING LAKE PLATFORM

A. Erosional surface beneath Redwater reef. Surface truncates both sediment (arrow) and stromatoporoid (A), indicating syndepositional lithification. Another small stromatoporoid encrusts the eroded stromatoporoid. Note that the walls of only the upper 2 cm of the burrow are sharply delineated (B). Walls along the lower portion of the burrow are fuzzy and gradational (C). Core slab. Bar length equals 3 cm. 5-36-56-21W4, 3924 feet.

B. Hardground in off-reef position. Note pyrite mineralized sharp surface (arrows). Burrow extends down from discontinuity surface. Note walls of only the upper 2 cm of the burrow are sharply delineated. Core slab. Bar length equals 3 cm. 4-1-59-21W4, 3912 feet.

C. Hardground in off-reef position. Arrows point to sharp, jagged nature of surface. Core slab. Bar length equals 3 cm. 10-33-55-23W4, 4764 feet.

D. Unlithified surface in off-reef position (arrow). Argillaceous wackestones overlie clean limestone. Core slab. Bar length equals 3 cm. 11-16-47-19W4, 5032 feet.

A
B
C
a
B
b
A
A
c
d

PLATE 10

A. Hardground at top of Cooking Lake platform in off-reef position. Surface truncates intraclast (A), bivalves (B) and cement (C). Thin section, plane light. Bar length equals 157 microns. 4-1-59-21W4, 3912 feet.

B. Hardground at top of Cooking Lake platform beneath reef. Possible boring (A) truncates pellets (B) and very fine spar cement. Photomicrograph of peel. Bar equals 392 microns. 12-28-57-21W4, 3915 feet.

C. Rugose coral beneath top surface of Cooking Lake (A) in off-reef position. Coral calyx is successively filled by: (1) marine bladed spar cement (B), (2) internal marine sediment (C) and (3) later equant calcite cement (D). Bladed cements have an asymmetric distribution in large open chambers (B). Cements line upper but not lower surfaces. Identical cements have an isopachous distribution in protected small chambers (below B). These occurrences indicate a competitive emplacement of both cements and sediment and hence imply syndepositional lithification in a water saturated (phreatic) marine environment. Thin section, plane light. Bar length equals 3.83 mm. 10-33-55-23W4, 4764 feet.

D. Hardground capping cycle in Middle Platform Member. Note truncation of brachiopods (A) along this surface. Core slab. Bar length equals 3 cm. 4-7-59-20W4, 3943 feet.

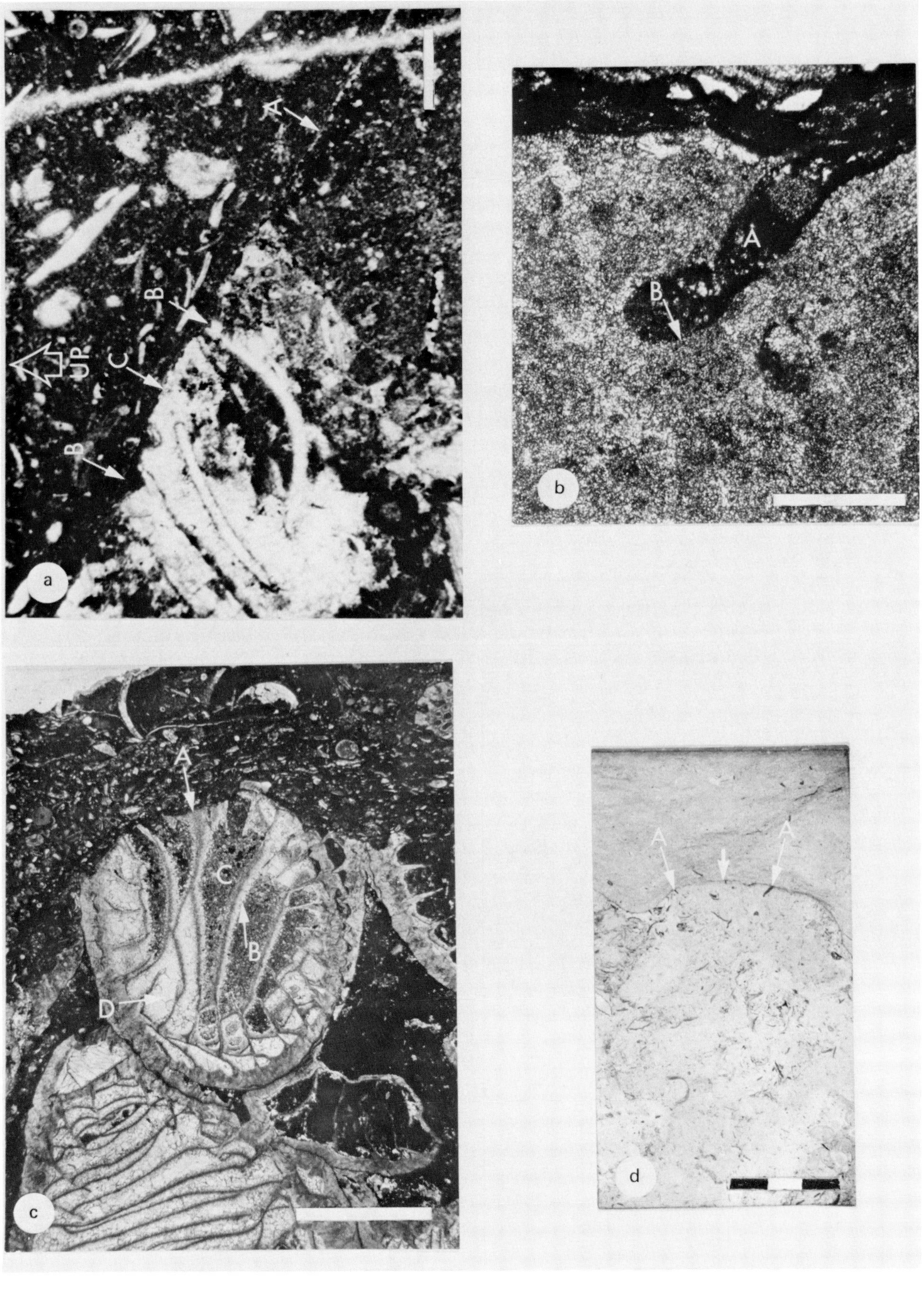
A
B
B
C
UP
a
A
B
b
A
C
B
D
c
A
A
d

## REFERENCES

ANDRICHUK, J.M., 1958a, Stratigraphy and facies analysis of Upper Devonian reefs in Leduc, Stettler, and Redwater areas, Alberta: American Association Petroleum Geologists Bulletin, v. 42, p. 1-93.

ANDRICHUK, J.M., 1958b, Cooking Lake and Duvernay (Late Devonian) sedimentation in Edmonton area of central Alberta, Canada: American Association Petroleum Geologists Bulletin, v. 42, p. 2189-2222.

ANDRICHUK, J.M., 1961, Stratigraphic evidence for tectonic and current control of Upper Devonian reef sedimentation, Duhamel area, Alberta, Canada: American Association Petroleum Geologists Bulletin, v. 45, p. 612-632.

COOK, H.E., 1972, Miette platform evolution and relation to overlying bank ("reef") localization, Upper Devonian, Alberta: Bulletin of Canadian Petroleum Geology, v. 20, p. 375-411.

DOLPHIN, D.R., AND KLOVAN, J.E., 1970, Stratigraphy and paleoecology of an Upper Devonian carbonate bank, Saskatchewan River crossing: Bulletin of Canadian Petroleum Geology, v. 18, p. 289-331.

DRAVIS, J., 1979, Rapid and widespread generation of Recent oolitic hardgrounds on a high energy Bahamian platform, Eleuthera Bank, Bahamas: Journal of Sedimentary Petrology, v. 49, p. 195-208.

DUNHAM, R.J., 1962, Classification of carbonate rocks according to depositional texture, *in* Ham, W.E., ed., Classification of carbonate rocks: American Association Petroleum Geologists, Memoir no. 1, Tulsa, Oklahoma, p. 108-121.

EMBRY, A.F., AND KLOVAN, J.E., 1971, A Late Devonian reef tract on northeastern Banks Island, Northwest Territories: Bulletin of Canadian Petroleum Geology, v. 19, p. 730-781.

KLOVAN, J.E., 1964, Facies analysis of the Redwater reef complex, Alberta, Canada: Bulletin of Canadian Petroleum Geology, v. 12, p. 1-100.

LAPORTE, L.F., 1967, Carbonate deposition near mean sea-level and resultant facies mosaic: Manlius Formation (Lower Devonian) of New York State: American Association Petroleum Geologists Bulletin, v. 51, p. 73-101.

MOUNTJOY, E.W., 1967, Factors governing the development of the Frasnian, Miette, and Ancient Wall reef complexes (banks and biostromes), *in* Oswald, D.H., ed., International Symposium on the Devonian System: Alberta Society Petroleum Geologists, v. 2, p. 387-408.

PURSER, B.H., 1969, Syn-sedimentary marine lithification of Middle Jurassic limestones in the Paris Basin: Sedimentology, v. 12, p. 205-230.

SHINN, E.A., 1968, Practical significance of birdseye structures in carbonate rocks: Journal Sedimentary Petrology, v. 38, p. 215-223.

TEBBUTT, G.E., CONLEY, C.D., AND BOYD, D.W., 1965, Lithogenesis of a distinctive Carbonate rock fabric: Contributions to Geology, University of Wyoming, Laramie, v. 4, No. 1, 13 p.

WALLS, R.A., MOUNTJOY, E.W., AND FRITZ, P., 1979, Isotopic composition and diagenetic history of carbonate cements in Devonian Golden Spike reef, Alberta, Canada: Geological Society of America Bulletin, v. 90, p. 963-982.

WENDTE, J.C., 1974, Sedimentation and diagenesis of the Cooking Lake platform and Lower Leduc reef facies, Upper Devonian, Redwater, Alberta: Unpublished PhD. Dissertation, University California at Santa Cruz, 222 p.

CHAPTER 4

# EVOLUTION OF THE JUDY CREEK COMPLEX, A LATE MIDDLE DEVONIAN ISOLATE PLATFORM-REEF COMPLEX IN WEST-CENTRAL ALBERTA

J.C. WENDTE

## INTRODUCTION

Reef complexes situated on carbonate platforms contain a disproportionate amount of hydrocarbon reserves. In Western Canada, approximately half of the oil and gas pools with recoverable reserves greater than 16 X $10^6 m^3$ (100 million barrels) of oil or its gas equivalent are Devonian reef complexes. Cores and wireline logs from producing and outlying wells from these pools provide perhaps an unparalleled opportunity to examine the evolution of reef complexes in all three dimensions.

Most of the hydrocarbon bearing Devonian reef complexes in the WCSB are isolate features situated on areally widespread carbonate platforms. They include the Middle Devonian (late Givetian) Swan Hills reef complexes in west-central Alberta and the Kee Scarp buildup at Norman Wells in the Northwest Territories; and the Late Devonian (Frasnian) Leduc reef complexes in both east-central and west-central Alberta. The previous chapter dealt with the evolution of the underlying carbonate platform and its relationship to reef inception and localization. This chapter focuses primarily on the evolution of isolate platform-reef complexes. The Swan Hills buildup at Judy Creek in west-central Alberta is used as an example to illustrate the evolutionary development of these complexes.

Despite the importance of these deposits, only a few papers dealt with the evolution of Swan Hills or time-equivalent buildups in the WCSB up until the 1980's. Of the pre-1980 papers, Fischbuch's (1968) paper was the most directly concerned with determining the evolution of platform-reef complexes. He studied six Swan Hills reef buildups and divided them into as many as nine divisions based on distribution and type of facies, fossils and occurrence of green shales. He interpreted each division to be a single stage of reef growth, separated from others by a period of submarine or subaerial erosion. In the opinion of this author, this paper marked a distinct advancement in the understanding of the evolution of reef complexes from previously published papers.

In 1982, Wendte and Stoakes published the results of their initial study of the Judy Creek reef complex. This paper differed from all previous ones in that evolutionary divisions were based solely on the recognition and delineation of depositional cycles. The "megacycles" defined in that paper did not correspond to growth stages of other papers concerned with the evolution of Swan Hills reefs. In retrospect, the significance of this paper in my opinion was not that it provided a comprehensive model of the evolution of the Judy Creek reef complex, but that it laid out a methodology that could be used to reconstruct the evolution of platform-reef complexes. After 1982, the author continued to examine the Judy Creek complex while employed by Esso Resources Canada. A three dimensional, sequence-facies model of the buildup was ultimately constructed. This model incorporated information gained from the use of wireline logs from approximately 250 wells and the examination of about 6,000 m of core from 100 of these wells. The evolution of the Judy Creek complex presented here is based on that study.

The incentive to develop a methodology to reconstruct the evolution of platform-reef complexes and to build three-dimensional models was not only academic. Because both Stoakes and I were employed by an oil company that operated many of the giant Devonian reef pools, it behooved us to develop a means to accurately portray and predict reservoir continuity in these reefal buildups. Because most of the porosity in the majority of Esso's reefal pools is primary interparticle and intraparticle and because the lithologies are (fortunately!) limestone, an overall excellent correspondence exists between depositional facies and reservoir quality. Shifts in depositional facies at cycle contacts matched corresponding shifts in porosity and permeability belts. This correspondence between depositional facies and reservoir quality was demonstrated in the 1982 paper.

Since the mid 1980's, Esso-employed geologists have constructed three-dimensional models of other Devonian limestone reef reservoirs. These include the Swan Hills Snipe Lake pool (Muir and others, 1990), the time-equivalent Norman Wells pool in the Northwest Territories (Muir and others, 1984), the Leduc Redwater pool (Wendte and Callow, 1987) and the Leduc Golden Spike pool (Heidgerken and others, 1989). These examples will be illustrated in the following chapters.

## GEOLOGICAL HISTORY AND FIELD SETTING

Judy Creek is one of many large Swan Hills atoll reef complexes in central Alberta which contain large petroleum reserves. These reef complexes are situated on a Swan Hills carbonate platform and intertongue with, and are overlain by, deeper water basinal limestones of the Waterways Formation (Figure 1). Platform carbonates overlie interbedded anhydrites and carbonates of the Fort Vermilion Formation, a succession about 10 m thick, which in turn overlies shales of the Watt Mountain Formation.

The underlying platform is an areally extensive, shallow-marine carbonate. In the immediate area surrounding the Judy Creek reef complex, the platform unit is approximately 35 m thick. As shown in Figure 2, it covers approximately 13,000 $km^2$ (5,000 square miles) along the western margin of the Alberta Basin, approximately 240 km (150 miles) northwest of Edmonton. Topographically high areas of the platform provided the sites for subsequent reef growth. This pattern of localization has been observed in the

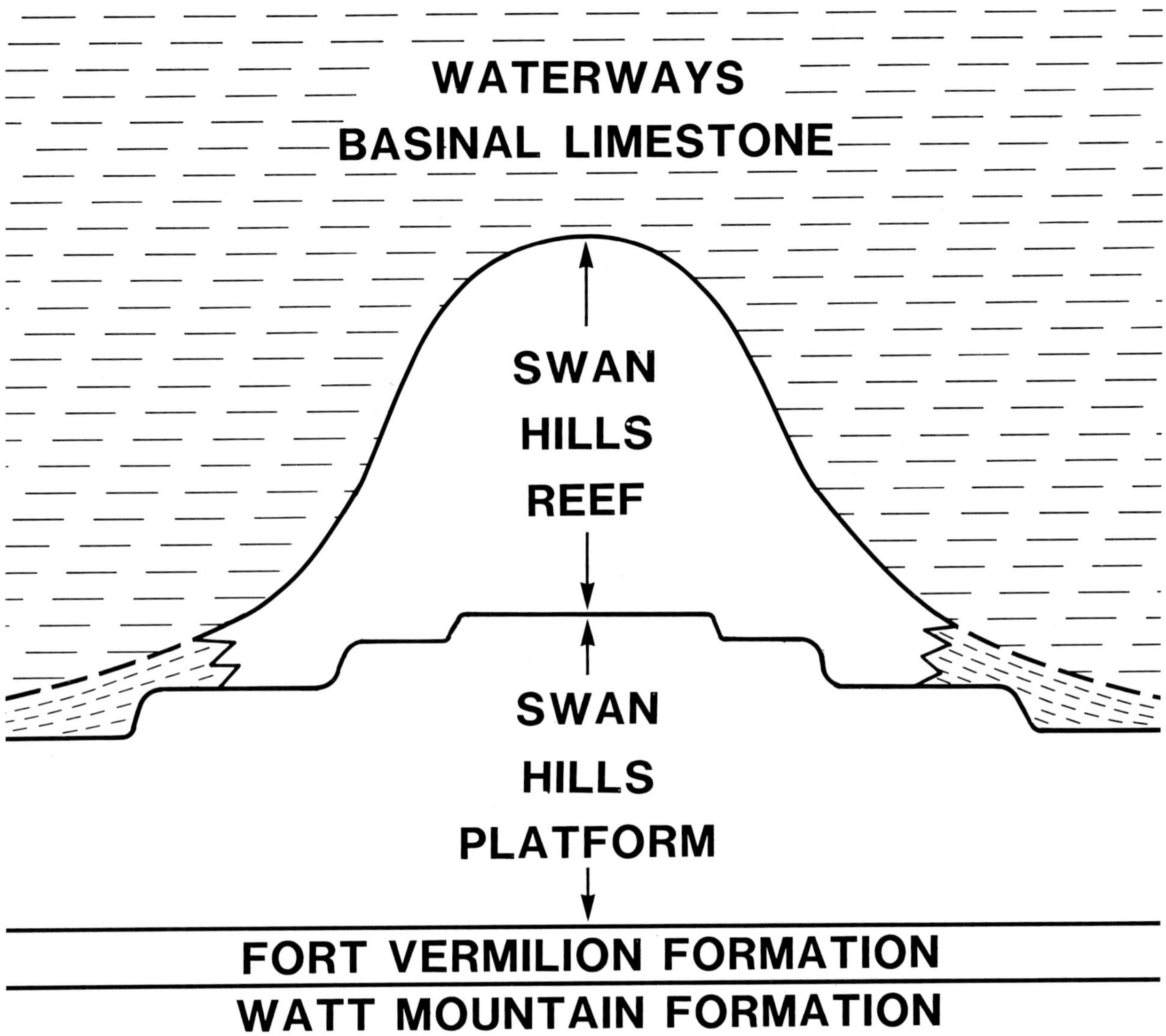


Figure 1. Swan Hills reef sequence.

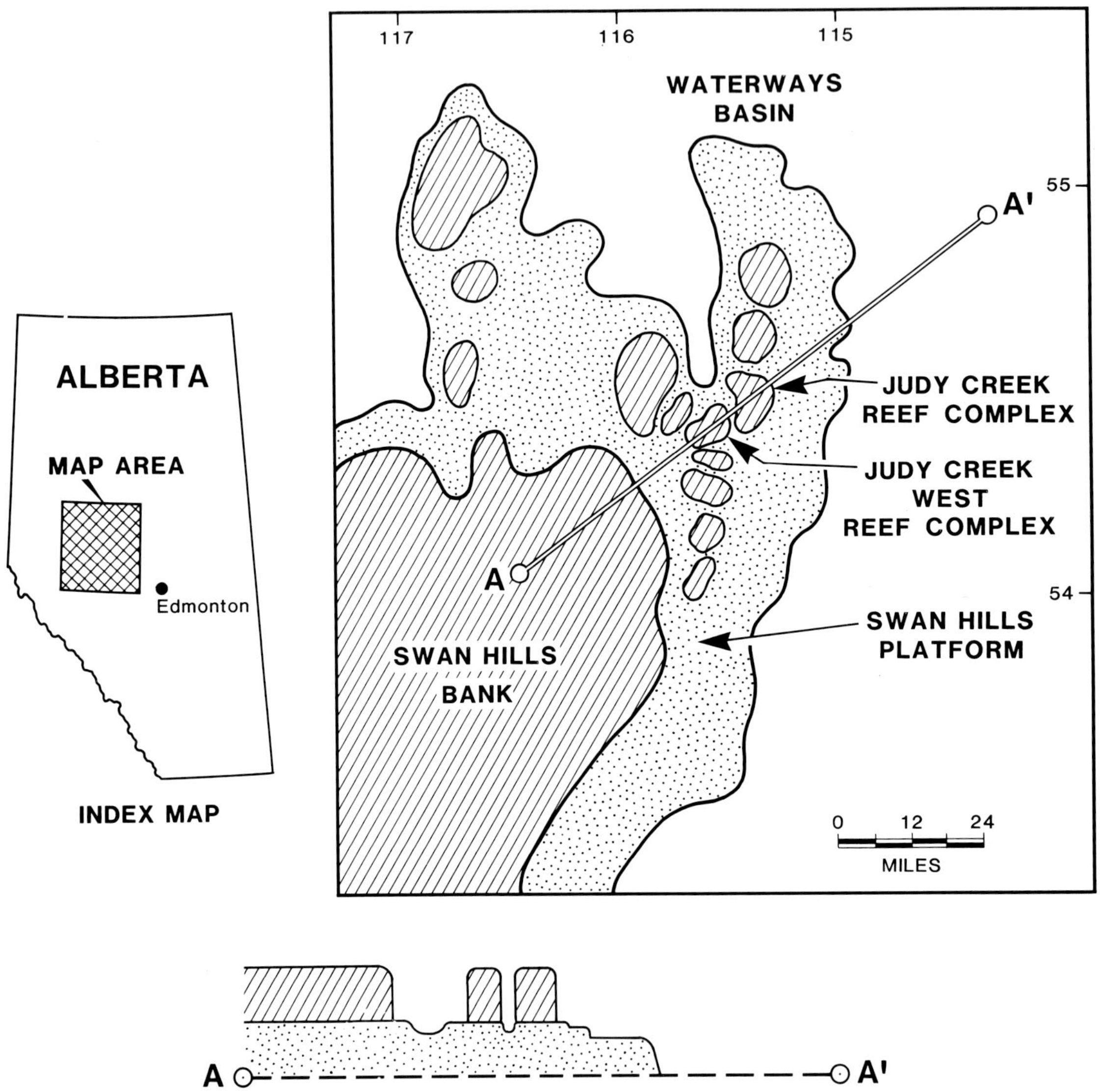


Figure 2. Location and paleogeography of the Swan Hills area and diagrammatic topographic profile.

Leduc reefs of central Alberta (see previous chapter). The Judy Creek reef complex occurs towards the eastern margin of this platform, the northeast side facing the prevailing northeasterly winds (Murray, 1966 and Leavitt, 1968). It is separated from the Judy Creek West complex by a deep-water channel extending from the basinal embayment to the northwest. Shallow-water platform sedimentation continued further to the southwest on a regional Swan Hills bank during the same time frame as reef growth (Figure 2).

The Judy Creek reef complex is up to 73 m thick and covers 122 $km^2$ (47 square miles). It lies at a present burial depth of around 2,600 m with a structural dip to the southwest of 8 m per km. About 75 percent of the reef complex was oil filled with free formation water occurring only in the downdip, southwestern part of the complex. Underlying platform carbonates are mainly tight. Jardine and others (1977) have summarized the production history of this field. The field was discovered and put on production by Imperial Oil in 1959. Due to ineffective aquifer support, reservoir pressures soon declined and a downdip, peripheral waterflood was initiated in 1962. As pool production increased from 4,700 $m^3$ (30,000 barrels) per day in the late sixties to 11,130 $m^3$ (70,000 barrels) per day in the early seventies, the ineffectiveness of the peripheral waterflood became apparent. Extreme pressure gradients developed. By early 1973, reservoir pressure close to the injection wells was about 1,000 psi above the original reservoir pressure, whereas, some internal areas of the pool were below the bubble point. Obviously, some permeability barriers prevented the peripheral waterflood from functioning as predicted. In order to gain a better model of reservoir continuity, integrated geological-engineering studies, described by Jardine and others (1977), were conducted. As a result of these studies, an infill pattern waterflood was initiated in 1974. By 1975, pressure gradients drastically declined. In 1979, Esso Resources (Imperial Oil) sought and received approval (in 1980) to put the field on a tertiary miscible $CO_2$ flood. This required the development of a highly sophisticated three-dimensional model of reservoir continuity resulting in the initiation and completion of the present study.

## ENVIRONMENTAL FACIES

Eleven environmental facies displaying distinctive textures, sedimentary structures, fossils and other constituents were recognized. Depositional interpretation of these facies is based on both their lateral and vertical facies position and on comparisons with Recent analogies as well as the before-mentioned compositional attributes. Ten of these facies are depicted on a paleobathymetric profile (Figure 3) and are discussed below, starting in the reef interior and progressing basinward:

1. *Tidal Flat/Beach Limestones* are deposited along the shorelines of islands in the interior reef lagoon. Tidal flat limestones formed along low energy shorelines, above mean low tide, and generally consist of pelletal mudstones to packstones with crinkly cryptalgal

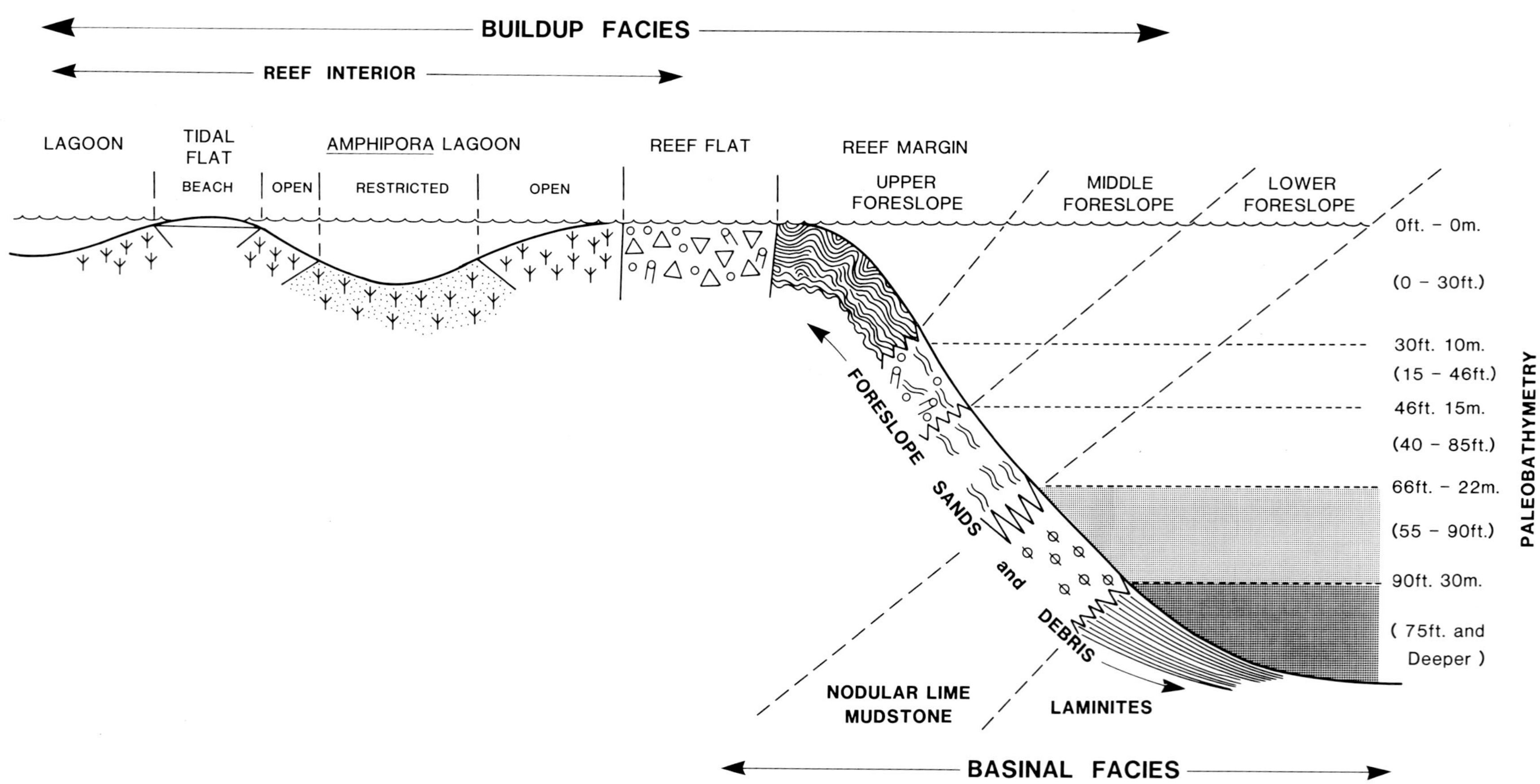


Figure 3. Paleobathymetric profile of Swan Hills reef facies.

laminations, laminar fenestrae and tabular lithoclasts (Plate 1A). Beach limestones formed along high energy shorelines and vary from lime grainstones to stromatoporoid rudstones with a grainstone matrix. Some coarse rudstones contain milky white microstalactitic beach rock cements.

2. *Amphipora Lagoon Limestones* are comprised of light brown to dark carbonaceous rudstones and floatstones containing the stick-like stromatoporoid *Amphipora* (Plate 1B). The dark coloured limestones contain abundant micrite and are interpreted as deeper restricted lagoonal sediments. These darker coloured limestones occur more abundantly toward the centre of the reef complex, where circulation was the most sluggish.

3. *Reef Flat And Shoal-Margin Limestones* consist of an assemblage of brecciated and/or well-washed *Amphipora*, cylindrical and more massive stromatoporoid rudstones and lesser amounts of lime grainstones. More irregular stromatoporoids commonly encrust abraded stromatoporoid clasts. Deposition occurred on shallow flats, immediately behind the reef margin, subject to intensive reworking by major storms.

   Energy conditions along "backstepped" margins and on the leeward western side of the complex were not as high as on the northeastern face of the complex. In these settings, upper foreslope, reef-margin and reef-flat facies exhibit only minor textural and faunal variations and are collectively termed *Shoal-Margin Limestones*. These limestones are predominantly rudstones with abundant branching cylindrical stromatoporoids and varying amounts of abraded thick-tabular and irregular-encrusting stromatoporoids. A light-coloured lime grainstone matrix reflects the moderately high energy conditions in these settings.

4. *Reef Margin/Upper Foreslope Limestones* formed in high energy, wave swept, turbulent conditions and consist of alternating *in situ* thick-tabular stromatoporoids (boundstones) up to a few tens of cm thick and broken-up stromatoporoid rubble (Plate 1C). The thick-tabular stromatoporoids were adapted to an encrusting habitat, generally comparable to that of the coralline red algae *Lithothamnion* in Recent reefs.

5. *Middle Foreslope Limestones* consist of rudstones containing abundant branching cylindrical stromatoporoids, significantly lesser amounts of *Amphipora* and a light to medium brown finer grained matrix (Plate 1D). Higher parts contain thick-tabular stromatoporoids, abraded stromatoporoid clasts and a carbonate sand matrix. Lower parts contain some thin-tabular (wafer) stromatoporoids and have a more micritic matrix. Deposition occurred in well-circulated but lower wave-energy environments similar to that for the staghorn coral *Acropora cervicornis* in Recent reefs.

6. *Lower Foreslope Limestones* comprise darker, carbonaceous and more micritic rudstones and floatstones and minor boundstones with thin-tabular or wafer stromatoporoids and common crinoids and brachiopods (Plate 2A). In most occurrences, the majority of the thin-tabular stromatoporoids are abraded, reflecting an allochthonous origin. Some thin-tabular stromatoporoids, especially in higher parts of the facies belt appear to be *in situ*, growing on a micritic substrate. These wafer stromatoporoids have a growth form similar to some platy corals in Recent deep forereefs. Unlike these corals, that appear to be adapted to reduced light conditions, the wafer stromatoporoids appear to have adapted to growing on a generally micritic substrate where sedimentation rates were low.
7. *Nodular Lime Mudstones* are composed of bioturbated micritic limestones with a characteristic nodular to banded aspect, containing a sparse fauna of crinoids and atrypid brachiopods (Plate 2D). The nodular texture appears to have resulted from a combination of burrowing and selective early cementation accentuated by later compaction. Deposition of these sediments is inferred to have taken place in deeper, low-energy dysaerobic to aerobic environments.
8. *Laminites* are dark unfossiliferous lime mudstones with essentially undisturbed alternating planar millimetre laminations of carbonate and carbonaceous material. These sediments may contain a sparse pelagic fauna of conodonts, stylolinids and tentaculitids. The sediments represent accumulation in deep-water, oxygen-starved euxinic conditions of the surrounding basin.
9. *Foreslope Debris* consists of rudstones and floatstones with a light to medium brown wackestone to commonly a grainstone matrix (Plate 2B). Large bioclasts are abraded, and in places encrusted. These constituents include thick-tabular, encrusting and branching cylindrical stromatoporoids, and branching thamnoporid and massive tabulate corals. These constituents were derived from higher parts of the foreslope and, thus, have an allochthonous origin. In some sections, the matrix becomes less micritic upwards.
10. *Foreslope Sands* are made up of micritized skeletal and pelloidal grainstones and packstones commonly interbedded with micritic basin sediments (Plate 2C). These lime sands were shed from upslope shallow reef environments where they were formed by the physical and biological breakdown of larger reef formers. Foreslope sands do not characterize all depositional cycles but appear to correspond to stages of low reef aggradation and more extensive reworking.
11. *Stromatoporoid Sand Shoal Limestones* include those limestones deposited on a ramp-bounded shoal during the last major phase of reef growth. These deposits include *Amphipora*, cylindrical, bulbous and encrusting

stromatoporoid rudstones and some skeletal and peloidal grainstones. The major difference between this phase and previous reef stages is not as much the composition but the disposition of the facies. Instead of facies corresponding to that of an atoll reef complex, with a well defined reef margin and an interior lagoon, the distribution corresponds to that along a ramp profile. The facies occur in much wider belts and transitions between facies generally are much more gradational. This is especially the case along the northeast face of the shoal where basinward slopes are very gentle.

Two subfacies are recognized on the cross sections in Figures 4 and 5. The *Stromatoporoid Shoal* subfacies corresponds to a higher energy deposit. These limestones are mainly rudstones and floatstones with a packstone to grainstone matrix. Organic constituents include cylindrical, irregular encrusting and bulbous stromatoporoids as well as *Amphipora*. These deposits typify high stratigraphic and marginal positions on the shoal. The second subfacies consists of *Amphipora* rudstones and floatstones and interbedded tidal-flat deposits, analogous to those described for the preceding growth stages.

Most of the facies outlined are depth-controlled and co-existed on distinct portions of the bathymetric profile as depicted in Figure 3. Attempts to quantify paleobathymetry have been made and are discussed further in a later section. Notable exceptions are the foreslope debris and sands, which because of their gravity-derived nature may occur anywhere downslope on the profile and the last facies outlined in the text which represents the final evolutionary stage of reef growth.

## GREEN SHALES

Thin green "shales", containing micrite and carbonate grains as well as clay minerals, occur in layers 1 mm to a few cm thick in parts of the reef complex. Even though these shaley intervals are volumetrically a very minor component of the reef, numerous workers have placed major significance on their occurrence. Wendte and Stoakes (1982) also contended that their occurrence was significant to understanding reef evolution but that previous workers misinterpreted their significance. Most of the following discussion is taken from that summary.

Most previous workers (Murray, 1966; Leavitt, 1968; Havard and Oldershaw, 1975 and others) regard these green shales as representing the residual soil of a former paleoexposure surface. Consequently many authors evoke sea-level drops and in some instances considerable erosion and solution accompanying green shale emplacement (Leavitt, 1968, p. 317). Others have used green shales to separate divisions in the reefs and as correlation surfaces.

Instead of the presently accepted solution-karstification origin for these green shales, we interpret them to be the products of storms, which transported this fine sediment from some external source to the reef complex. The mere presence of green shales in a reef complex implies an external source, for no reefs in the Recent are known to produce illitic green shales. The following observations support the depositional mode of formation (Wendte and Stoakes, 1982).

1. The occurrence of green shales in subtidal lagoonal and reef flat environments implying deposition in other than an exposed tidal flat setting.
2. The interlamination of thin, millimetre-thick green shales with carbonaceous (algal mat?) laminae implying a physical rather than chemical process of sedimentation.
3. The geopetal green shale infill of parts of large inter- and intraparticle pores in reef margin and reef flat sediments further suggesting a physical process of sedimentation.
4. The compositional similarities of green shales in the reefs and basinal argillaceous sediments to the northeast.
5. The repeated occurrence of green shales implying some episodic process. Green shales generally characterize the uppermost portions of shoaling-upward cycles in the reef, a stage when sea level was considered relatively stationary and shale progradation had become re-established (Stoakes, 1980).
6. Sharp, erosional bases of some thicker green shales with incorporated angular lithoclasts, intraclasts and bioclasts implying rapid emplacement with only minor erosion of the reef surface.

Contemporaneous shale banks infilling the basinal area to the northeast, during relatively stable stands of sea level, appear to be likely sources for these green shales. The model of reciprocal reef growth and basin filling has been described from the Leduc and Ireton formations of Alberta (Stoakes, 1980) and appears to have application here.

Immediately following a sea-level rise terrigenous influx to the basin is reduced and reefs grow vertically forming an upward-shoaling sequence. Prolonged stillstand allows terrigenous sedimentation to become re-established and shale banks again prograde into the basin. Occasional major storm rework shale from these approaching banks and "dusts" the reef complex. The patchy distribution of the green shales is related to both their irregular pattern of deposition and to their preservability in the various environments.

## CYCLICITY

Major shifts and changes of facies belts occur during the evolution of the Judy Creek reef complex and are illustrated on the cross section in Figure 4. This cross section extends across the reef complex from the windward northeast face to the leeward southwest side. Five major depositional sequences, whose bases are denoted by $R_0$, $R_1$, $R_2$, $R_3$ and $R_4$, are traced across the complex. In foreslope settings, each of the lower four units consists of an overall upward-shoaling succession of facies and therefore is said to represent a single "megacycle" of reef sedimentation. Two other distinct "megacycle" levels, $R_{00}$ and $R_{000}$, precede reef inception ($R_0$). All these "megacycles" are equivalent levels to the second-order Devonian cycles described in Chapter 2. The change from one megasequence to another is marked by a shift to a deeper water or more basinward facies.

Megacycles are defined along the margins of the complex. In many peripheral positions, the top of each megasequence is marked by an abrupt change from shallow-water facies deposited at or near sea level to substantially deeper water, basinal limestones. These abrupt deepenings record the "backstepping" nature of the reef in these locations. Once defined along the periphery of the reef, each sequence boundary was correlated throughout the reef complex.

The tops of pre-reefal and reefal megacycles vary from sharp to gradational. Where reefal facies of a megasequence are overlain by basinal or lower-foreslope facies of a succeeding cycle, the contacts are mainly sharp (Plate 3A). Along the peripheral parts of the complex, pre-reef contacts $R_{00}$ and $R_{000}$ are predominantly sharp and are commonly encrusted by thin-tabular stromatoporoids (Plate 3B). The top of the reef is always sharp and is marked by truncation of underlying stromatoporoids, *Trypanites* borings and small digitate stromatolite encrustation (Plate 3C). These occurrences imply lithification of the top of the reef prior to the deposition of the overlying Waterways Formation. Because evidence indicating subaerial exposure is lacking, the surfaces are interpreted as submarine hardgrounds. Similar hardgrounds cap portions of other reefal megacycles (Plate 3A).

The recognition of these megacycles is particularly important to understanding the evolution of the Judy Creek reef complex. As discussed before, megasequence contacts in certain positions in the reef complex are marked by pronounced shifts in facies belts. Facies intermediate on a depositional profile between those immediately below and above the contacts are typically missing. The cyclic nature is interpreted to reflect an episodically rising sea level. The initiation of each megacycle corresponds to a marked increase in the rate of sea-level rise. During these times, reef growth along certain positions of the reef periphery was outpaced by a rapidly rising sea level resulting in a submarine-cemented hardground surface. At times of slower relative sea-level rise (often termed stillstand), reef growth kept pace or exceeded the rate of sea-level rise, producing the characteristic upward-shoaling sequences.

The $R_4$ cycle contact is different from all other megasequence contacts. The contact is ubiquitously sharp and, with the exception of sections along the periphery of the complex, is marked by the superposition of shallow-water over deeper water facies.

Major changes in sea level for any given megacycle are made up of minor increments most accurately recorded by the response of the interior lagoonal sediments. Sediments in the lagoon accumulated

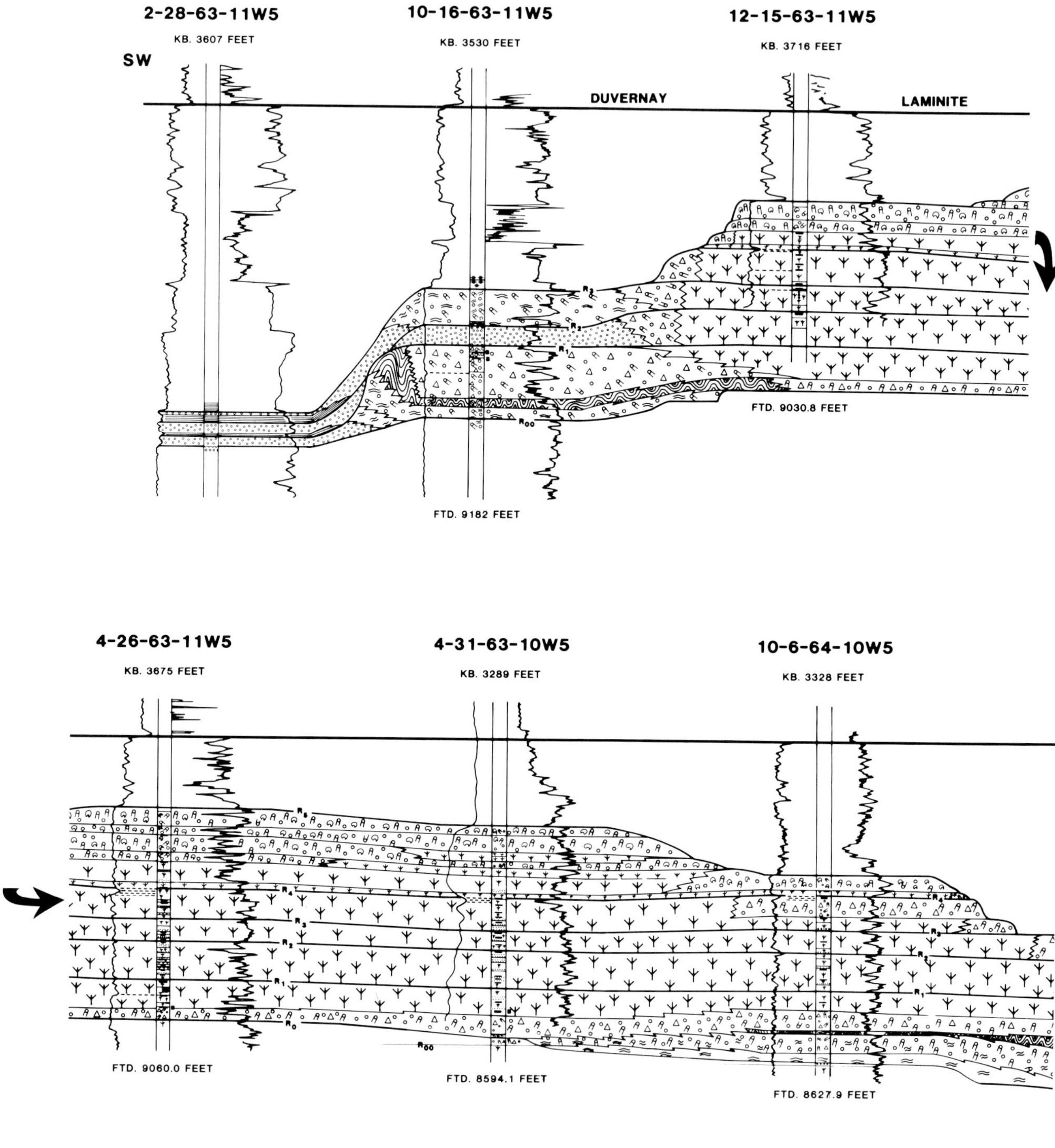
2-28-63-11W5
KB. 3607 FEET
10-16-63-11W5
KB. 3530 FEET
12-15-63-11W5
KB. 3716 FEET
SW
DUVERNAY
LAMINITE
FTD. 9030.8 FEET
FTD. 9182 FEET
4-26-63-11W5
KB. 3675 FEET
4-31-63-10W5
KB. 3289 FEET
10-6-64-10W5
KB. 3328 FEET
FTD. 9060.0 FEET
FTD. 8594.1 FEET
FTD. 8627.9 FEET

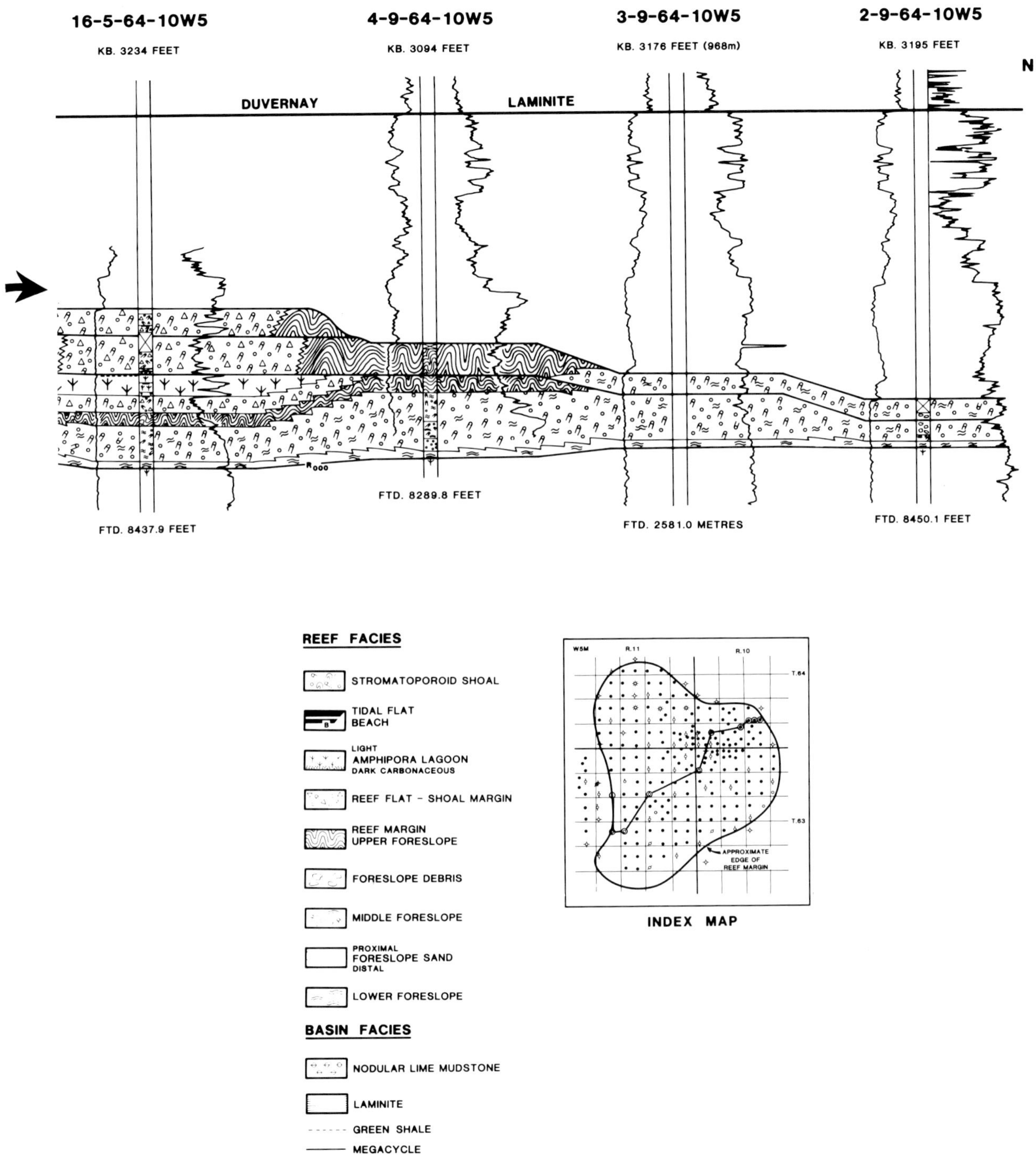


Figure 4. Sequence-facies cross section across the Judy Creek reef complex.

at much lower rates than those in the more actively growing reef margin. Consequently these sediments were outpaced by relative sea-level rises that had little or no effect on the more competent margin. These minor rises of sea level produced the cyclic pattern of lagoonal sedimentation shown on the cross section in Figure 4. Where completely developed, each lagoonal cycle grades up from dark carbonaceous and micritic *Amphipora* rudstones to light brown, grainy *Amphipora* rudstones into tidal flat or beach deposits. Due to the complexity of these cycles, they were not correlated in a three-dimensional manner throughout the reef complex. However, megacycle contacts from the reef margin were correlated to pronounced lagoonal cycles bases. The bases of these cycles were then correlated throughout the reef complex. Correlations of this nature argue against the autocyclic model of lagoonal deposition proposed for the Kaybob Swan Hills reef by Wong and Oldershaw (1980).

The use of cycles defined in the Judy Creek reef enables true growth stages to be defined. The correlation of each megacycle boundary may be double-checked by the reference to the facies immediately below each contact. A cycle boundary reflecting the operation of an external control, in this case sea-level rise, should preserve a normal lateral sequence of depositional facies. Furthermore, knowing that tidal flats in the interior and the reef flat at the margin approximate sea level at that depositional stage, the approximate paleobathymetry of the various facies can be measured. This has been done and the various values are presented on Figure 3 along with the average observed variations. These variations appear to result from the evolutionary stage of the reef at that time. This calculation ignores any differential compaction that may have subsequently taken place. It is notable that the cross section in Figure 4, as reconstructed, shows a little difference in thickness between muddy lagoonal sediments of the interior and reef-margin sediments within the same depositional megacycle.

The reefal stages, based on the recognition of natural cyclicity in the buildup, differ from those of previous workers in the Swan Hills. For example, some of Fischbuch's (1968) divisions do not appear to be separate growth stages. In the nearby Goose River reef, the change from his Division V to Division VI expressed in part by a perceptible southwesterly shift in the loci of reef growth (op. cit. Fig. 8a, p. 462-463) could be explained more readily as the lateral progradation of shallow-water reef facies over deep-water "limy shales" within the same growth stage.

The recognition by Fischbuch (1968) and others of a number of comparable "divisions" within various Swan Hills buildups suggests that the present approach of defining individual growth stages could be attempted basinwide to produce fairly exact paleogeographic reconstructions. Their widespread nature further suggests that the model of relative sea-level control of the various stages is indeed valid. However, based on our work in the Judy Creek reef complex, we see no evidence for and would disagree strongly with, the postulation of <u>major</u> sea-level drops either forming or modifying the various growth stages.

## REEF INITIATION

The onset of reefing at Judy Creek corresponds to a time of rapid sea-level rise. Similar to the localization of Leduc reefs, Swan Hills reefs developed on underlying platform highs. The formation of this high at Judy Creek resulted from a phase of backstepping platformal successions. The evolution of this phase is illustrated on the reef transect in Figure 4 and on the cross section along the north side of the complex in Figure 5.

This phase of backstepping resulted in the shift of shallow-water platform facies toward the centre of the complex. During this phase, downslope settings in front of the backstepping platform margins were generally areas of sediment bypass where little or no sediment accumulated. In the immediate area around the Judy Creek reef complex, this phase of backstepping records approximately 18 m of stratigraphic accommodation or growth from $R_{000}$ to $R_0$. Near the centre of the complex, the platform attains a thickness of approximately 54 m. The configuration of this platform high and the disposition of facies immediately prior to $R_0$ are illustrated in Figure 6.

The nature of the backstepping in the platformal succession varies areally. Shifts in facies belts indicate that backstepping occurred in major discrete steps along the north and east sides of the complex. These steps correspond to the $R_{000}$, $R_{00}$ and $R_0$ levels. In an embayment on the northeast side of the complex, backstepping occurred not only at major steps, but also at intermediate intervals. This style of backstepping is demonstrated on Figure 4, where an intermediate step between $R_{000}$ and $R_{00}$ is evident at the 10-6 location. Core control on the western side of the complex is insufficient to document a specific style of backstepping.

Backstepping prior to the onset of reefing has been beautifully and fully documented in the study of the Swan Hills Snipe Lake complex by Muir and others (1990). Muir and his co-workers demonstrated the occurrence of five major backstepping sequences from the top of the Fort Vermilion Anhydrite to the base of the reef complex, equivalent to $R_0$ at Judy Creek. The reef at Snipe Lake is localized on the shallow-water portion of their final platform succession.

The nature of the contact at the base of the reef, $R_0$, differs from that of $R_{000}$ and $R_{00}$. $R_0$ is commonly transitional and is never marked by the sharp, lithified nature inherent in $R_{000}$ and $R_{00}$. *In situ* deposits of the basal reef cycle overlie this surface. The absence of synlithification is interpreted to imply that the rate of relative sea-level rise marking the change from platform to reef was not as rapid as preceding ones initiating platformal successions at the $R_{000}$ and $R_{00}$ levels.

## REEF EVOLUTION FROM INCEPTION ($R_0$) TO $R_4$

The initial cycle of reef growth, from $R_0$ to $R_1$, records an early phase of reefal progradation followed by aggradation. This style is evident on the cross section in Figure 4. During the progradational phase, the reef and its foreslope built out over the

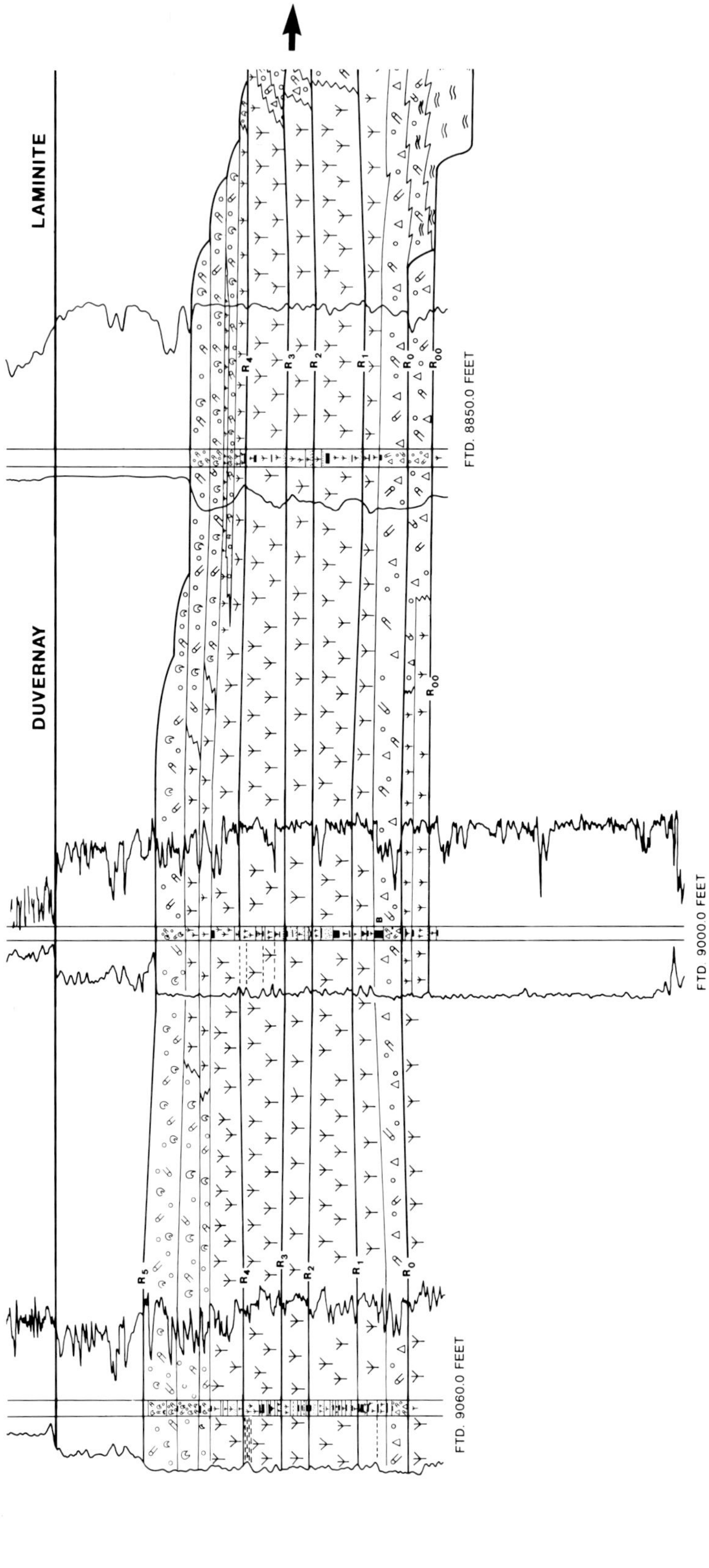

4-26-63-11W5
KB. 3675.0 FEET
FTD. 9060.0 FEET
4-35-63-11W5
KB. 3447.0 FEET
FTD. 9000.0 FEET
4-12-64-11W5
KB. 3529.0 FEET
FTD. 8850.0 FEET
DUVERNAY
LAMINITE
R5
R4
R3
R2
R1
R0
R00
B

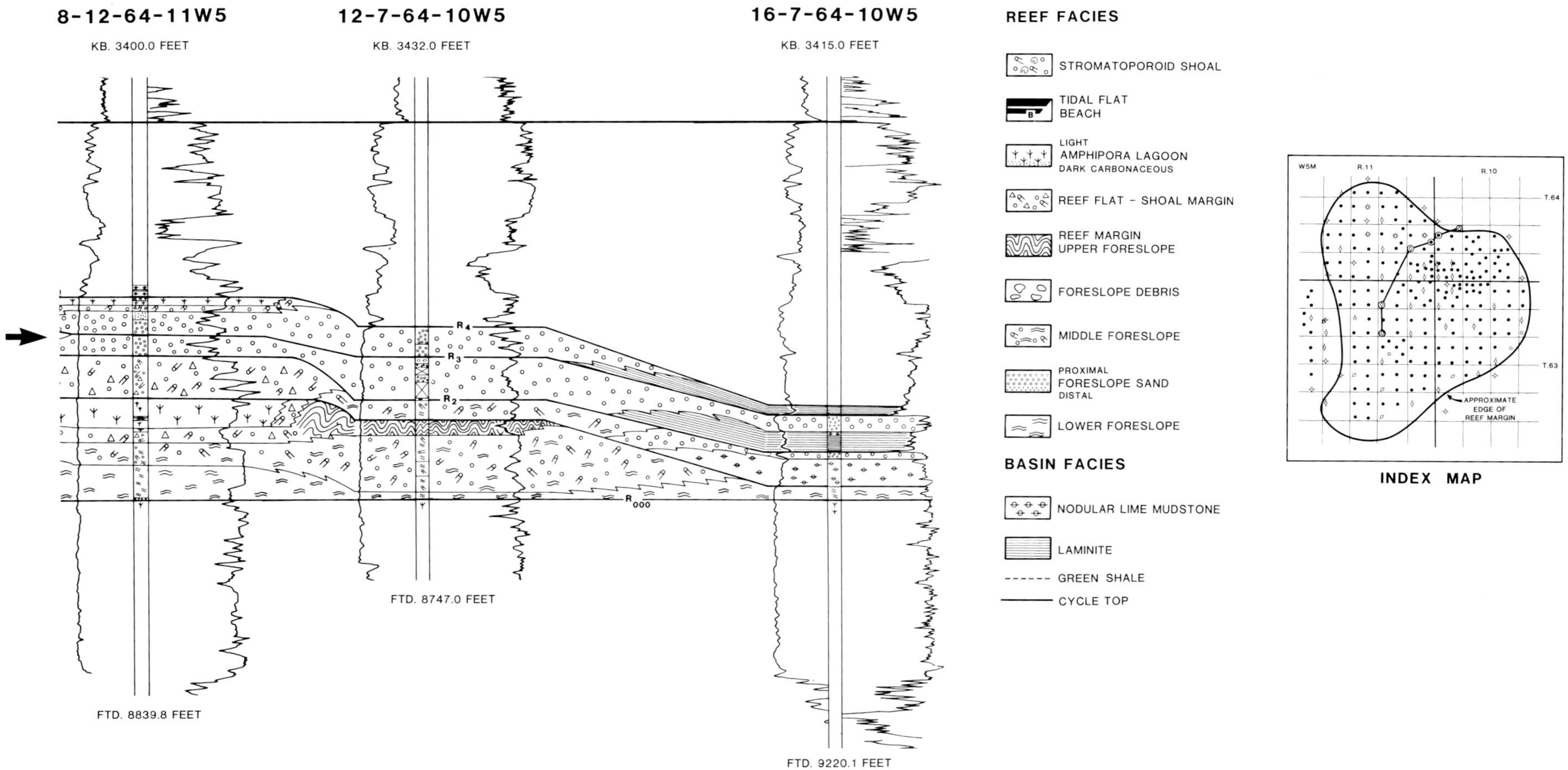


Figure 5. Sequence-facies cross section from the centre to the north edge of the Judy Creek complex.

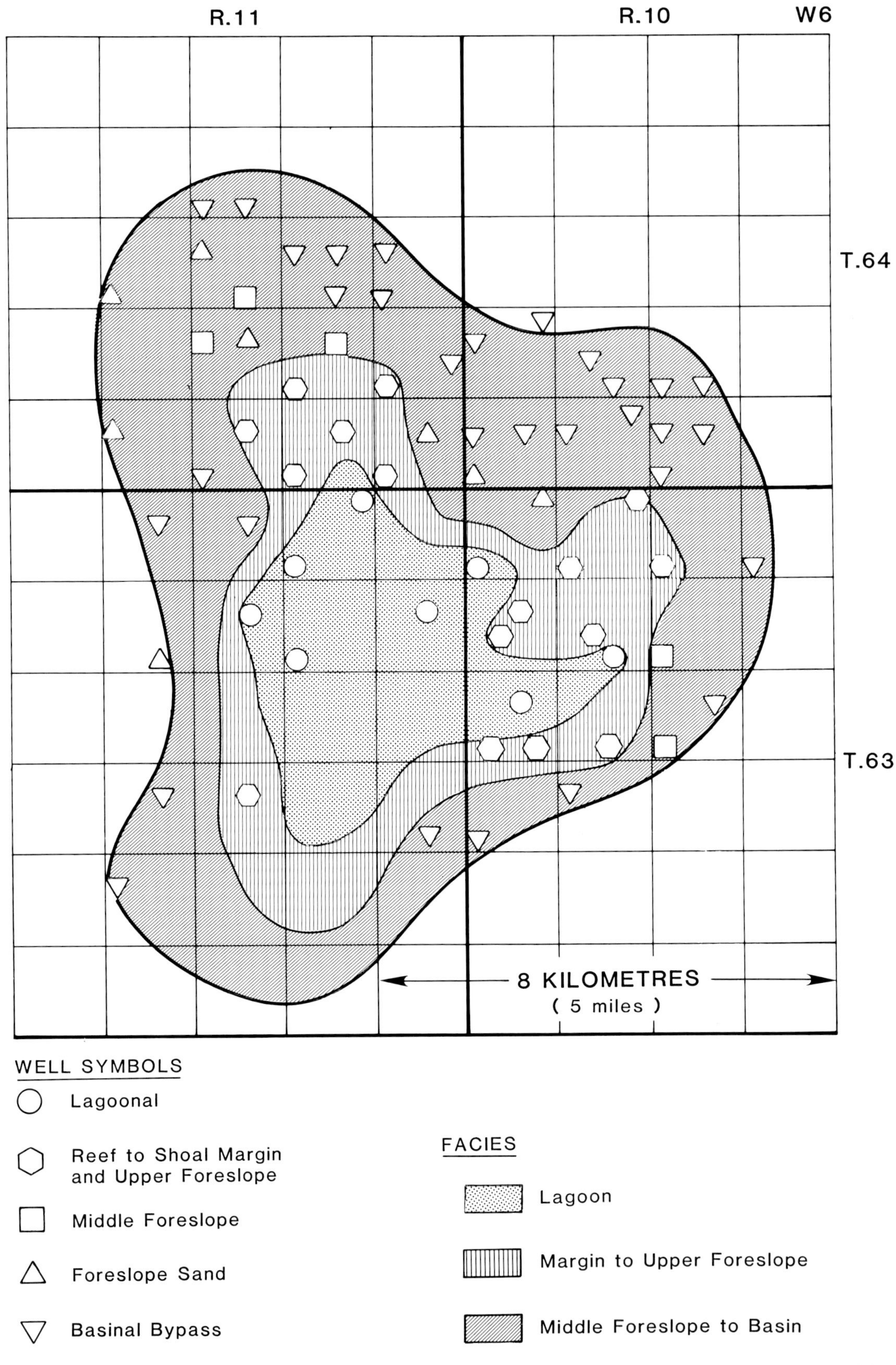


Figure 6. Disposition of platform facies immediately beneath $R_0$. The configuration of the platform high corresponds to the distribution of lagoonal, reef or shoal margin, and upper foreslope facies.

previously drowned platform steps. The amount of progradation varies from as little as approximately 1 km on the sides of the complex to as much as 4 km along the northeast periphery of the pool. Where the reef prograded out beyond the edge of the platform succession capped by $R_{00}$, lower-foreslope deposits of the basal reef cycle typically sit directly on the $R_{000}$ level. The overlying foreslope and reef-margin deposits record a remarkably uniform shoaling-upward sequence of facies, as illustrated in the 16-5 well in Figure 4.

The final phase of growth in the initial reef cycle is aggradational. This style is demonstrated in Figure 4 between the locations at 16-5 and 4-9. The phase of aggradation results in a very thick development of highly porous and permeable upper-foreslope and reef-margin facies.

The style of reef growth from the top of the first reef cycle ($R_1$) to $R_4$ varies laterally. This variability is summarized on Figure 7. This figure shows the position of the reefal crests corresponding to the cycles capped by $R_1$, $R_2$, $R_3$ and $R_4$. All variations in stacking from backstepping, retreating, upbuilding to (very minor) forestepping are documented.

The close to virtual coincidence of the lines marking the position of the reefal crests along the northeast margin of the complex depicts an upbuilding to retreating style of reef growth. This style is illustrated on the cross section in Figure 8. Upper foreslope and reef-margin facies in the 2-8 well are especially thick (40 m). This abnormally thick development records the upbuilding to retreating style of reef growth from $R_1$ to $R_3$ in this area. This style of reef growth is interpreted to reflect essentially continuous reef growth along the most favourable, windward northeast "buttress" of the complex.

The positions of the reefal crests from $R_1$ to $R_4$ are more separated along the north and especially the east sides of the complex. This separation records more of a backstepping style of reef growth. As illustrated on the cross section in Figure 9, this imparts a "terrace" morphology to the reef complex. Poorly developed reef margins, termed shoal margins, backstep away from the periphery of the reef complex.

This backstepping morphology is interpreted to reflect less than optimum conditions on the "sides" of the reef complex. Incipient drownings mark times of rapid sea-level rise. Increments of sea-level rise outpaced reef growth. During subsequent times of slower sea-level rise, poorly developed reef or shoal margins became re-established in more hospitable, lower-energy settings, further removed the previous break in slope.

The nature along the leeward, western margin of the complex is not as well known. The relationships between reefal stages indicate more of an overall upbuilding to retreating style of reef growth. This indicated style of reef growth is interpreted to be the consequence of a deep-water embayment rooted in the platform succession, that separates Judy Creek from the Judy Creek West complex. The presence of this embayment is evident in the 2-28 well on the west end of the transect in Figure 4.

A cross section along the western margin of the pool is shown in Figure 10. The $R_3$ to $R_4$ cycle exhibits basinward progradation

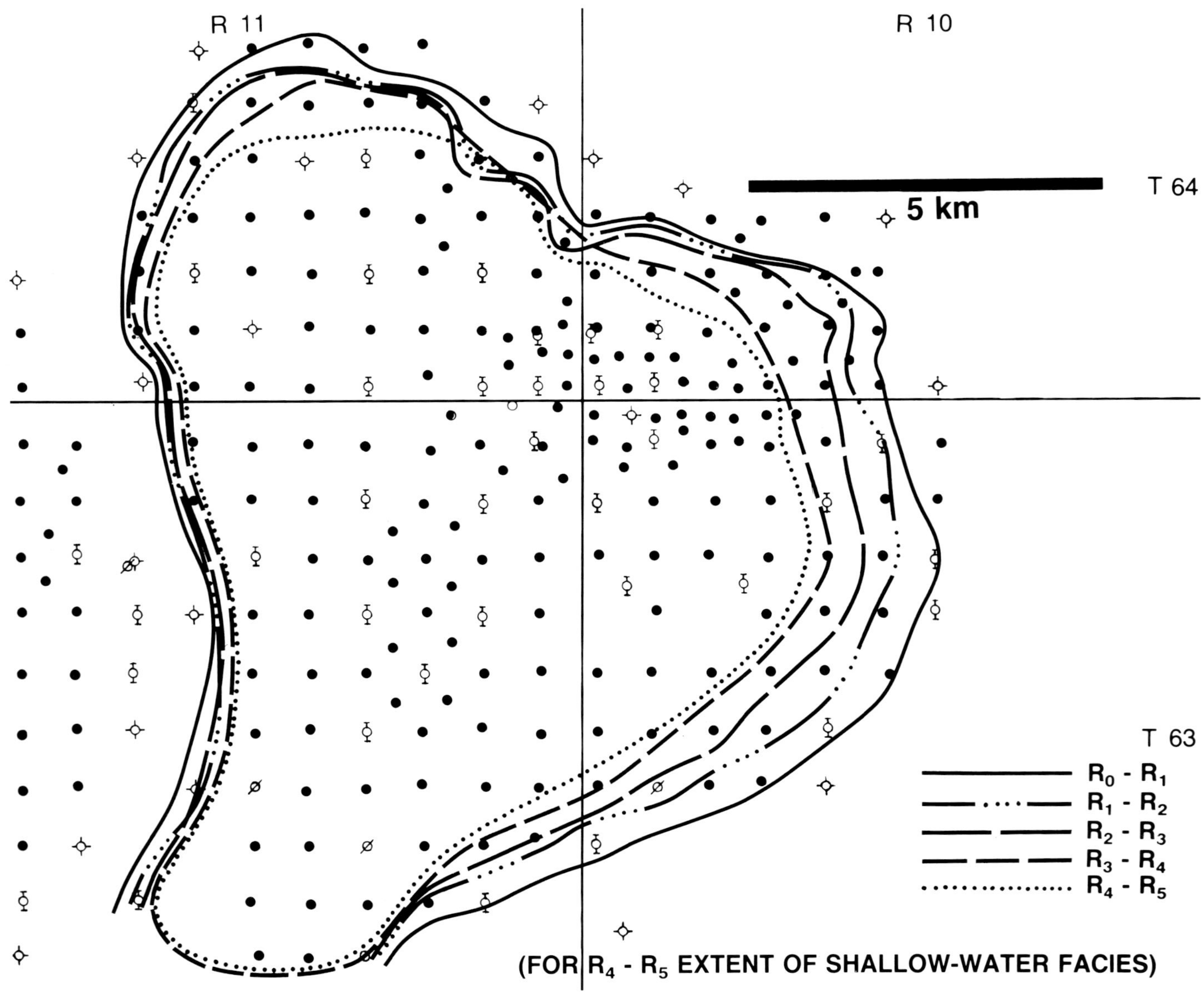


Figure 7. Position of reef or shoal margins from $R_0$ to $R_4$. The outline of the $R_4$-$R_5$ shoal complex is also shown.

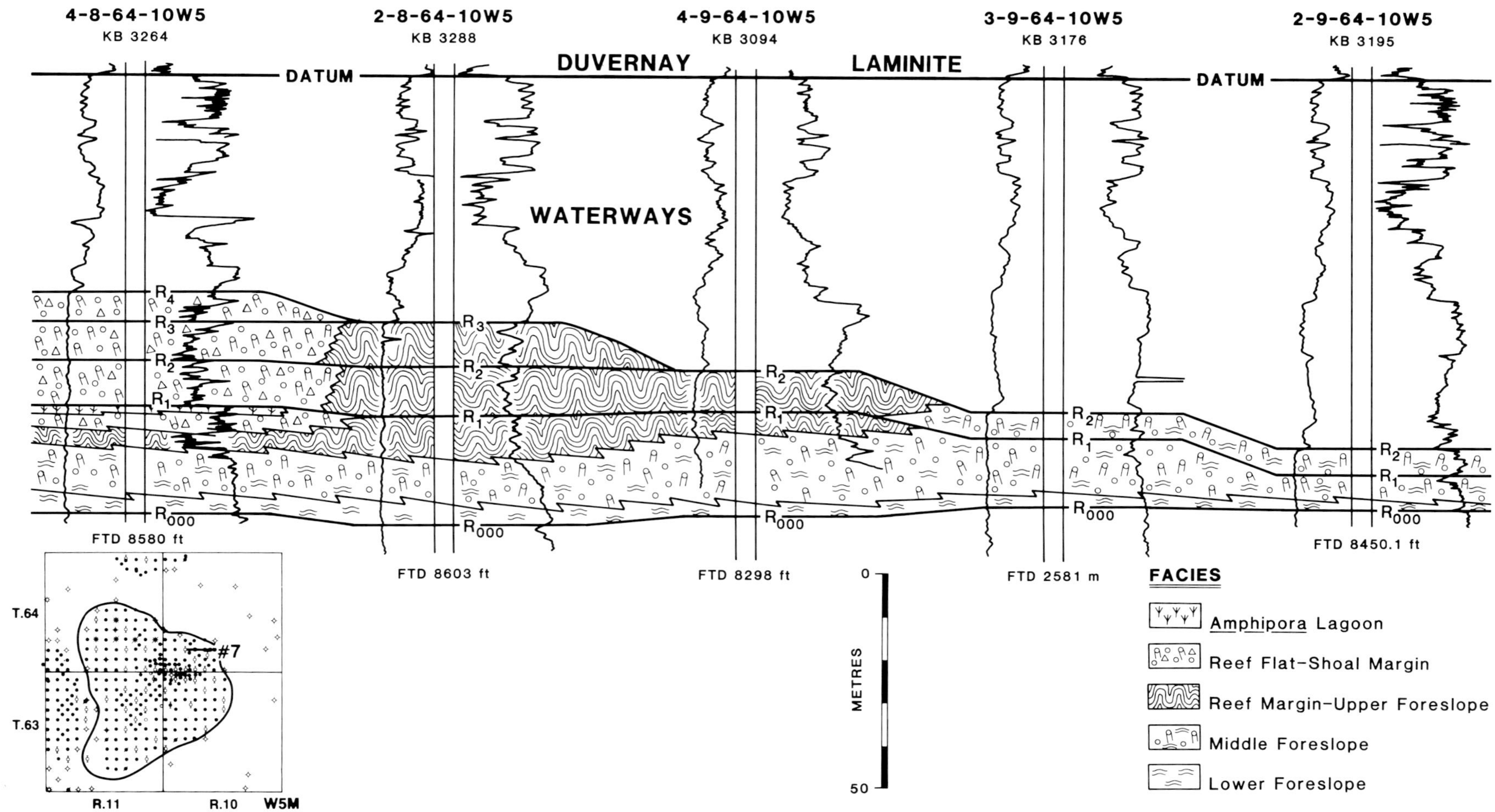


Figure 8. Upbuilding to retreating style of reef growth along the northeast margin of the reef complex.

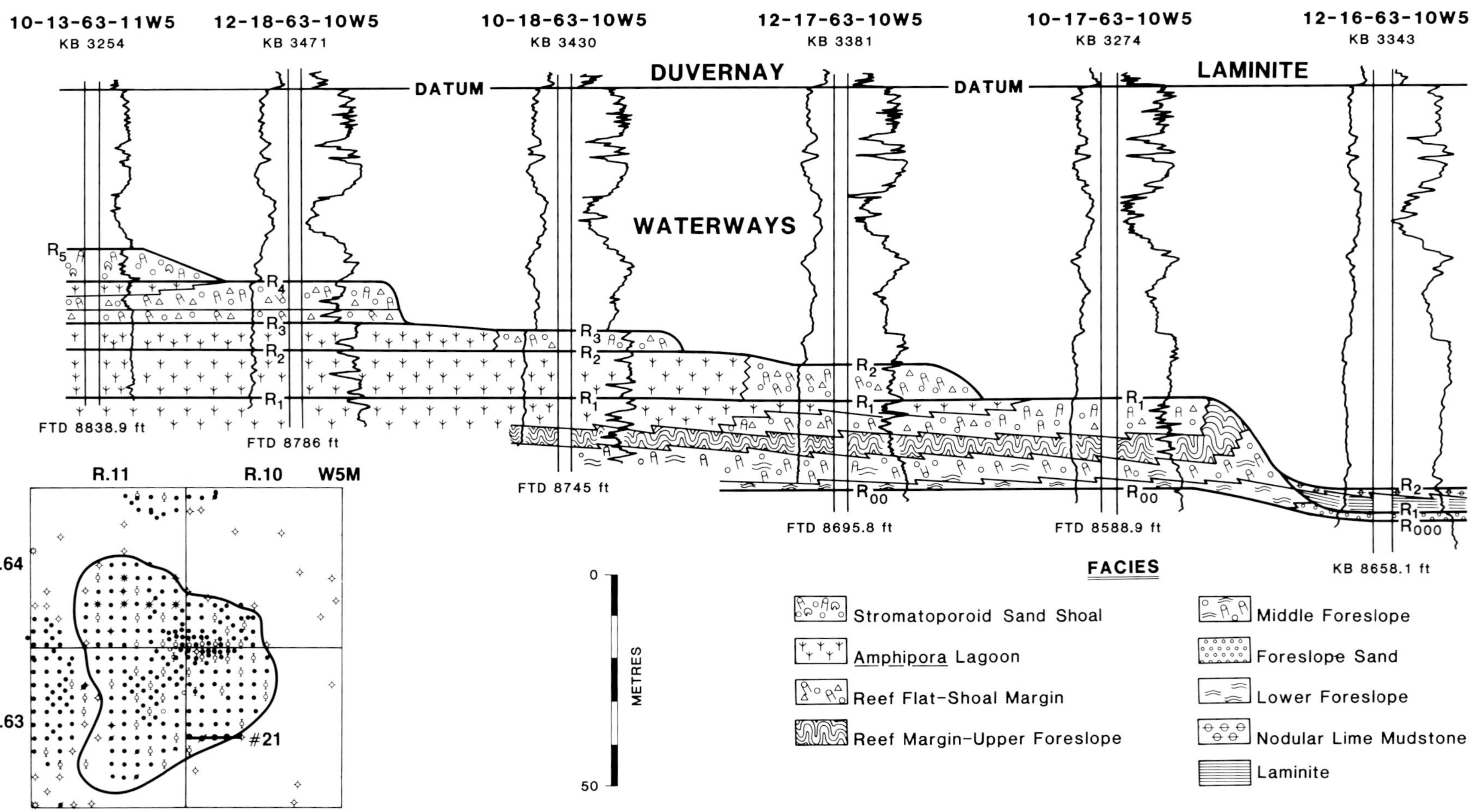


Figure 9. Backstepping style of reef growth along the east margin of the reef complex.

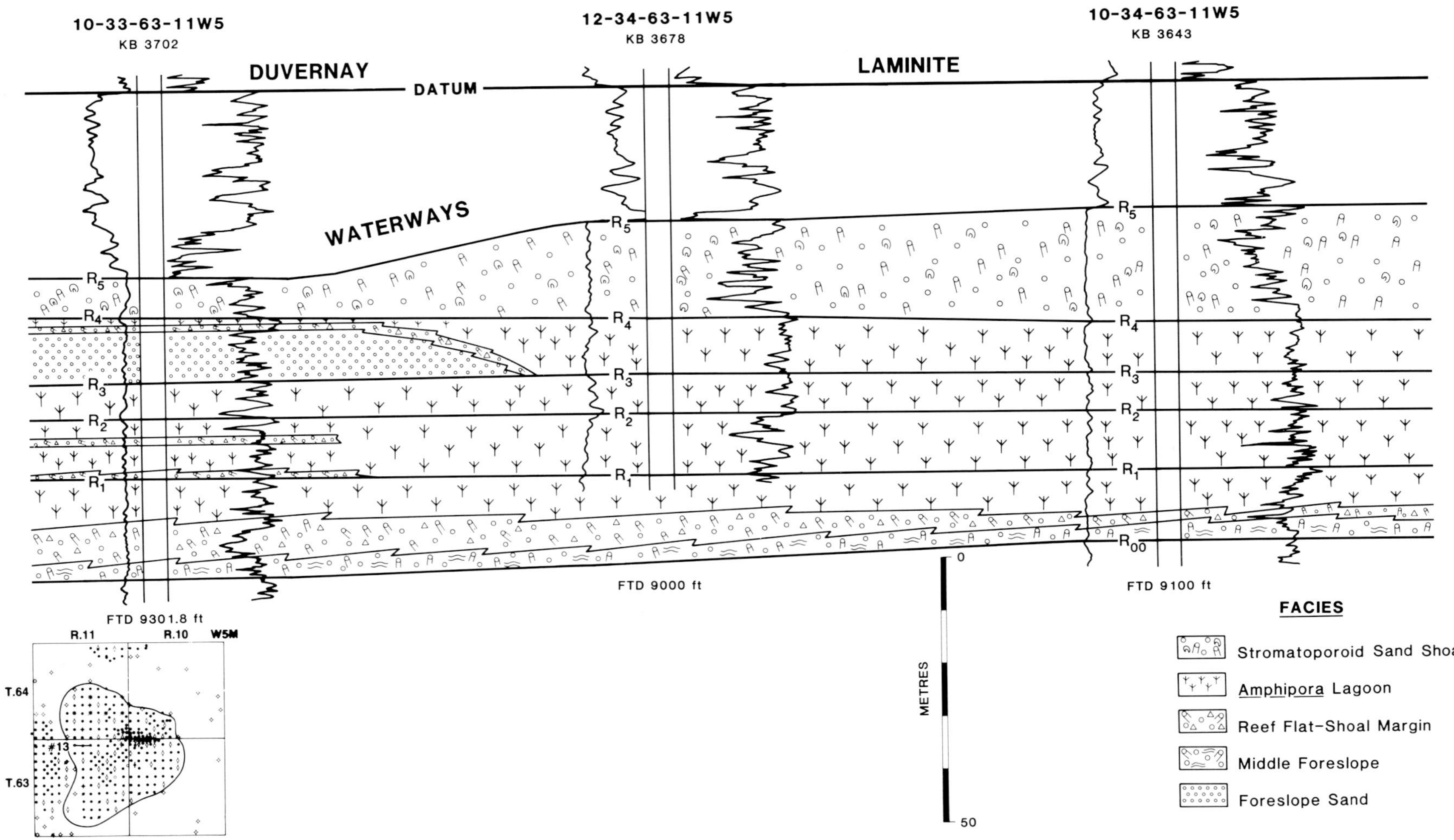


Figure 10. Progradational nature of $R_3$ to $R_4$ cycle along the leeward west margin of the reef complex.

of a very poorly developed shoal margin over a shallow-foreslope sand. Within well control, the outer limit of this shoal margin approximates that of the margins of the previous megacycles (see Figure 7). This coincidence may reflect the presence of a deeper hole to be filled. Obviously, a deeper depression requires more sediment to fill it and, as a consequence, to allow for progradation across it. A deeper water embayment can therefore act as an impediment to progradation. Because of this impediment, the basinward extent of reefal progradation in all megacycles may be approximately the same.

In summary, the evolution from $R_1$ to $R_4$ records an overall transgressive phase of sedimentation. An overall greater shift of facies at $R_3$, especially on the windward northeast side of the complex, may suggest a greater rate of relative sea-level rise at this level than at $R_1$ and $R_2$.

The upper part of the $R_3$ to $R_4$ megacycle also records a significant change in the depositional style of the lagoonal cycles. Preceding lagoonal cycles, especially toward the centre of the reef complex, commonly are fully developed: dark carbonaceous subtidal sediments grading up into shallower, lighter coloured lagoonal sediments and into tidal-flat and beach deposits. Reef interior cycles immediately prior to $R_4$ <u>totally</u> lack the darker, carbonaceous lagoonal facies, typical of a deeper water lagoon. Instead, sediments in these cycles are ubiquitously cream to light brown. In addition, these cycles contain abundant tidal-flat deposits. The nature of these cycles indicates that the rate of accommodation or aggradation was declining. As a consequence, holes sufficiently deep enough to accumulate carbonaceous rich sediment were never created.

The interval prior to $R_4$ also contains the greatest number and abundance of green shales in the reef complex. This occurrence is evident in lagoonal wells on the cross section in Figure 4. As previously proposed (Wendte and Stoakes, 1982), this correspondence is interpreted to reflect very slow rates of reef aggradation.

## EVOLUTION FROM $R_4$ TO REEF TERMINATION

The $R_4$ surface marks an unique event in the evolution of the Judy Creek reef complex. This contact is always sharp and deposits below this surface show evidence of lithification prior to the deposition of the overlying sediments (Plate 3D). No where along this contact do facies below and above this surface intertongue. $R_4$ is thus a cycle contact.

This cycle contact differs in one fundamental way from <u>all</u> other cycle contacts in the Judy Creek reef complex: shallow-water facies are superimposed over deeper water facies. As illustrated in Plate 3D, cryptalgal mats, inferred to represent tidal-flat conditions, overlie bioturbated shallow lagoonal sediments.

These observations support an interpretation that cessation of sedimentation at this level was caused by a lowering of sea level. Because no "low-stand" reefs could be found superimposed on deeper water foreslope deposits, the amount of lowering is considered to be small, perhaps only a few metres. The following sequence is postulated:

1. Gradually declining rates of sea-level rise or accommodation following $R_3$.
2. A small increment of sea-level fall, terminating growth along the top of the reef complex at the $R_4$ level.
3. A subsequent slow rate of rise of sea level, allowing onlap of tidal-flat sediments onto the previously exposed and lithified reef top.

The distribution of facies immediately overlying the $R_4$ surface supports an interpretation of a gradual rise of sea level following exposure. These facies occur in almost a concentric pattern, as illustrated in Figure 11. Dark coloured cryptalgal mats occur in the centre of the reef complex, and are surrounded by belts, in an outward direction, of light coloured cryptalgal mats and subtidal deposits. The dark coloured tidal-flat sediments are interpreted to reflect the more restricted circulation conditions in the interior of the complex. These conditions permitted carbonaceous matter to accumulate and to be preserved. Improved circulation and exchange of waters around the periphery of the complex resulted in oxidation of the carbonaceous matter. The outer belt merely records the permanently submerged setting around the entire complex.

The overlying facies exhibit the anatomy of a retreating and backstepping, ramp-bounded shoal complex. As illustrated on the cross section in Figure 4, the more restricted *Amphipora* lagoonal facies characterize the lower and more central positions of this complex. The less restricted Stromatoporoid Shoal Limestones increase in abundance upwards and toward the periphery of the complex. The superposition of the Stromatoporoid Shoal deposits over the *Amphipora* lagoonal facies records shifts in facies belts toward the centre of the reef. These shifts are interpreted to represent times of more rapid sea-level rise.

The profile across the complex shows an inherent difference between the windward and leeward sides. This difference is depicted on the isopach map of the shoal complex shown in Figure 12. The windward northeast side has a gently sloping ramp profile, whereas, the leeward westward flank has more of a steeper, aggradational aspect.

The retreating and backstepping, ramp-bounded style of evolution is interpreted as a consequence of both an episodically rising sea level and the configuration of the initial profile across the shoal complex. Prior to initial onlap, the $R_4$ surface was essentially flat over the entire reef complex. Initial onlap resulted in retreat of the tidal-flat sediments toward the reef centre. Succeeding stages of deposition resulted in some degree of sediment aggradation. But because of the retreating and backstepping style, only minor relief was created along the shoal margins, especially on the windward northeast flank. The relief was insufficient to create a pronounced break in slope and to localize reef-margin facies. Instead, ramp-bounded shoal sequences occur right to the top of the reef complex. The final stages of reef growth are localized along northeast-southwest trending shoals as depicted on Figure 12. The orientation of these shoals match

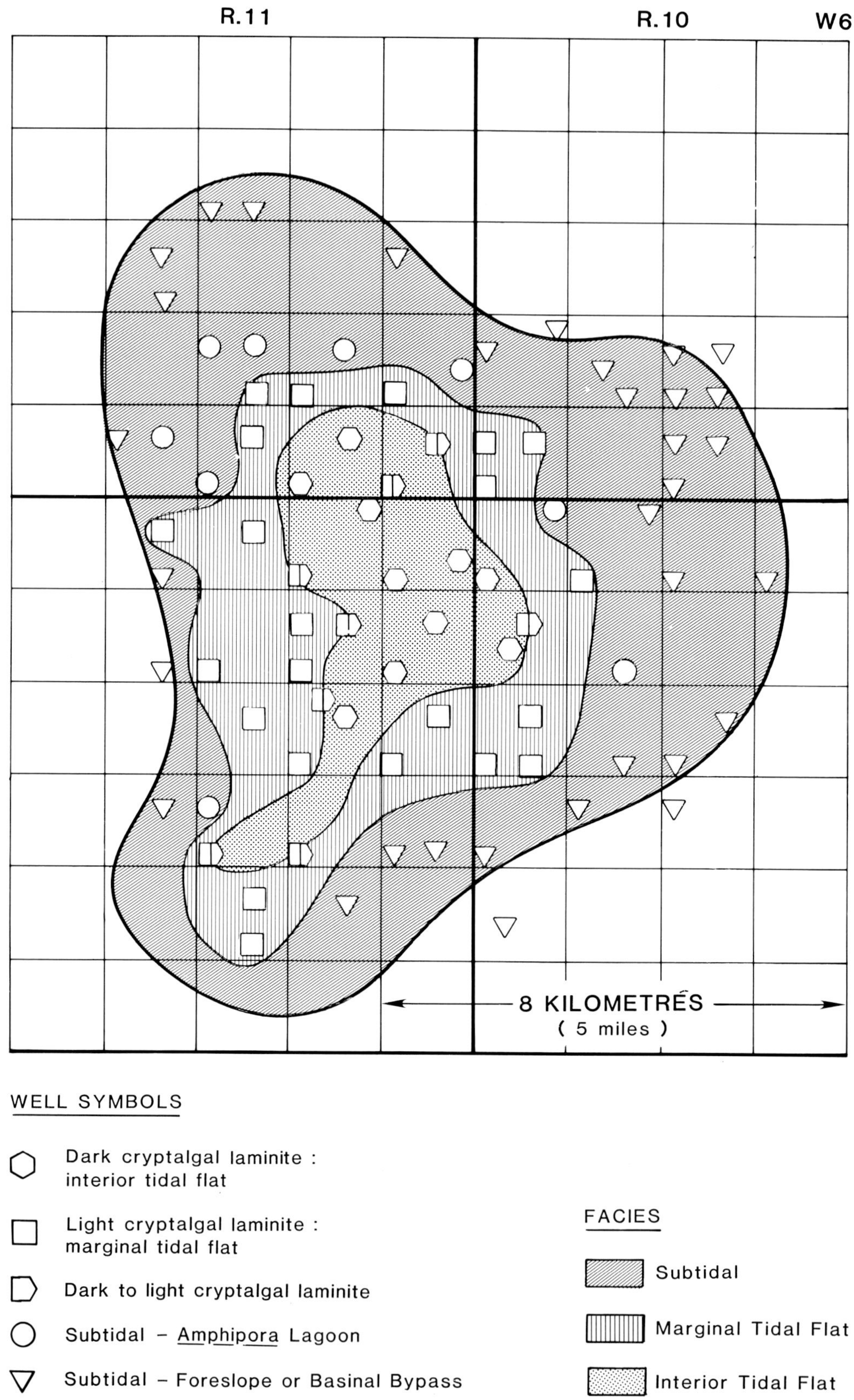


Figure 11. Distribution of facies immediately above $R_4$ surface.

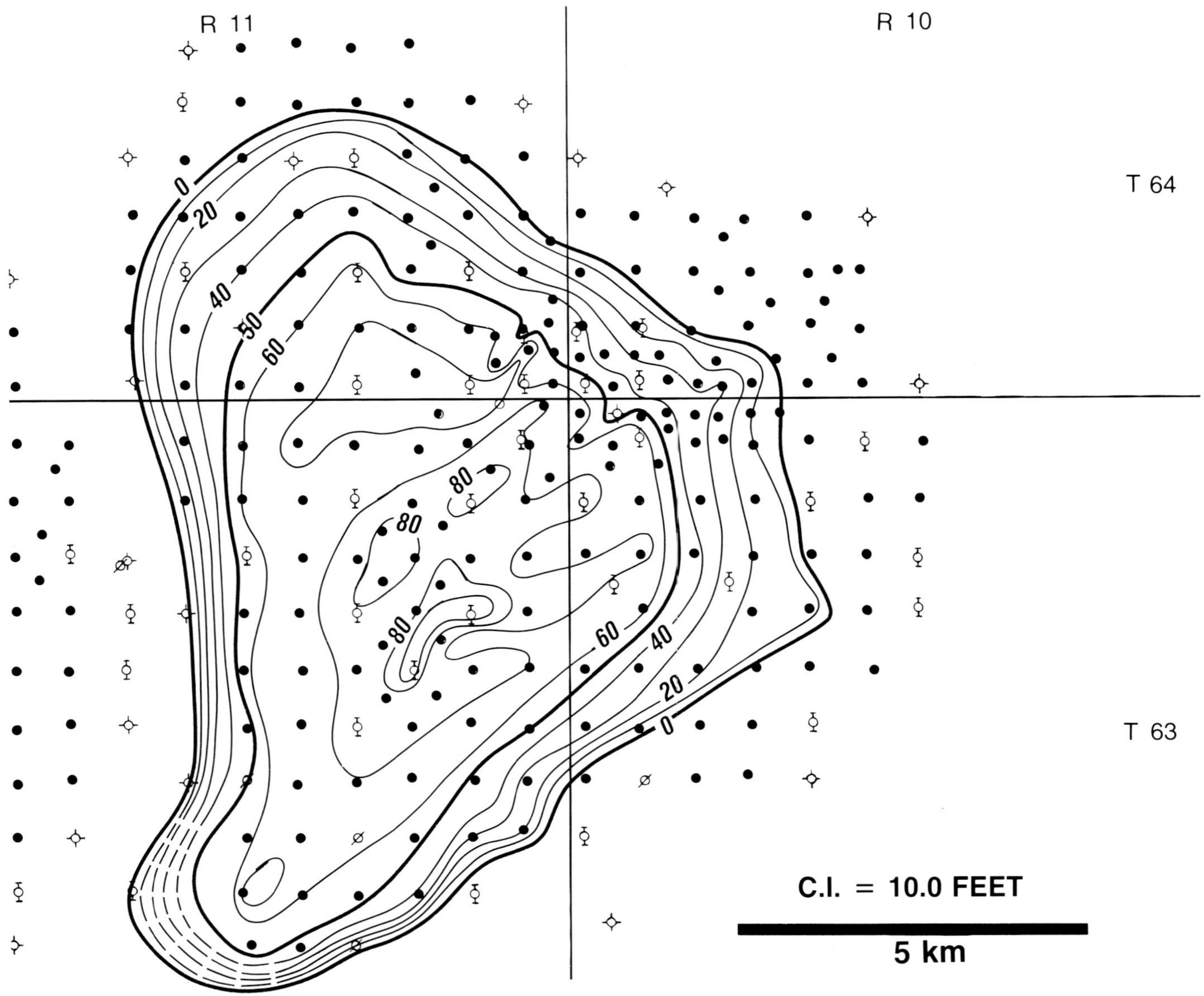


Figure 12. Isopach map of $R_4$ to $R_5$ shoal complex. Thicks exceeding 80 feet correspond to isolate shoals with a northeast-southwest orientation. These isolate shoals are the final stages of a backstepping, ramp-bounded shoal complex.

the direction of the prevailing trade wind circulation belt described in Chapter 1.

The style of growth supports an interpretation that the reef complex simply "backstepped" itself out of existence. The ultimate drowning of the complex is not then due to one episode of an accelerated rate of sea-level rise, but records the results of a continued phase of backstepping and shrinking resulting from a number of periods of accelerated rates of sea-level rise. Thus the ultimate drowning of the complex not only attests to the effects of a rapidly rising sea level, but can also be attributed to the progressive shrinkage in areal size. One conclusion drawn from this study is that the areal extent of complexes exert a pronounced control on "drowning" episodes.

## SUMMARY

The evolution of the Judy Creek reef complex records the interaction of both external relative sea-level and internal depositional controls. The initial localization of the isolate reef is the culmination of progressive backstepping of the underlying carbonate platform succession. Stages of reef growth record various relative sea-level signatures and the interaction of sea level and depositional controls, such as circulation direction and depositional topography. Ultimate drowning is simply the final act in a backstepping phase of growth.

## PLATE 1

A. Cryptalgal laminite. 4-24-63-11W5, 8812 feet. Note laminar and irregular fenestrae.

B. Lagoonal cycle contact. 16-31-63-10W5, 8432 feet. Dark coloured, carbonaceous *Amphipora* floatstone overlying light brown *Amphipora* floatstone.

C. Upper Foreslope facies. 4-9-64-10W5, 8174 feet. Highly porous stromatoporoid debris overlying thick-tabular stromatoporoid, which may be in growth position.

D. Middle Foreslope facies. 4-21-63-10W5, 8646 feet. *Stachyodes* branching stromatoporoids with fine grainstone-packstone matrix. Minor tabular and encrusting stromatoporoids.

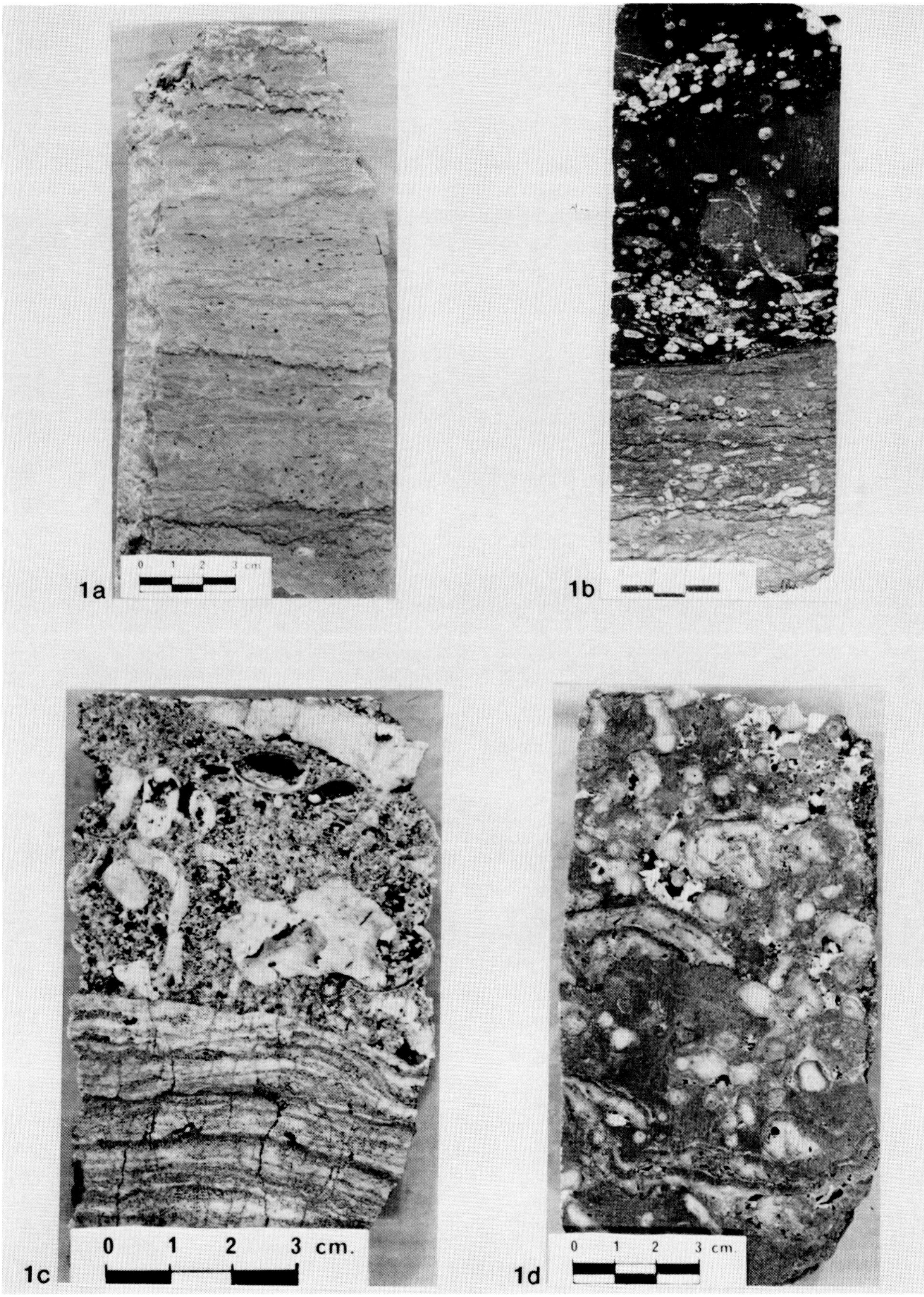
1a
0 1 2 3 cm
1b
1c
0 1 2 3 cm.
1d
0 1 2 3 cm.

PLATE 2

A. Lower Foreslope facies. 12-13-64-11W5, 8559 feet. Thin-tabular stromatoporoid rudstone to boundstone (?) with dark carbonaceous packstone and wackestone matrix.

B. Foreslope debris. 12-7-64-10W5, 8552 feet. Abraded stromatoporoid rudstone with a skeletal grainstone matrix.

C. Foreslope sand. 16-7-64-10W5, 8563 feet. Peloidal grainstone with some abraded brachiopod fragments and crinoid columnals.

D. Nodular lime mudstone. 16-7-64-10W5, 8605.5 feet.

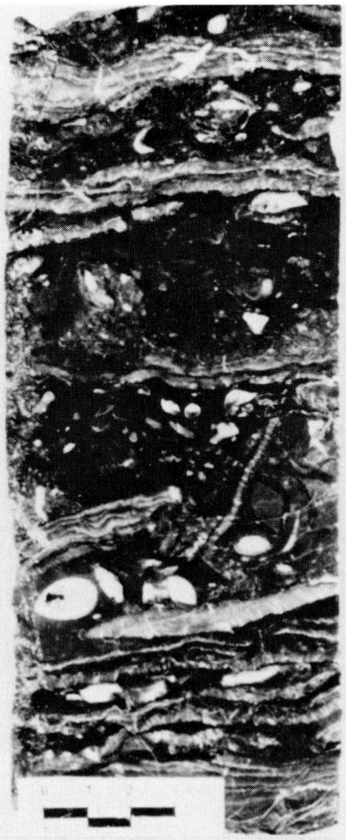

2a

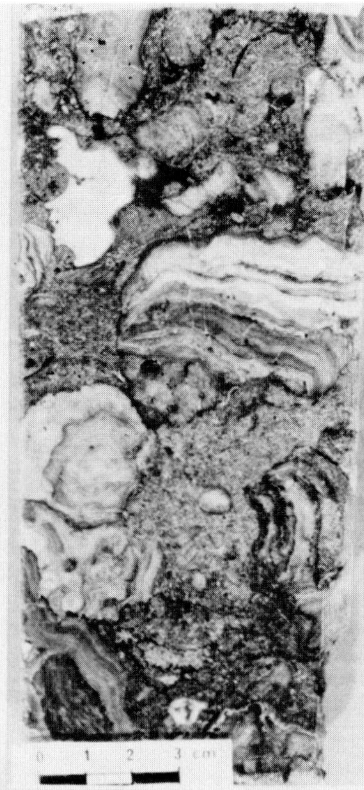


2b

2c

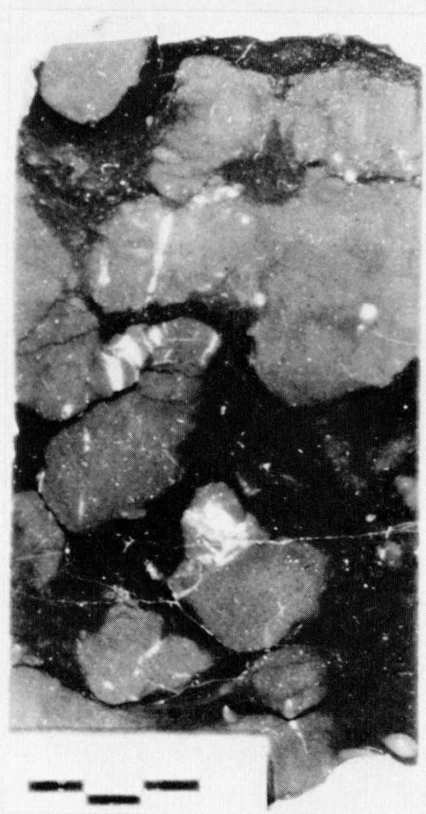

2d

PLATE 3

A. Megacycle contact on the foreslope. 16-7-64-10W5, 8625 feet. Stromatoporoid debris overlain by dark basinal limestone. Surface shows minor erosion and truncation of wafer stromatoporoid and other grains. Lack of differential topography between hard stromatoporoids and enclosing matrix indicates lithification of deposits below the contact prior to deposition of basinal limestones.

B. $R_{000}$ contact. 16-7-64-10W5, 8636 feet. Lower Foreslope facies overlying light coloured *Amphipora* floatstone. Contact is marked by tabular stromatoporoid encrusting the surface, suggesting syndepositional lithification of the underlying sediments.

C. Reef top. 2-8-64-10W5, 8340 feet. Abruptly truncated thick-tabular stromatoporoid abruptly overlain by more open-marine basinal sediments. The truncated stromatoporoid indicates lithification of the top of the reef prior to deposition of the basinal sediments.

D. $R_4$ contact. 10-3-64-11W5, 8803.5 feet. Crinkly cryptalgal mat (arrow) overlying light coloured, bioturbated lagoonal facies.

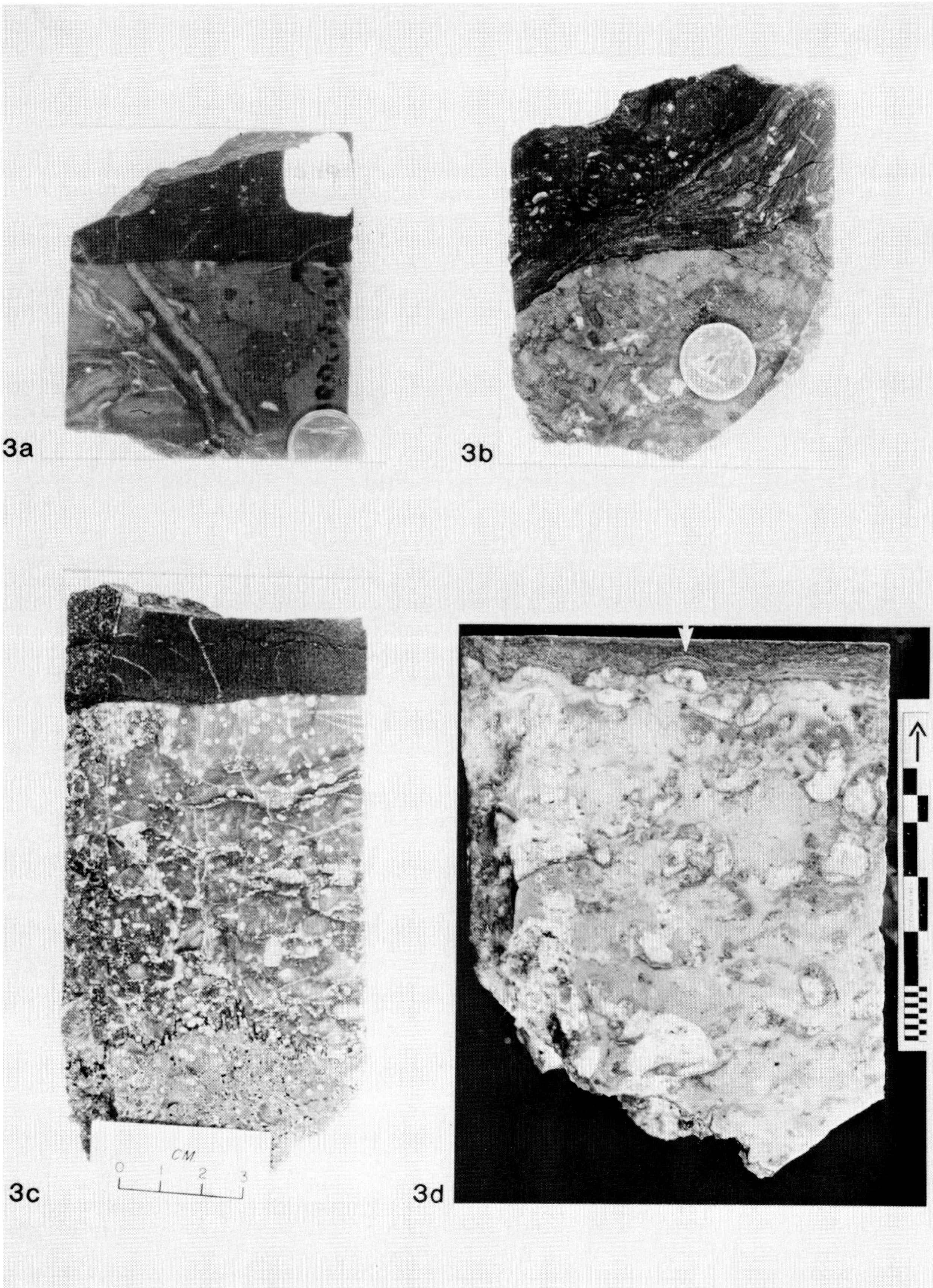


3a 3b 3c 3d

## REFERENCES


FISHCBUCH, N.R., 1968, Stratigraphy, Devonian Swan Hills reef complexes of central Alberta: Bulletin of Canadian Petroleum Geology, v. 16, p. 446-587.

HAVARD, C., AND OLDERSHAW, A.E., 1976, Early diagenesis in back-reef sedimentary cycles, Snipe Lake reef complex, Alberta: Bulletin of Canadian Petroleum Geology, v. 24, p. 27-69.

HEIDGERKEN, L.D., WENDTE, J.C., AND STOAKES, F.A., 1989, Stratigraphic evolution and its control on reservoir facies of a Leduc reef complex, Golden Spike, Alberta, *in* Exploration Update, Program and Abstracts booklet for annual CSPG-CSEG convention, Calgary, Alberta, p. 67.

JARDINE, D., ANDREWS, D.P., WISHART, J.W., AND YOUNG, J.W., 1976, Distribution and continuity of carbonate reservoirs: Journal of Petroleum Technology, p. 873-885.

LEAVITT, E.M., 1968, Petrology, palaeontology, Carson Creek North reef complex, Alberta: Bulletin of Canadian Petroleum Geology, v. 16, p. 298-413.

MUIR, I.D., SPRINGATE, G.L., AND MAWDSLEY, J.R., 1990, Geological model of the Middle-Upper Devonian Snipe Lake BHL "A" pool platform -reef complex, Western Canada: American Association of Petroleum Geologists Bulletin, v. 74, p. 726-727.

MUIR, I.D., WONG, P.K., AND WENDTE, J.C., 1984, Devonian Hare Indian-Ramparts (Kee Scarp) evolution, McKenzie Mountains and subsurface Norman Wells, N.W.T.: basin-fill and platform-reef development, *in* Eliuk, L., Kaldi, J., and Watts, N., eds., Carbonates in subsurface and outcrop: Canadian Society of Petroleum Geology Core Conference, p. 82-102.

MURRAY, J.W., 1966, An oil-producing reef-fringed carbonate bank in the Upper Devonian Swan Hills Member, Judy Creek, Alberta: Bulletin of Canadian Petroleum Geology, v. 14, p. 1-103.

PURSER, B.H., 1972, Subdivision et interpretion des sequences carbonates: Memoir British Royal Geological Museum, v. 77, p. 678-698.

STOAKES, F.A., 1980, Nature and control of shale basin fill and its effect on reef growth and termination: Upper Devonian Duvernay and Ireton Formations of Alberta, Canada: Bulletin of Canadian Petroleum Geology, v. 28, p. 345-410.

WENDTE, J.C., AND CALLOW, G.O., 1987, Upper Devonian Woodbend platform and reef evolution; Redwater, Alberta, 1987, *in* Canadian Reef Inventory Project Symposium, Program and Abstracts booklet, Banff, Alberta, p. 110.

WENDTE, J.C., AND STOAKES, F.A., 1982, Evolution and corresponding porosity distribution of the Judy Creek reef complex, Upper Devonian , Central Alberta, *in* Cutler, W.G., ed., Canada's giant hydrocarbon reserves: Canadian Society of Petroleum Geologists, Core Conference Manual at CSPG-AAPG Convention, Calgary, Alberta, p. 63-81.

WILSON, J.L., 1967, Carbonate-evaporite cycles in lower Duperow Formation of Williston Basin: Bulletin of Canadian Petroleum Geology, v. 15, p. 230-312.

WONG, P.K., AND OLDERSHAW, A.E., 1980, Causes of cyclicity in reef interior sediments Kaybob reef, Alberta: Bulletin of Canadian Petroleum Geology, v. 28, p. 411-424.

# CHAPTER 5

# NATURE AND SUCCESSION OF BASIN FILL STRATA

F.A. STOAKES

## INTRODUCTION

Thick basinal sequences comprising very fine-grained sediments are common in the geological record. For the most part they present a monotonous succession of poorly defined lithologies. Although not the direct focus of exploration efforts these intervals are important in terms of permeability seals and more importantly the accumulation of hydrocarbon source sections.

This article examines the nature and controls on basin filling during the Upper Devonian - Early Mississippian of the WCSB. It focuses primarily on the Ireton and Duvernay intervals of the Woodbend sequence (see Chapter 8). However, conclusions and general inferences are then related to other basin-fill successions, namely the Waterways Formation of the Beaverhill Lake Group (Chapter 7), the Winterburn shales of the Winterburn Group (Chapter 9) and finally the Exshaw - Banff shales of the early Mississippian (Chapter 11).

Volumetrically basin-fill successions comprise a significant portion of the Upper Devonian to Lower Mississippian section in the WCSB. They may be grouped into three types based on gross lithology:

1) Evaporitic
2) Carbonate-rich
3) Shale-rich

The first category reflects the regressive component of Upper Elk Point Group stratigraphy (Chapter 6) and is not dealt with further here. The last two reflect the interplay of two sedimentary contributions to basin fill. These are: a) A fine-grained carbonate component produced through the biological and physical erosion of shallow-water carbonates and reworked to the basin of deposition (intrabasinal source). It should be noted that Paleozoic carbonates lack the significant planktonic carbonate component of Mesozoic and Cenozoic systems so that although basinal sedimentation reflects settling of fine carbonate material from suspension in the water column, it's ultimate origin is primarily the shallow-water carbonate system around the basin edge. b) A fine-grained clay-shale component sourced by erosion of distant terrains (extrabasinal source) and brought into the basin of deposition by oceanic circulation.

## MODES OF BASIN FILL: THE IRETON AND DUVERNAY FORMATIONS

The stratigraphy and stratal relations of these two units are

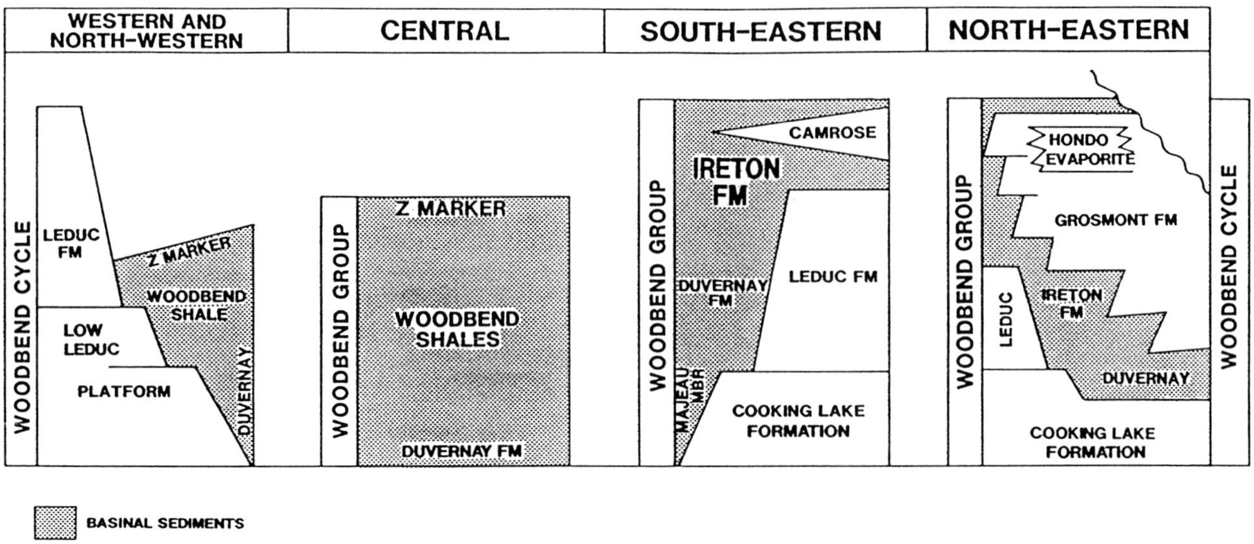


Figure 1. Stratigraphic nomenclature of the Woodbend Group.

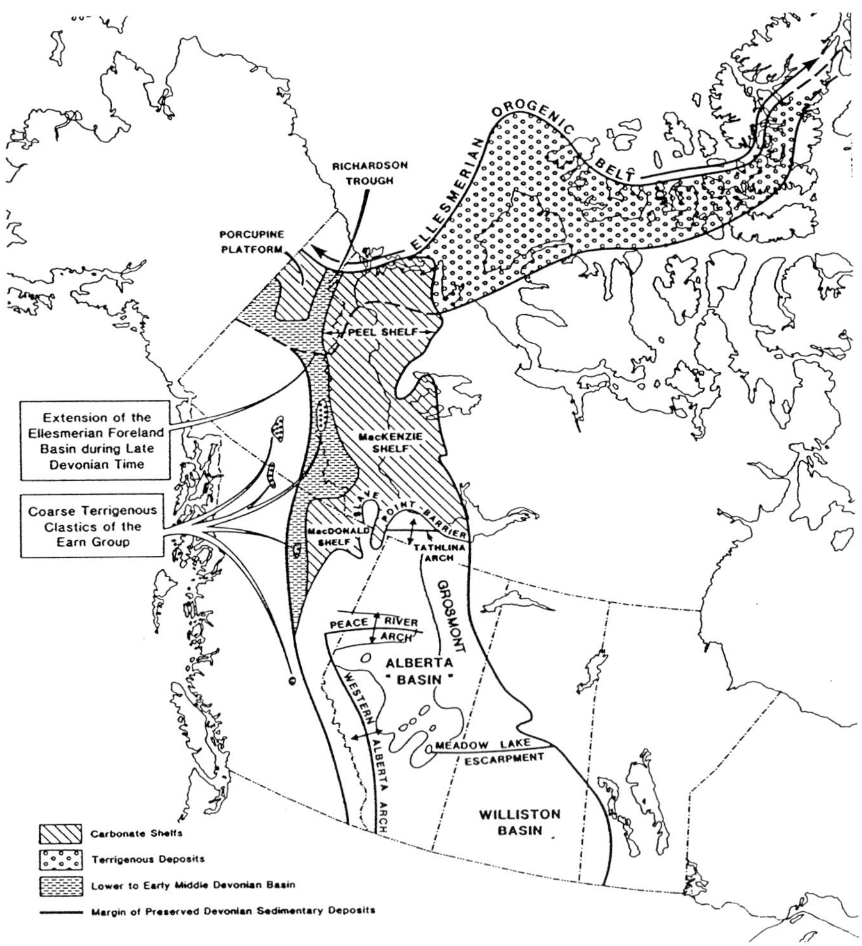


Figure 2. Physiographic and tectonic elements of the Western Canada Sedimentary Basin (from Wendte Chapter 1, this volume).

dealt with in Chapter 8. These two formations comprise the regressive basin-filling portion of the Woodbend succession (Figure 1) and mark the first significant influx, into central Alberta, of fine terrigenous material from a distant extrabasinal source. Compositionally the Ireton is comprised of varying amounts of illite and chlorite (Campbell and Oliver, 1968) which give it a distinctly greenish colour. The source of the shale component is thought to be the Ellesmerian Fold Belt in the Canadian Arctic Archipelago (Stoakes, 1979). It gained access to the basin via a seaway that existed between the Peace River Arch landmass and the Grosmont Platform Complex (Stoakes 1980 and Figure 2).

*Composition of The Basin Fill* - The basin-fill succession exhibits systematic compositional variations in a number of respects:

a) Proximity to regional input
b) More local sources of carbonate supply
c) Bathymetric depth

The first two factors reflect the relative contribution of material on different scales. On Figure 9 north and eastern portions of the Ireton basin contain more argillaceous material whereas the southwestern part is more carbonate-rich ("Westerdale facies" of Chapter 8). This reflects the direction of input of the terrigenous material. The more distant (or distal) southwestern portion of the basin receiving volumetrically more material from the encircling reef masses. This mechanism also operates on a more local scale with individual reef masses or shallow-water platforms contributing more carbonate material to adjacent coeval basinal settings. McCrossan (1961) refers to carbonate "haloes" around reefs reflected by resistivity log response. He used this analysis to interpret current patterns in the basin especially as it related to the "clastic" (i.e. reef-derived) carbonate component. Finally, mention may be made of the bathymetric variation in composition. Distinct marker surfaces (discussed later) reflect the bathymetric profile at a number of stages during basin filling (Oliver and Cowper 1963, Stoakes 1980). These allow for an assessment of the variation in composition at various "paleowater" depths in the basin. In part this must reflect the ability of the fine suspended particles to settle out from the water column. This compositional variation is illustrated in Figure 3. Major sedimentation of the clay-shale component takes place on the inclined portion of the profile. Further basinwards away from this influx, starved basin conditions existed with settling out of only fine carbonate material. This produces the carbonate-rich bottom-set beds of the Duvernay Formation. On the upper undathem portion of the profile, local carbonate enrichment takes place through *in situ* skeletal productivity (see later discussion of facies).

*Facies of The Basin Fill* - The basin-fill succession contains a number of distinct facies related primarily to water depth (Figure 4). Lithofacies characterizing the undathem portion contain evidence of an impoverished though normal-marine calcareous fauna. Further basinward the profile is characterized by development of

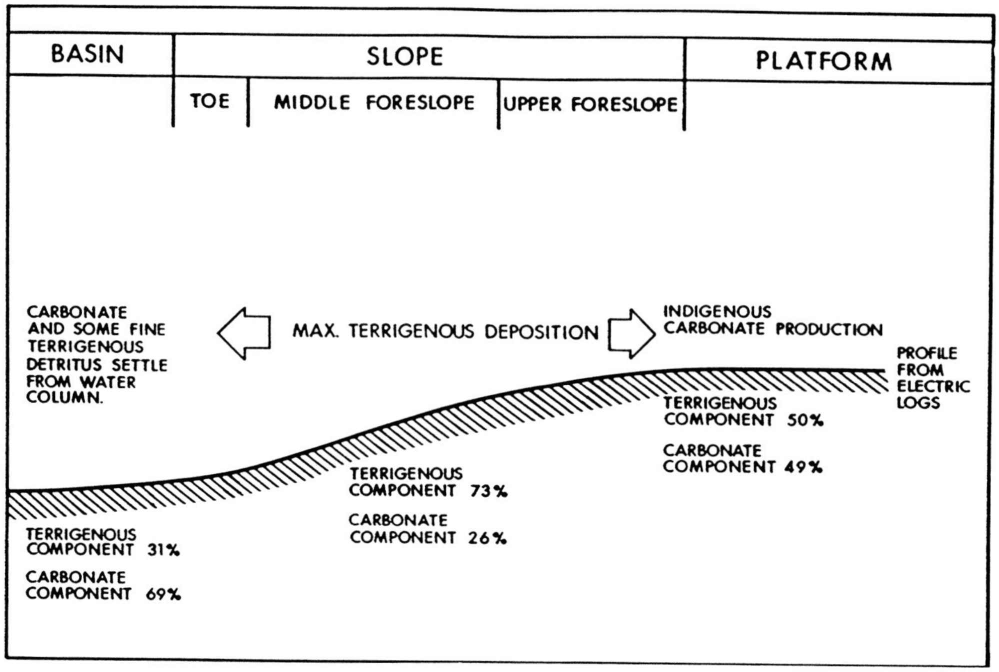


Figure 3. Compositional variation across depositional topography (from Stoakes, 1980).

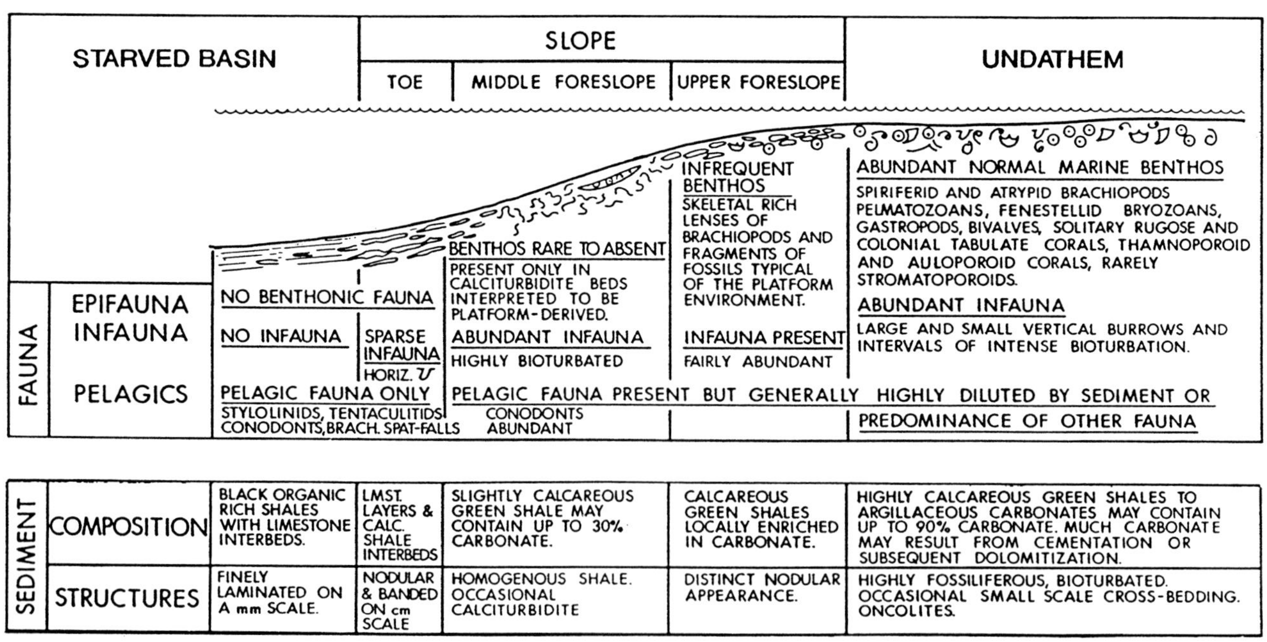


Figure 4. Facies of the Duvernay-Ireton basin fill (from Stoakes, 1980).

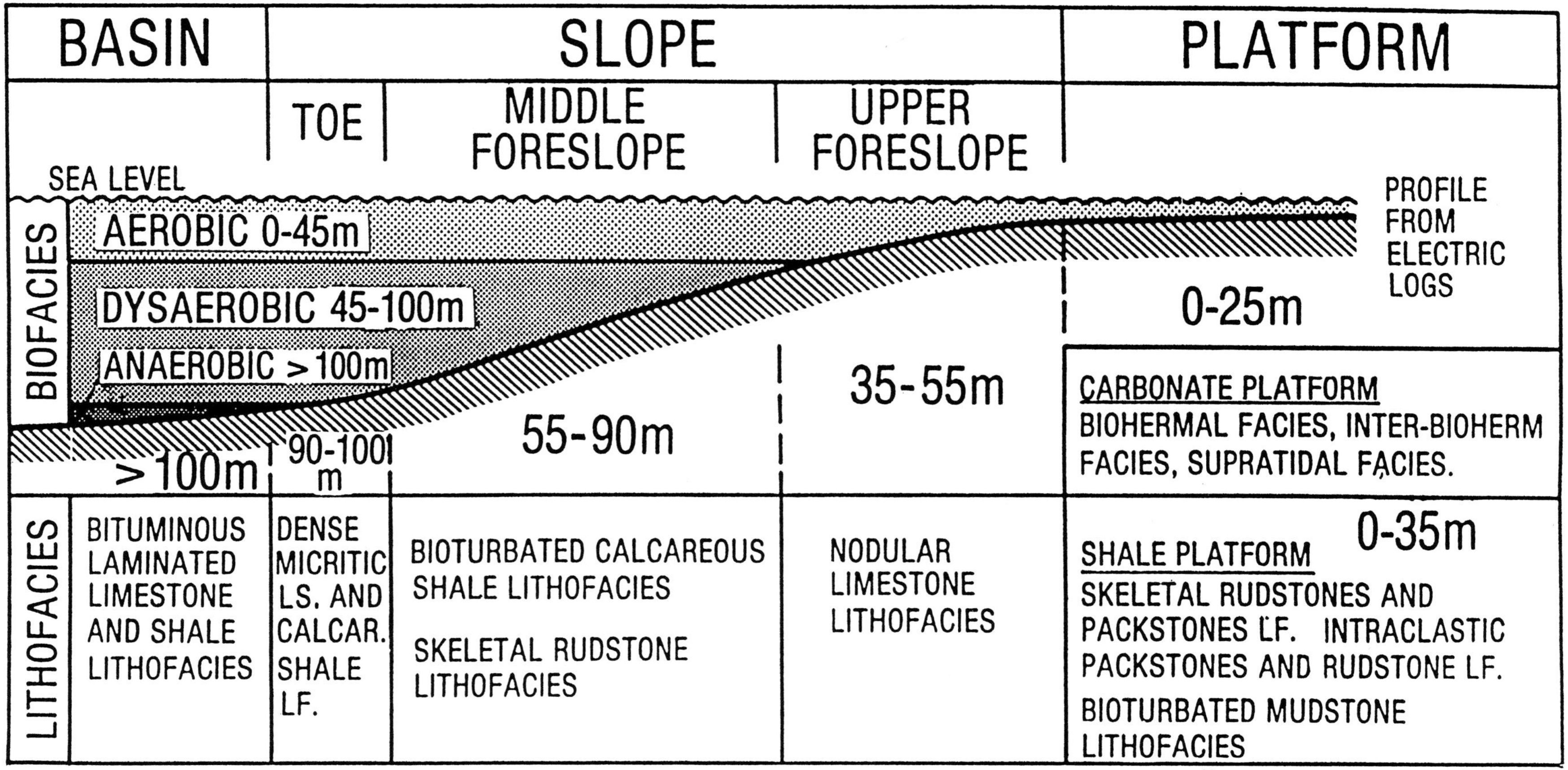


(from STOAKES, 1980)

Figure 5. Duvernay-Ireton paleobathymetry depicting idealized stratification of basin waters.

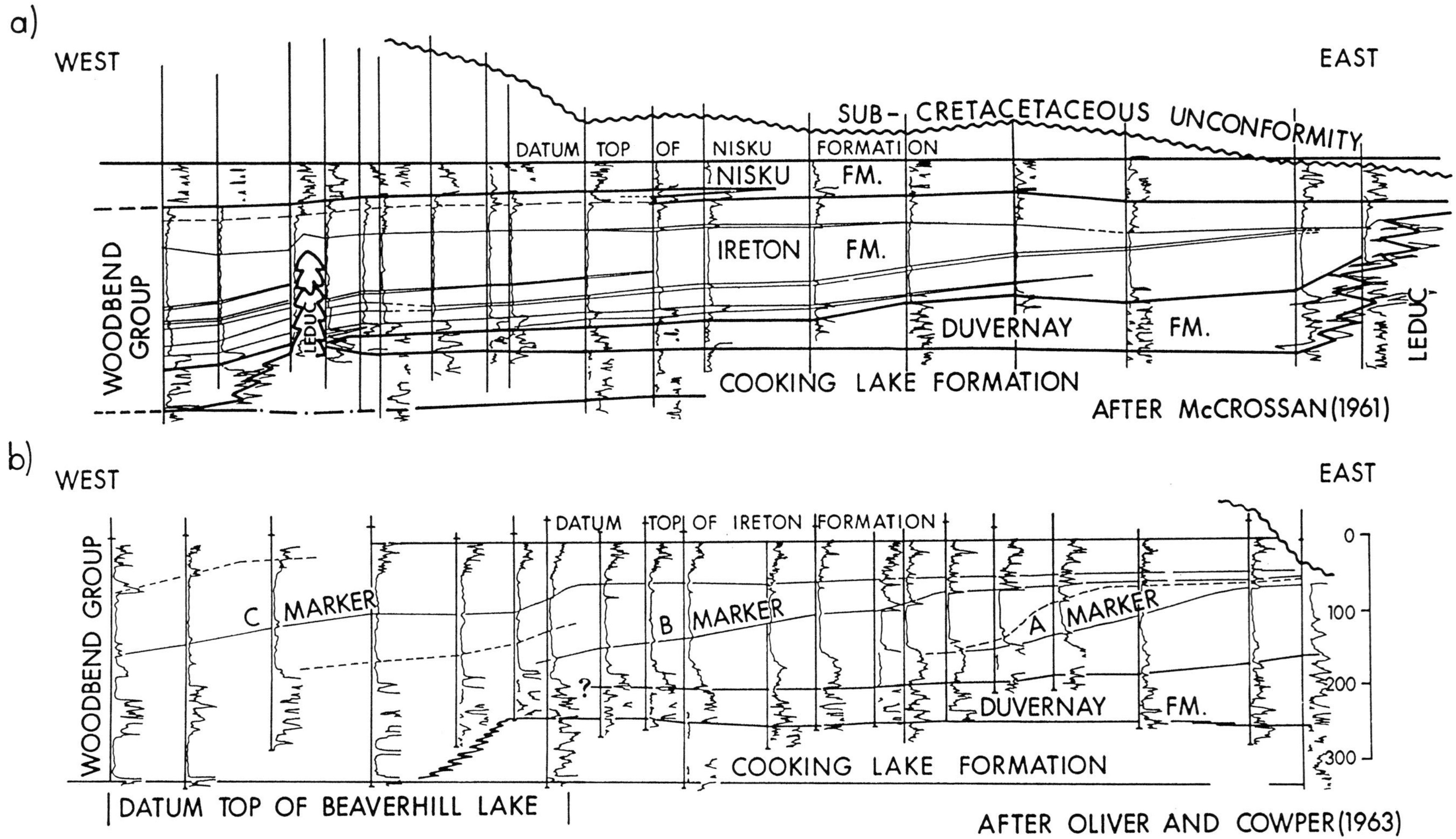


Figure 6. Marker correlations with the basin fill after various authors.

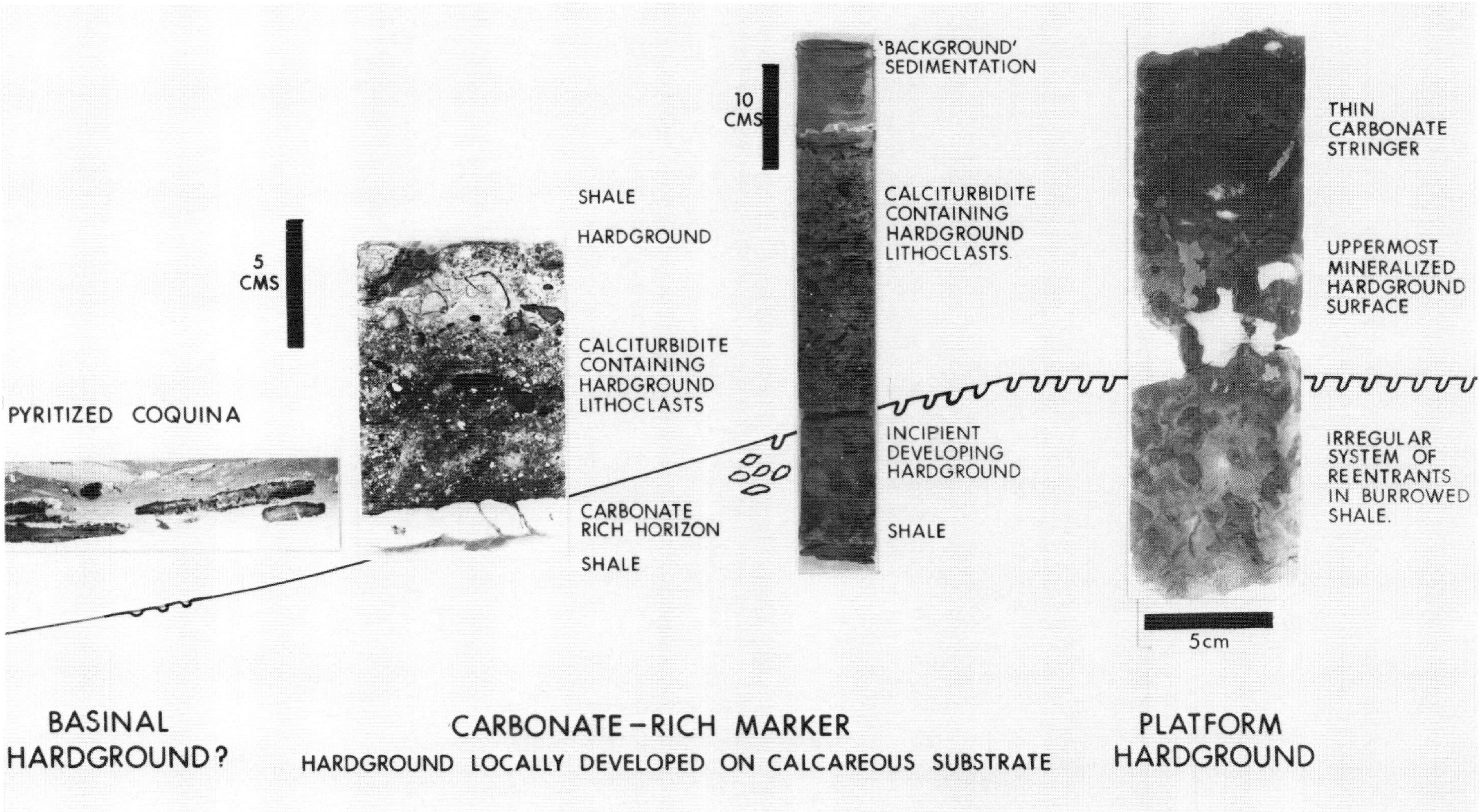


Figure 7. Lithologic components of a log marker surface.

bioturbated lime mudstones which contain fewer faunal elements and are characterized by a nodular, probably early diagenetic, texture. They are dominated by infaunal burrowers. On the toe of the slope and basinal portions sediments contain only a sparse pelagic fauna. These sediments are finely laminated and often organic-rich suggesting inhospitable probably anoxic bottom conditions. Indeed the succession of lithofacies reflects a tripartite layering of the water column similar to the model proposed by Byers (1977) and this is depicted in Figure 5. Faced with this monotonous succession of lithofacies it is easy to understand how patterns within the basin fill were long ignored or misunderstood.

*Stages of Basin Fill* - As increased control in basinal successions became available, a number of authors (McCrossan 1961; Oliver and Cowper, 1963) began to recognize the significance of correlatable surfaces within the basin fill (Figure 6). Stoakes (1980) first detailed a model for their occurrence and significance in understanding patterns of basin fill and reef growth in central Alberta. These beds have expression as resistivity "kicks" or clean gamma zones that are widely correlatable as surfaces in the basin fill. Examination in core revealed them to be either submarine hardground surfaces or calciturbidites (Figure 7). This suggested that they resulted from a pause in sedimentation that allowed the bathymetric profile to become lithified (Stoakes, 1980). Examination of the lateral disposition of lithofacies beneath a log marker surface revealed a normal distribution, related to depth, suggesting that the log marker depicted original submarine topography. Associated with the formation of this submarine-cemented log marker was a compositional change from more argillaceous below to more carbonate-rich above. Stoakes (1980) argued that this reflected control by relative sea-level rise which shut-off the extrabasinal source but permitted carbonate deposition to persist. This model is summarized in Figure 8.

Following relative sea-level rises, at certain stages of basin filling portions of the clinoform were in water depths suitable for carbonate platform or ramp deposition. This gave rise to a thin extensive unit that reflected the bathymetry of the pre-existing shale clinoform. It extends out into the basin and reflects a return to shallow-water carbonate growth. In this account it is termed a "piggyback" carbonate to reflect it's relation to shale deposition. The Camrose Member of the Ireton Formation (Stoakes, 1977) is one such "piggyback" carbonate.

*Reef Growth And Basin Fill* - Deposition of the basinal Duvernay and Ireton successions is assigned primarily to the regressive phase of the Woodbend megasequence (Chapter 8). The occurrence of a number of marker surfaces reflecting the basin configuration at various stages during basin filling allows for a more detailed evaluation of this assignation. Leduc reef growth was initiated by "drowning" of the more areally extensive Cooking Lake platform (Chapter 3). Initial sedimentation in the basin consists of micritic downslope-ramp lithologies that pass upwards into dark-coloured, organic-rich laminites of the Duvernay Formation. Subsequent markers in the Ireton downlap into this "condensed" interval. This suggests that

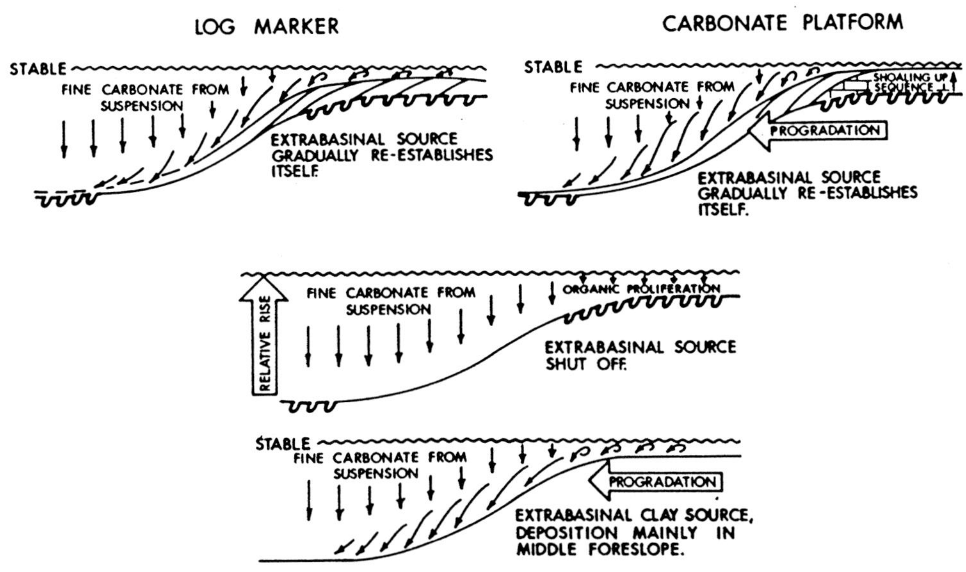


Figure 8. Sequential development of a log marker surface or "piggyback" carbonate ramp.

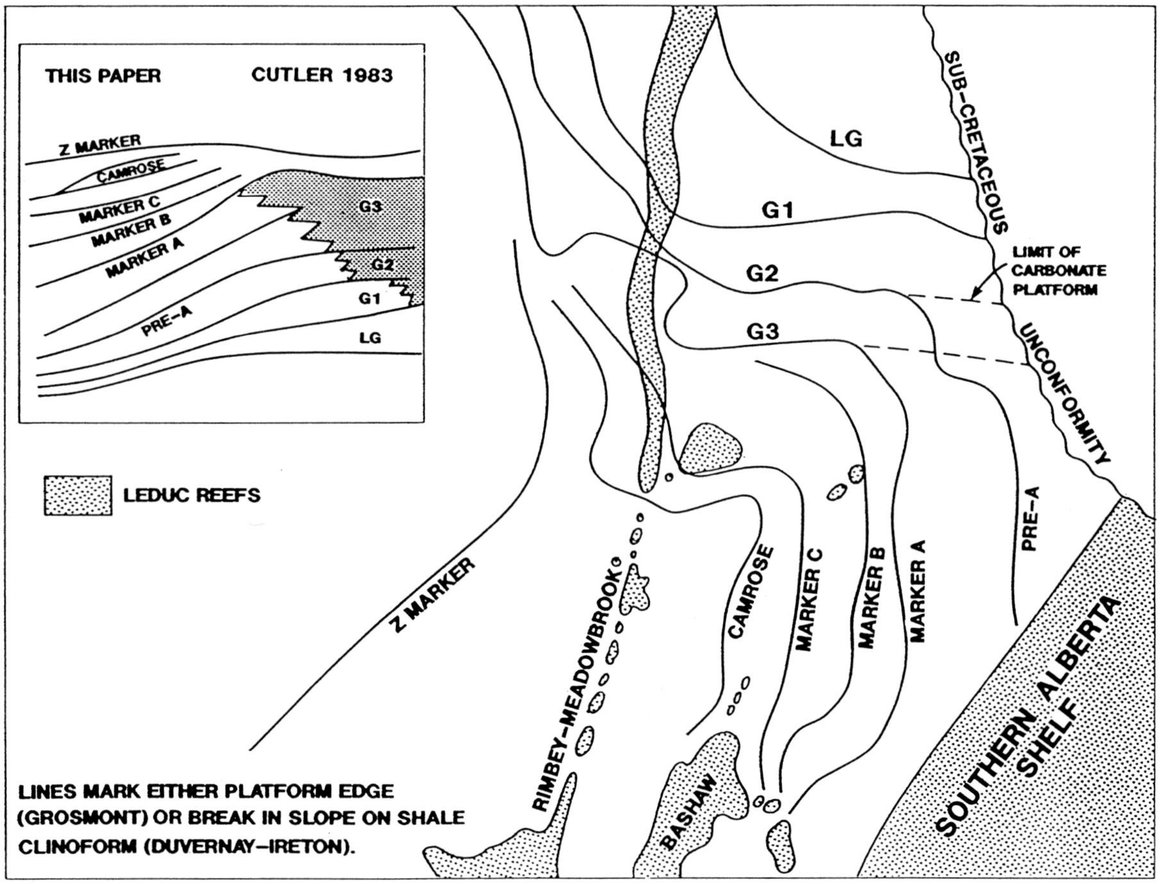


Figure 9. Map showing progressive basin filling of the Woodbend shales.

at least the initial stages of reef growth equate to only a small percentage of equivalent basinal sedimentation in central Alberta. This can be seen through the detailed correlations at Redwater (Chapter 8, Figure 4) where individual stages of reef growth are related to basin fill in the adjacent embayment. Following deposition of the uppermost middle Leduc, basin filling becomes dominant with individual clinoforms gradually encroaching upon and finally terminating areas of reef growth. This pattern is apparent first in the Grosmont Platform Complex which progressively terminated reef growth along the northerly extent of the Rimbey-Meadowbrook trend, and then in the Ireton of the East Shale Basin (Figure 9). It is apparent that deposition of the regressive phase was also characterized by an intermittently rising sea-level history with reef growth and basin filling taking place in a reciprocal fashion. However, following deposition of the upper middle Leduc overall slower rates of sea-level rise combined with the delivery of fine terrigenous sediment to the basin permitted material to progressively infill deeper-water areas, gradually encroaching upon and finally terminating areas of reef growth.

The Z Marker is a clinoform surface that reflects basin configuration at the end of the Woodbend megasequence. It indicates that Leduc reef growth on the east side of the Alberta Basin had ceased while those on the west continued to exist in open marine conditions. This marker provides a significant division in the basin-fill succession between Woodbend and Winterburn shale deposition (Stoakes and Wendte, 1987). However in all other respects (overall composition, lithofacies, biofacies etc.) the two basin-fill units are virtually indistinguishable.

## COMPARISON TO OTHER BASIN-FILL SUCCESSIONS

Deposition of The Ireton - Duvernay basin fill provides a model that can be applied to other clay-shale basin-fill successions in the Upper Devonian - Early Mississippian portion of the WCSB (Figure 10).

The Waterways Formation of the Beaverhill Lake Group (Chapter 7) differs the most of any of these basin-fill successions. In overall geometry and depositional patterns it is similar and contains a number of log marker-defined units revealing the internal depositional patterns (Sheasby, 1971; Keith, 1990). However compositionally the unit is dominantly comprised of fine-grained carbonate lacking any significant clay-shale content. The reason for this difference is thought to be a function of both the shallow bathymetry of this initial basin-flooding unit combined with the large areas of the basin succession that existed as shallow carbonate-dominated "banks" that shed material to the adjacent basin.

The Winterburn shale is similar in many respects to the Ireton model but appears to reflect more local variations as the basin of deposition sub-divides and fills (Chapter 9). Following Z Marker deposition a wedge of shale closes the entryway to the central Alberta Basin and for the first time the effects of the extrabasinal source area diminish. At this stage more localized source areas may have become important (i.e. Peace River Arch

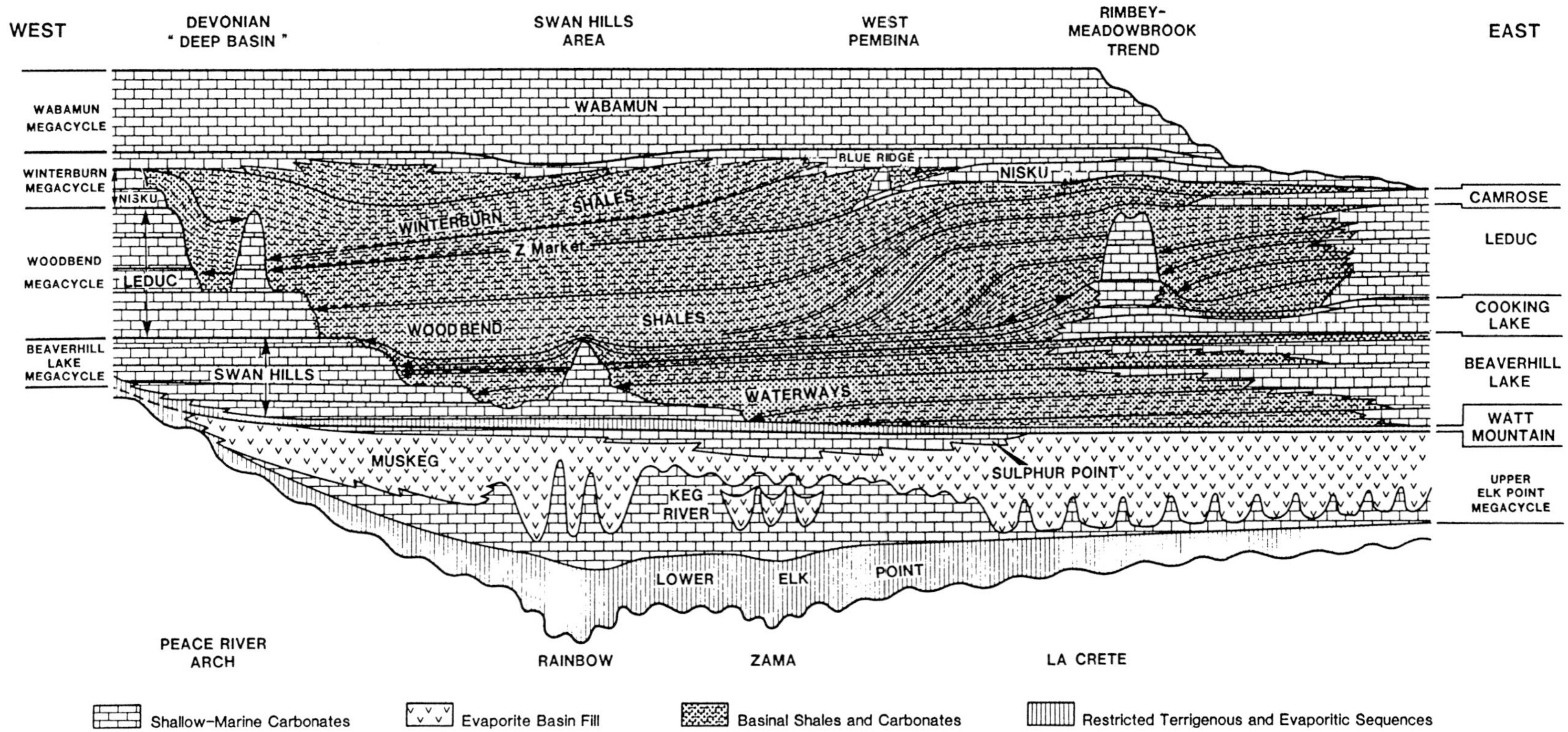


Figure 10. Composite schematic cross section illustrating the cyclicity of Devonian sequences.

landmass) and later stages of the "upper" Winterburn basin fill became more anhydritic. Deposition of shale from the extrabasinal source continues to be deposited north of the Peace River Arch infilling the basin out to a longitude of approximately $122^0$ degrees in northeast British Columbia and contiguous areas of the Northwest Territories.

Details of the Wabamun basin-fill succession are unclear due primarily to it's relatively limited areal extent in the WCSB (see Chapter 10).

Finally the Early Mississippian, Exshaw - Banff basin-fill exhibits a return to significantly siliciclastic basin filling (Chapter 11). It comprises a series of westerly prograding ramp systems (O'Connell, 1990) that infill and shallow the basin. The Exshaw Formation is an organic-rich unit that comprises the condensed interval and bottom-set beds of this prograding package.

*Hydrocarbon Source Rock Potential* - Appreciation of the nature and patterns of basin-fill sequences has many applications but perhaps the most important is an appreciation of hydrocarbon source potential. This section attempts to establish a model of source rock accumulation through detailed consideration of the Duvernay Formation of central Alberta and then applies this model to other potential source horizons.

*Duvernay Formation* - The Duvernay Formation of east-central Alberta is a typical marine, oil-prone source rock (Stoakes and Creaney, 1985). It accumulated as the deep-water distal toes of a number of individual Ireton clinoforms. Sediments reflect accumulation in water depths of greater than 100 m based on detailed paleobathymetric reconstructions (Stoakes, 1980).

Accumulation of organic-rich source intervals appears to be governed by the presence of euxinic (oxygen depleted) conditions at or below the sediment-water interface, a process well documented by Demaison and Moore (1980). This has the effect of inhibiting the burrowing infauna and allowing the accumulation and preferential preservation of organic matter. Stoakes (1980) has proposed a tripartite layering of the water column based on degree of oxygenation within the Duvernay - Ireton basin (Figure 5).

The Duvernay Formation comprises a number of interbedded lithofacies:

Nodular lime mudstones - generally less than 0.5 % Total Organic Carbon (TOC) by weight.

Laminites - dark organic-rich sediments that may contain up to 10 % TOC by weight.

The values of paleo-water depth given in Figure 5 serve as a general guide and may be expected to vary with the stage of basin evolution and circulation patterns. Many of the smaller scale variations in source quality and amount reflect short-term local perturbations of the system.

This phase of Woodbend sedimentation marks the maximum incursion of the Upper Devonian seas onto the craton. As such the Duvernay Formation marks the maximum paleobathymetry of the system

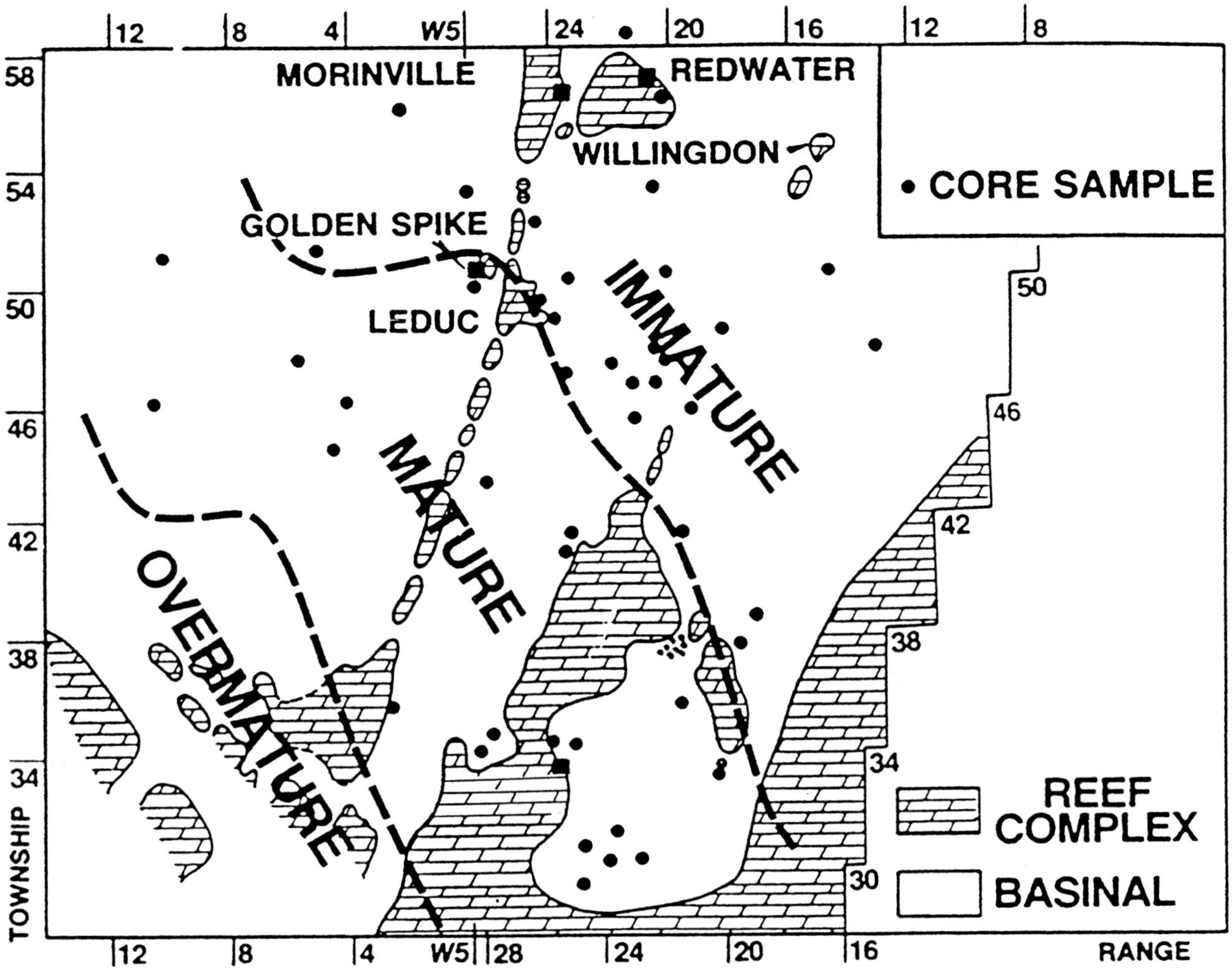


Figure 11. Map of Duvernay source rock maturity (from Stoakes and Creaney, 1985).

and this explains why it is so rich and laterally extensive (Figure 10). In east-central Alberta in the vicinity of the Rimbey - Meadowbrook reef trend the Duvernay has attained the oil window (Figure 11). Within this area alone an estimated 5 billion barrels of oil and 17 Tcf of gas, sourced from the Duvernay, are trapped in Leduc Formation reefs. This example represents a relatively "closed" system of maturation and migration. In other areas of Alberta oil from this prolific source rock has migrated stratigraphically upwards and downwards but structurally updip. In this way it has come to fill reservoirs in other megasequences, the Swan Hills Formation reefs of the Swan Hills area, the Keg River Formation platform of the Senex-Panny area and even the basal Devonian Sands (Watt Mountain) in a series of stratigraphic pinch-outs furthest updip (Creaney and Allan, 1990). If basin bathymetry combined with the presence of anoxia and slow sedimentation rates govern the accumulation of source rocks then this is favoured at a number of times during basin evolution. In particular it occurs during the early transgressive deepening phase of individual megasequences when deposition is predominantly concentrated around the perimeter of the basin system. In the Upper Devonian - Early Mississippian of the WCSB this is reflected at a number of separate levels (Figure 10).

*Beaverhill Lake Group* - To date no significant hydrocarbon source section has been identified in the Beaverhill Lake Group. This is not due to the absence of deeper-water sediments which can be documented in this sequence. Rather it appears to be more related to the overall shallow nature of the Waterways Basin which perhaps permitted more frequent overturning of any apparent "layering" in the water column.

*Winterburn Group* - The transgressive event that initiated Winterburn sedimentation over the Z Marker surface eventually led to the deposition of hydrocarbon source rocks in the Cynthia shale of the West Pembina area (Exploration Staff, Chevron Standard Limited, 1979). However this source interval is generally thin and rarely exhibits values of over 2 % TOC. It's occurrence is largely postulated because oils in the West Pembina Zeta Lake reefs show a progressive increase in API gravity with increasing burial that reflects the maturation of an adjacent source rock (Creaney and Allan, 1990). This combined with the lack of any effective conduit suggests they are locally sourced from an intimately associated source interval. The impoverished nature of this source interval may be explained because it is still relatively shallow in terms of absolute bathymetry. In addition the time duration of sedimentation is relatively brief with the Cynthia platform margin recovering and rapidly prograding the short distance basinwards to the area of the reefs. What perhaps is less readily explained is the absence of a significant source interval overlying the Z Marker in more basinward locations. If the Z Marker does indeed reflect basin bathymetry it is much greater in areas to the west which should have favoured source rock accumulation. Cored evidence, it is true, is unavailable but $\Delta$ log R (Passey and others, 1990) does not reveal any source potential on all the available wireline log control.

*Wabamun Group* - As mentioned previously, little is known of the basinal succession equivalent to the Wabamun ramp carbonates of the WCSB. However examination of the distal part of this ramp sequence in the Parkland Field area of northeast British Columbia (Chapter 10) reveals a potential source interval in the middle of the unit presumably equating to the change from transgressive to regressive sedimentation. It's exact distribution is still largely undefined at this time.

*Exshaw Formation* - The transgressive deepening that initiates the Banff cycle is marked by the occurrence of another prolific source section, the Exshaw Formation. This unit contains Type 1 organic material with TOC's up to 12.5 % (Creaney and Stoakes, 1987) and reaches up to several metres in thickness. In northern areas of the WCSB it overlies tight distal carbonate ramps of the uppermost Wabamun (Chapter 11) and may be considered a "closed" hydrocarbon system. In these areas it acts as the local source for fault-related dolomitization traps (Stoakes,1987). Further south it encounters porosity in the underlying Wabamun and is involved with the erosional unconformity.

*Pekisko Formation* - Finally mention may be made of the base of the Pekisko-Shunda cycle detailed in Chapter 11. Here transgressive deepening over a basal Pekisko ramp sequence gave rise to isolate pinnacle reef growth concomitant with backstepping of the basin margin. Here $\Delta$ log R mapping has revealed the presence of source intervals that accumulated in deeper water adjacent to areas of reef growth (Stoakes and others, 1989). In this respect the situation bears marked similarity to the Cynthia shale occurrence at West Pembina. It differs in one significant respect, namely that the area of the presently known Pekisko pinnacles occurs in a source rock section that has not yet reached thermal maturity.

## SUMMARY

Within the Upper Devonian - Early Mississippian of the WCSB a number of thick basin-fill successions occurs. They largely relate to deposition during the regressive phase of sedimentation of successive megasequences. The thickest and most areally extensive package belongs to the Duvernay and Ireton formations of the Woodbend megasequence. This marks the regression following the point of maximum incursion of the Upper Devonian seas onto the craton. It also coincides with the delivery of the first significant amounts of terrigenous clay-shale sediments into central Alberta. Mapping of a number of synchronous log marker surfaces reveals the pattern and progression of basin fill and it's relation to reef growth. Log markers formed in response to relative sea-level rises that shut-off terrigenous influx to the basin but did not affect fine carbonate sedimentation. The ultimate source of the extrabasinal terrigenous sediment is thought to be the Ellesmerian Orogeny in the Canadian Arctic Archipelago. It gained access to the basin vis a seaway between the Peace River

Arch landmass and Grosmont Platform Complex. This funnelling to eastern portions of the basin gave rise to an asymmetry of fill that is most marked in Woodbend deposition and continued into the Winterburn shale interval. Differences existing between this model succession and other basin-fill sequences are apparent. Early in Winterburn "time" the entry point to the basin became closed and the nature of Winterburn shale fill changed from shale to more evaporitic in nature. The Waterways Formation of the Beaverhill Lake Group is significantly more carbonate-rich than the Woodbend or Winterburn successions reflecting the shallow bathymetry of the basin and presence of large shale banks, the shallow portions of which shed significant quantities of fine carbonate material.

Finally basin-fill sequences contain significant hydrocarbon source rocks. The most prolific of these is the Duvernay Formation of the Woodbend Group. It marks the maximum transgression and consequently paleobathymetry of the Upper Devonian Basin. It accumulated as a deep-water deposit under anoxic conditions that favoured the preferential preservation of organic material sourced in the overlying water column.

Hydrocarbon source intervals characterize other transgressive intervals within basinal areas. Consequently then the first-order predication of the nature and occurrence of hydrocarbon source intervals relates to sea-level history and stage of basin evolution. Secondary controls such as absolute bathymetry, circulation, degree of anoxia and time duration will affect the more local presence and thickness of these intervals.

## REFERENCES

BYERS, C.W., 1977, Biofacies patterns in euxinic basins: a general model, *in* Cook, H.E., and Enos, P., eds., Deep Water Carbonate Environments: Society of Economic Paleontologists and Mineralogists, Special Publication no. 25, p.5-17.

CAMPBELL, F.A., AND OLIVER, T.A., 1968, Mineralogic and chemical composition of Ireton and Duvernay formations, central Alberta: Bulletin of Canadian Petroleum Geology, v. 16, p.40-63.

CREANEY, S., AND ALLAN, J., 1990, Hydrocarbon generation and migration in the Western Canada Sedimentary Basin, *in* Brooks, J., ed., Classic Petroleum Provinces: Geological Society of London, Special Publication no. 50, p.189-202.

CREANEY, S., AND STOAKES, F.A., 1987, The Exshaw Formation Laminites in the Peace River Arch area: Pattern of maturation and migration with exploration implications for the Wabamun Group: Canadian Society of Petroleum Geologists, Second International Symposium on the Devonian System, Program Abstract, p. 64.

DEMAISON, G.J., AND MOORE, G.T., 1980, Anoxic environments and oil source bed genesis: American Association of Petroleum Geologists Bulletin, v. 64, p. 1179-1209.

EXPLORATION STAFF, CHEVRON STANDARD LIMITED, 1979, The geology, geophysics and significance of the Nisku reef discoveries, west Pembina area, Alberta, Canada: Bulletin of Canadian Petroleum Geology, v. 27, p. 326-359.

KEITH, J.W., 1990, The influence of the Peace River Arch on Beaverhill Lake sedimentation: Bulletin of Canadian Petroleum Geology, v. 38A, p. 55-65.

McCROSSAN, R.G., 1961, Resistivity mapping and petrophysical study of Upper Devonian inter-reef calcareous shales of central Alberta, Canada: American Association of Petroleum Geologists Bulletin, v. 45, p.441-470.

O'CONNELL, S.C., 1990, The development of the Lower Carboniferous Peace River Embayment as determined from Banff and Pekisko Formation depositional patterns: Bulletin of Canadian Petroleum Geology, v. 38A, p.93-114.

OLIVER, T.A., AND COWPER, N.W., 1963, Depositional environments of the Ireton Formation, central Alberta: Bulletin of Canadian Petroleum Geology, v. 11, p. 183-202.

PASSEY, Q.R., CREANEY, S., KULLA, J.B., MORETTI, F.J. AND STROUD, J.D., 1990, A practical model for organic richness from porosity and resistivity logs: American Association of Petroleum Geologists Bulletin, v. 74, p. 1777-1794.

RICH, J.L., 1951, Three critical environments of deposition and criteria for recognition of rocks deposited in each of them: Bulletin of the Geological Society of America, v. 62, p. 1-20.

SHEASBY, N.M., 1971, Depositional patterns of the Upper Devonian Waterways Formation, Swan Hills area, Alberta: Bulletin of Canadian Petroleum Geology, v. 19, p. 377-404.

STOAKES, F.A., 1977, The Camrose Member of the Ireton Formation in the Stettler area of central Alberta, *in* McIlreath, I.A., and Harrison, R.S., eds., The Geology of Selected Oil, Gas and Lead-Zinc Reservoirs in Western Canada: Fifth Canadian Society of Petroleum Geologists Core Conference, Calgary, Alberta, p.89-106.

STOAKES, F.A., 1979, Sea level control of carbonate deposition during progradational basin-filling: The Upper Devonian Duvernay and Ireton formations of Alberta, Canada: Unpublished Ph.D. thesis, University of Calgary, Calgary, Alberta, 346 p.

STOAKES, F.A., 1980, Nature and control of shale basin fill and its effect on reef growth and termination: Upper Devonian Duvernay and Ireton formations of Alberta, Canada: Bulletin of Canadian Petroleum Geology, v. 28, p. 234-410.

STOAKES, F.A., 1987, Fault controlled dolomitization of the Wabamun Group, Tangent Field, Peace River Arch, Alberta, *in* Krause F.F., and Burrowes, O.G., eds., Devonian Lithofacies and Reservoir Styles in Alberta: 13th CSPG Core Conference and Display and Second International Symposium on the Devonian System, p. 73-85.

STOAKES, F.A., AND CREANEY, S., 1985, Sedimentology of a carbonate source rock: The Duvernay Formation of Alberta, Canada, *in* Longman, M.W., Shanley, K.W., Lindsay, R.F., and Eby, D.E., eds., Rocky Mountain Carbonate Reservoirs - a core workshop: Society of Economic Paleontologists and Mineralogists Core Workshop no.7, Golden, Colorado, p.343-375.

STOAKES, F.A., HASSLER, G.T., AND FOELLMER, K.E.H., 1989, Stratigraphy and Hydrocarbon potential of the Pekisko Waulsortian mound play, Peace River Arch, Alberta: CSPG - CSEG Exploration Update, Programs and Abstracts, p.172.

STOAKES, F.A., AND WENDTE, J.C., 1987, The Woodbend Group, *in* Krause, F.F., and Burrowes, O.G., eds., Devonian Lithofacies and Reservoir Styles in Alberta: 13th CSPG Core Conference and Second International Symposium on the Devonian System, p. 153-170.

# CHAPTER 6

# UPPER ELK POINT MEGASEQUENCE

C.V. CAMPBELL

## INTRODUCTION

The Upper Elk Point Subgroup is comprised of a succession of shallow-water carbonates, (mainly basin-filling) evaporites and some siliciclastics up to over 400 m in thickness. This succession represents the initial open-marine inundation of the Alberta "Basin". Upper Elk Point strata overlie a more restricted assemblage of shallow-water carbonates, evaporites and siliciclastics of the Lower Elk Point Subgroup and are overlain by deposits of the Beaverhill Lake Group (discussed in Chapter 7). It is particularly well known for two reasons. Firstly, isolate reefs of the Keg River and more recently Winnipegosis formations provide significant hydrocarbon reservoirs sealed by evaporitic successions. Keg River reefs of the Rainbow area in northern Alberta are particularly well known. Secondly, deposits of the Muskeg and Prairie formations comprise one of the most widespread and thick evaporite successions in the world. Evaporites vary in composition from calcium carbonate to potash salts.

Most of the succession, up to the base of the Watt Mountain Formation, is interpreted to comprise part of a Devonian first-order cycle, as discussed in Chapter 2. The position of the base of this megacycle is unknown, but questionably occurs at the base of the Cold Lake salt. Anhydrites and restricted carbonates of the Chinchaga Formation, overlying the Cold Lake salt, would then be part of the transgressive phase of the Upper Elk Point megacycle. The overlying Watt Mountain Formation, a succession of dominantly siliciclastics, is the uppermost unit of the Upper Elk Point Subgroup, but is interpreted by us to be part of the Beaverhill Lake megasequence. This succession, therefore, is discussed in the ensuing Beaverhill Lake chapter.

Figure 1 depicts the general paleogeography of the Upper Elk Point megasequence. Carbonates and evaporites extend from the Presqu'ile barrier reef in northeast British Columbia, northwest Alberta, and the southern Northwest Territories down through Alberta and Saskatchewan into Manitoba and the contiguous parts of the northern United States. In central and southern Alberta, the "basin" is bounded by the Peace River and Western Alberta arches. Terrigenous deposits fringe these positive features and grade laterally into a widespread shelfal carbonate. Further basinward, thick evaporite successions occur in the Saskatchewan portion of the Elk Point Basin, in the La Crete Basin of northeast Alberta, and in small sub-basins, such as Rainbow, Zama and Shekilie north of the Peace River Arch. These basins and sub-basins also contain the hydrocarbon-bearing, isolate reefs previously alluded to.

Both Williams (1984) and Campbell (1987) emphasize that the

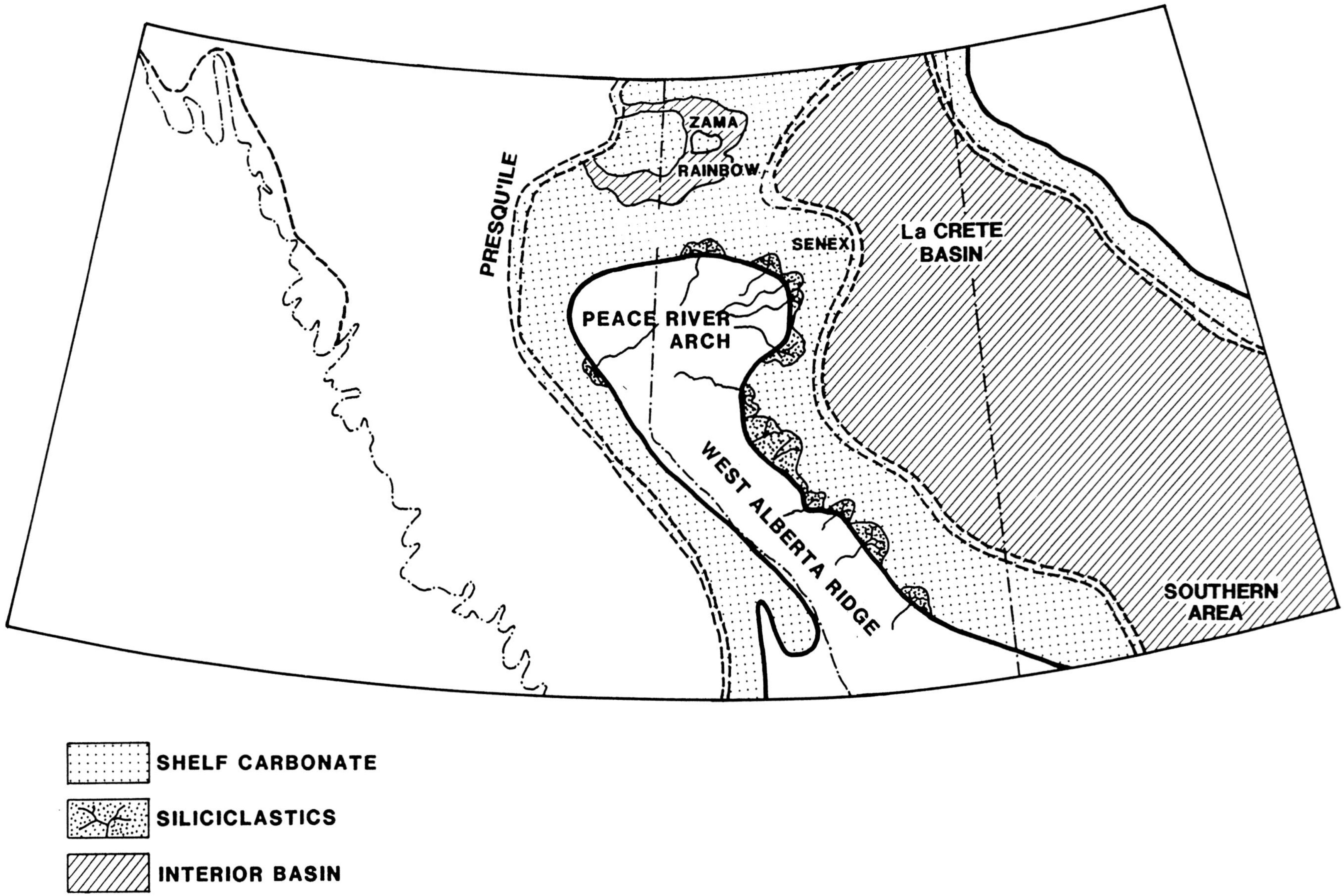


FIGURE 1. Paleogeographic Reconstruction of Elk Point Basin

present outline of the Elk Point Basin is largely the result of post-Devonian tectonics and erosion. A large segment of what was the Elk Point Basin is gone. The northeastern and eastern boundary of the basin is an erosional edge. Much of what is now the Canadian Shield may have been covered by Elk Point sediments. Because of the northeastern erosional edge, the projection of the Presqu'ile barrier to the northeast beyond its subcrop limits can only be hypothesized. As pointed out by Williams (1984), the present popular view that the Presqu'ile barrier streched across the northern depositional margin of the Elk Point Basin is only one of several possible configurations, as illustrated in Figure 2. What is clear, however, is that the Saskatchewan portion of the Elk Point Basin, the La Crete Basin in northeastern Alberta and the smaller sub-basins in northern Alberta were connected to an open-marine seaway during periods of isolate reef growth.

## STRATIGRAPHIC NOMENCLATURE

Stratigraphic nomenclature in this chapter generally follows that of previous authors (Law, 1955; Grayston and others, 1964; Hriskevich, 1966; and McCamis and Griffith, 1967; Reinson and Wardlaw, 1972 and Kendall, 1975). Figure 3 shows the nomenclature of the Elk Point Group for the various paleogeographic domains discussed in this chapter. Upper Elk Point deposits overlie anhydrites of the Chinchaga Formation and terrigenous clastics of the Ashern Formation.

Widespread, open-marine ramp carbonates of the Lower Keg River Member and the lower portion of the Winnipegosis Formation comprise the basal unit of the Upper Elk Point Subgroup throughout most of northeast British Columbia, Alberta and Saskatchewan. This unit is overlain by either widespread shelfal (bank) carbonates or by isolate reef and bank carbonates of the Upper Keg River Member and the upper part of the Winnipegosis Formation. In Saskatchewan, organic-rich carbonate laminites of the Brightholme Member (informal) and light coloured (non-organic) carbonate laminites of the Ratner Member occur between the crest of the isolate reefs and banks and are included as part of the Winnipegosis Formation. The Brightholme Member is coeval to shelfal, bank, and reef carbonates of the Upper Member of the Winnipegosis Formation. The Ratner Member postdates the growth of the adjacent Winnipegosis shelf, banks and reefs and represents a transitional, mesosaline phase. Evaporite deposits of the Muskeg and Prairie formations overlie the carbonate successions.

In the Rainbow and Zama sub-basins of northern Alberta, the Muskeg Formation is divided into three units: lower and upper evaporite successions and an intermediate carbonate interval. The lower evaporite succession corresponds to the Black Creek Salt Member in the Rainbow sub-basin and the Lower Anhydrite Member in the Zama sub-basin. Near the centre of the Zama sub-basin, the anhydrite in the latter member is largely a relict or residual deposit, the remains from wholesale dissolution of a Black Creek-equivalent salt. The overlying carbonate interval corresponds to the widespread, sheet carbonates of the Zama Member in the Zama sub-basin and a poorly developed equivalent in the Rainbow sub-

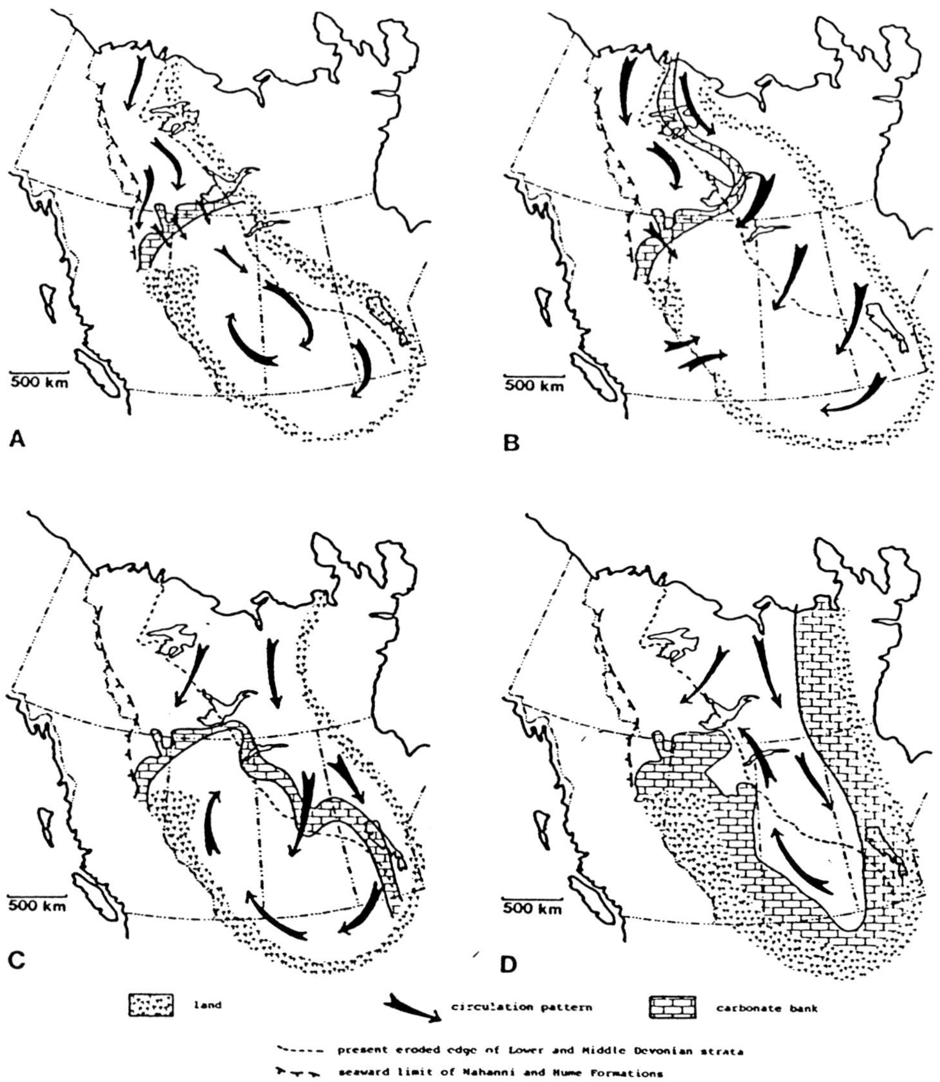


FIGURE 2. Four Possible Paleogeographic Reconstructions of the Elk Point Basin (Williams, 1984)

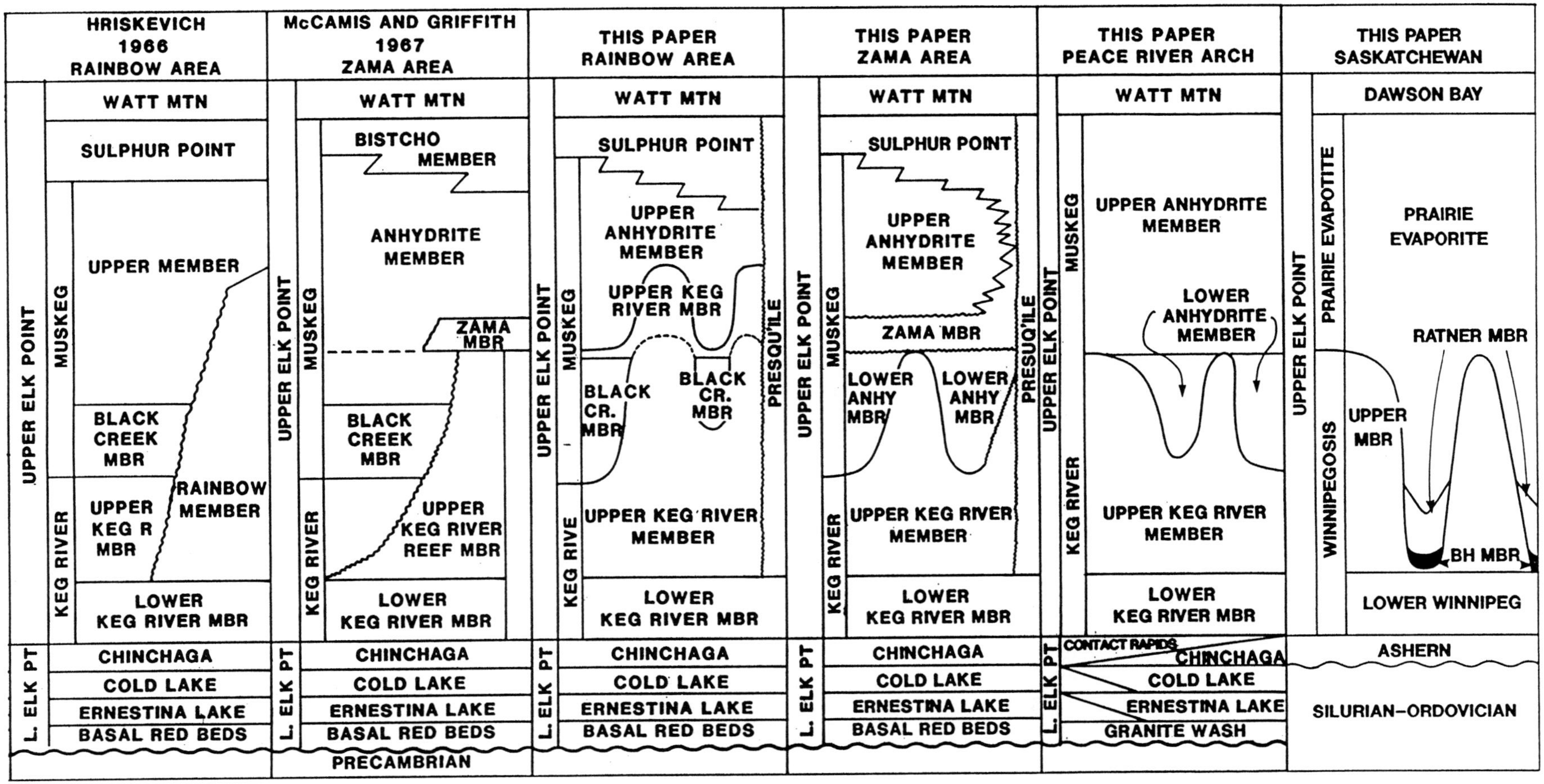


FIGURE 3. Elk Point Group Stratigraphic Nomenclature

basin that is difficult to identify. The upper evaporite succession corresponds to the Upper Anhydrite Member in both the Zama and Rainbow sub-basins. The subdivision of lower and upper evaporite successions also applies to the Peace River Arch.

The uppermost unit of the Upper Elk Point megasequence in northern Alberta and northeastern British Columbia is the Sulphur Point Formation. This is a wedge-shaped carbonate succession that roots directly on shallow-water carbonates (Presqu'ile Formation) along the Presqu'ile barrier margin complex and progressively thins into the Elk Point Basin to the south and east. In these directions, it changes facies into anhydrites of the Upper Anhydrite Member of the Muskeg Formation and is restricted in occurrence to the area north of the Peace River Arch.

## MODEL FOR UPPER ELK POINT MEGASEQUENCE

A general model depicting the evolution of the Upper Elk Point megasequence is shown on a generalized cross section extending from the Presqu'ile barrier complex, southeast across the Zama and Rainbow sub-basins, into the La Crete Basin and up onto the fringing shelf (Senex-Panny Platform) around the Peace River Arch (Figure 4). The evolution of the megasequence in the La Crete area closely approximates the overall evolution of the sequence in southern Saskatchewan.

The position of the actual base of the Upper Elk Point megacycle is conjectural, but lies somewhere in the Lower Elk Point Subgroup. What is known is that thick successions of Lower Elk Point strata are confined to lows in the Precambrian basement. These strata are dominantly evaporites and record the initial incursion of Devonian seas into the Alberta Basin.

The overlying Upper Elk Point succession generally conforms to the overall transgressive-regressive nature of first-order Devonian cycles. An important cycle level occurs near the middle of this succession and corresponds to a cycle magnitude somewhere between that of our first-order and normal second-order cycles, as discussed in Chapter 2.

The deposition of Lower Keg River carbonates marks the initial incursion of open-marine waters into the Alberta Basin. This interval consists of a succession of open-marine downslope ramp carbonates that shoal up into shallower facies over basement highs, such as the Hay River Fault zone along the southeast flank of the Rainbow sub-basin. Another increment of sea-level rise localized shelfal carbonate growth on a platform fringing the Peace River Arch and on banks overlying either pronounced basement highs (Hay River Fault Zone) or more subtle highs (corresponding to the position of the Hay River Bank). These banks were not isolate features, but were joined to the platform fringing the Peace River Arch in a configuration outside that depicted on the cross section. At this time, slightly lower lying areas of the Lower Keg River were drowned, and became deep-water embayments. Various segments of these embayments are referred to as sub-basins. Isolate pinnacle reefs formed on underlying Lower Keg River shoals that occur on the rampal succession in the sub-basins.

Shallow-water carbonate growth was halted by a sea-level drop.

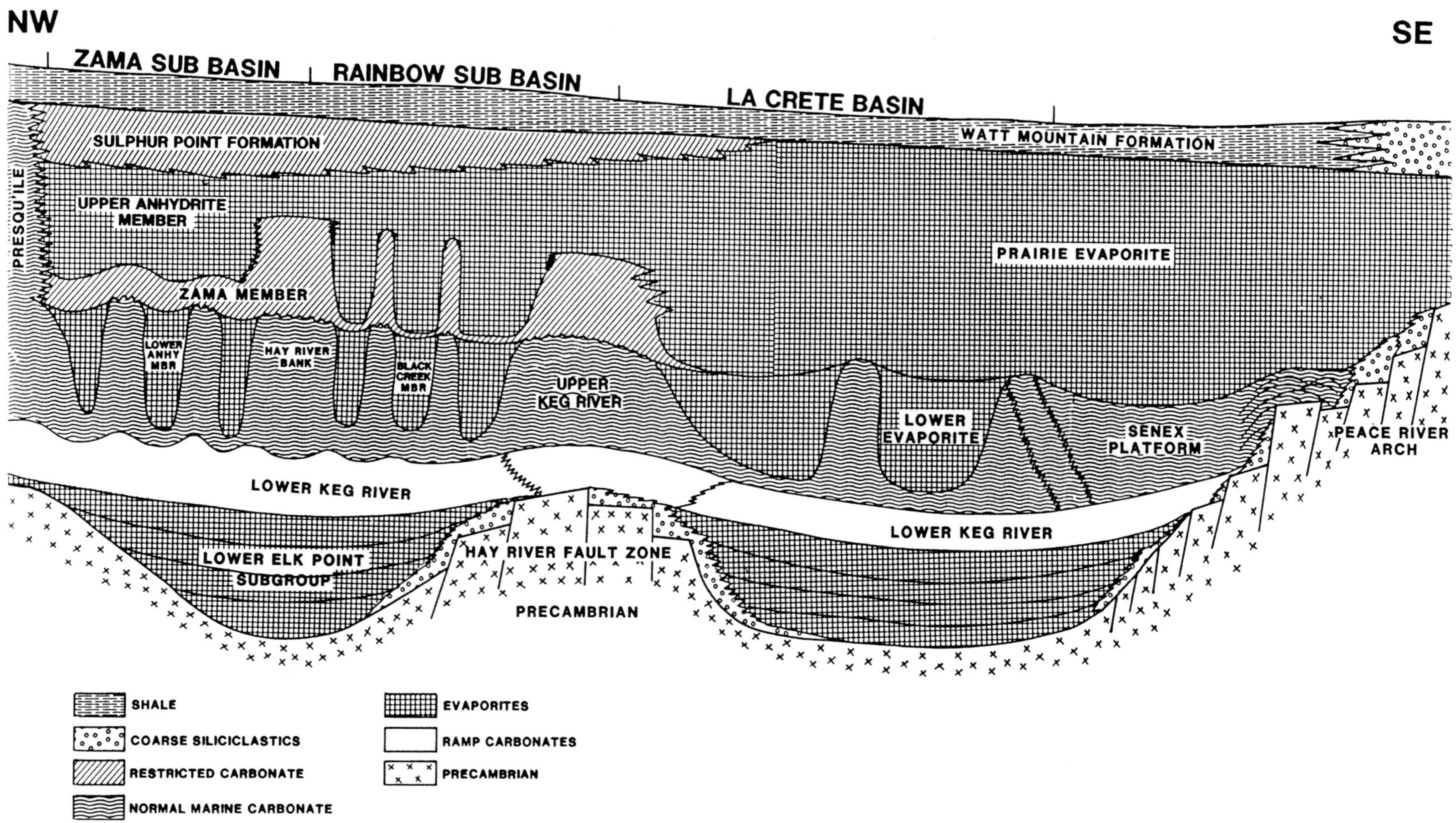


FIGURE 4. Generalized Cross Section from Northwestern Elk Point Basins to the Peace River Arch

Whether this drop resulted from a regional lowering of sea level or from evaporitic drawdown of marine water in the Elk Point Basin is not known. This drop led to exposure of shallow-water banks and reefs and to hypersaline conditions in the sub-basins. The hypersaline conditions caused the demise of reef growth, on their submerged flanks as well as on the emergent crests, and initiated precipitation of evaporites (mainly salts) that largely filled in the sub-basins and basins.

A gradual rise in sea level led to the reflooding of the shallow-water banks and reef crests in the Elk Point Basin. In this phase, a relatively restricted carbonate was added to the bank and reef tops. In the Zama sub-basin, where water depths in inter-reef areas were shallow, a shallow ramp-style carbonate was deposited on the reefs and in the intervening basinal areas. In the Rainbow sub-basin, where water depths between the reefs positions were apparently deeper, a shallow-water carbonate was deposited only on the reefs. Hypersaline conditions precluded carbonate growth on the Senex-Panny Platform and in Saskatchewan.

Circulation was sufficient for continued carbonate growth on both the relatively restricted reefs in the Rainbow sub-basin and on adjoining parts of the fringing banks or platforms. This shallow-water growth was partly contemporaneous with alternating gypsum (later converted to anhydrite) and carbonate deposition in the Rainbow and Zama sub-basins and on more restricted parts of the adjoining banks. Gradually circulation in the Rainbow sub-basin became more restricted and the waters more hypersaline. This led to cessation of shallow-water carbonate deposition on the adjoining banks. Shallow-water carbonate deposition only continued on the better circulated barrier platform complex.

Hypersalinity existed throughout the Elk Point Basin behind the barrier platform complex. Shallow-water carbonates were deposited continuously only on this more normal-marine barrier platform complex. Interbedded gypsum (later converted to anhydrite) and carbonates continued to be deposited behind the barrier. This successsion of evaporites marks the "regressive" infilling of the basins and sub-basins.

Toward the end of this megasequence, a gradually rising sea level caused incursion of more normal-marine waters into the Elk Point Basin from northwest to southeast. In the area north of the Peace River Arch, normal-marine waters gradually replaced hypersaline waters. This condition led to the deposition of shallow-water carbonates of the Sulphur Point Formation whereas Muskeg evaporites continued to precipitate from hypersaline waters to the south and east.

A gradual lowering of sea level in a later phase of Sulphur Point deposition led to subaerial exposure along the Peace River Arch and cessation of shallow-water carbonate deposition in basinward positions. The occurrence of "green shales" in the upper part of the Sulphur Point Formation in northern Alberta corresponds to the episodic shedding of fine siliciclastic sediments from the Peace River Arch. These beds may mark the initial phases of widespread subaerial exposure on the Arch.

More detailed accounts of Upper Elk Point strata in northern Alberta (northerly realm), along the Peace River Arch and in

Saskatchewan (southern region) are provided in the following sections.

## NORTHERLY REALM

The northerly realm comprises the area with small sub-basins behind the barrier margin complex in northwestern Alberta. The overall evolution of this realm was presented in the previous section. This section provides some additional detail on the makeup and evolution of reef complexes in the Rainbow, Zama and Shekilie sub-basins. Figure 5 shows a schematic cross section from the Shekilie sub-basin to the Rainbow sub-basin.

Isolate reef complexes in the Zama, Shekilie and Rainbow sub-basins overlie the Lower Keg River Member. This unit is a blanket-like biostromal carbonate up to approximately 60 m thick that sharply, but conformably overlies bedded anhydrites of the Chinchaga Formation. This carbonate comprises a succession of open-marine mudstones and wackestones deposited along a ramp. Crinoidal shoals developed during the latest phase of this succession and provided the topographic highs that localized Upper Keg River reef growth (Langton and Chin, 1968).

The isolate reefs in the Zama sub-basin are true "pinnacle" reefs. They are almost always less than a half section (0.65 $km^2$) in size, are approximately 100 m thick and have steeply dipping foreslopes ($30^0$-$40^0$). In vertical section they comprise one mega-shoaling upward cycle of deposition, a composite of the second-order cycles described in Chapter 2. The volumetrically most dominant facies in these complexes is allochthonous foreslope debris consisting of abraded stromatoporoid and coral bioclasts and peloidal and skeletal carbonate sands. These deposits either grade basinward into lime mudstones and wackestones or up into a narrow and rather discontinuous reef-margin facies. The reef-margin facies typically grades back into abraded stromatoporoid debris and peloidal sands deposited along a reef flat. The crest of these buildups consist of cyclic subtidal carbonate sands and muds and intertidal cryptalgal mat deposits. These deposits are restricted in occurrence to the upper part of the reef complexes, marking a phase where the reefs had finally built to sea level.

Ramp carbonates of the Zama Member (Muskeg Formation) overlie both the reefs and the off-reef evaporitic basinal-fill succession (the Lower Anhydrite Member). This succession varies in thickness from as thin as 16 m over the reefs to as much as 31 m between the reefs. Lithologically, the lower part of this unit comprises a succession of pellet carbonate sands, mudstones and wackestones. Some stromatoporoids and corals are present. The upper part of this member consists of pellet grainstones and mudstones with a much more restrictive fauna. These deposits grade up into bedded nodular anhydrites of the Upper Anhydrite Member. Subtle but notable variations from reef to off-reef positions indicate the presence of shoals corresponding to positions over the reefs.

The isolate reefs in the Shekilie sub-basin are generally similar to those in the Zama sub-basin. The Zama carbonate in this sub-basin, however, is thinner and has a deeper water aspect.

The isolate reefs in the Rainbow Sub-basin are thicker (up to

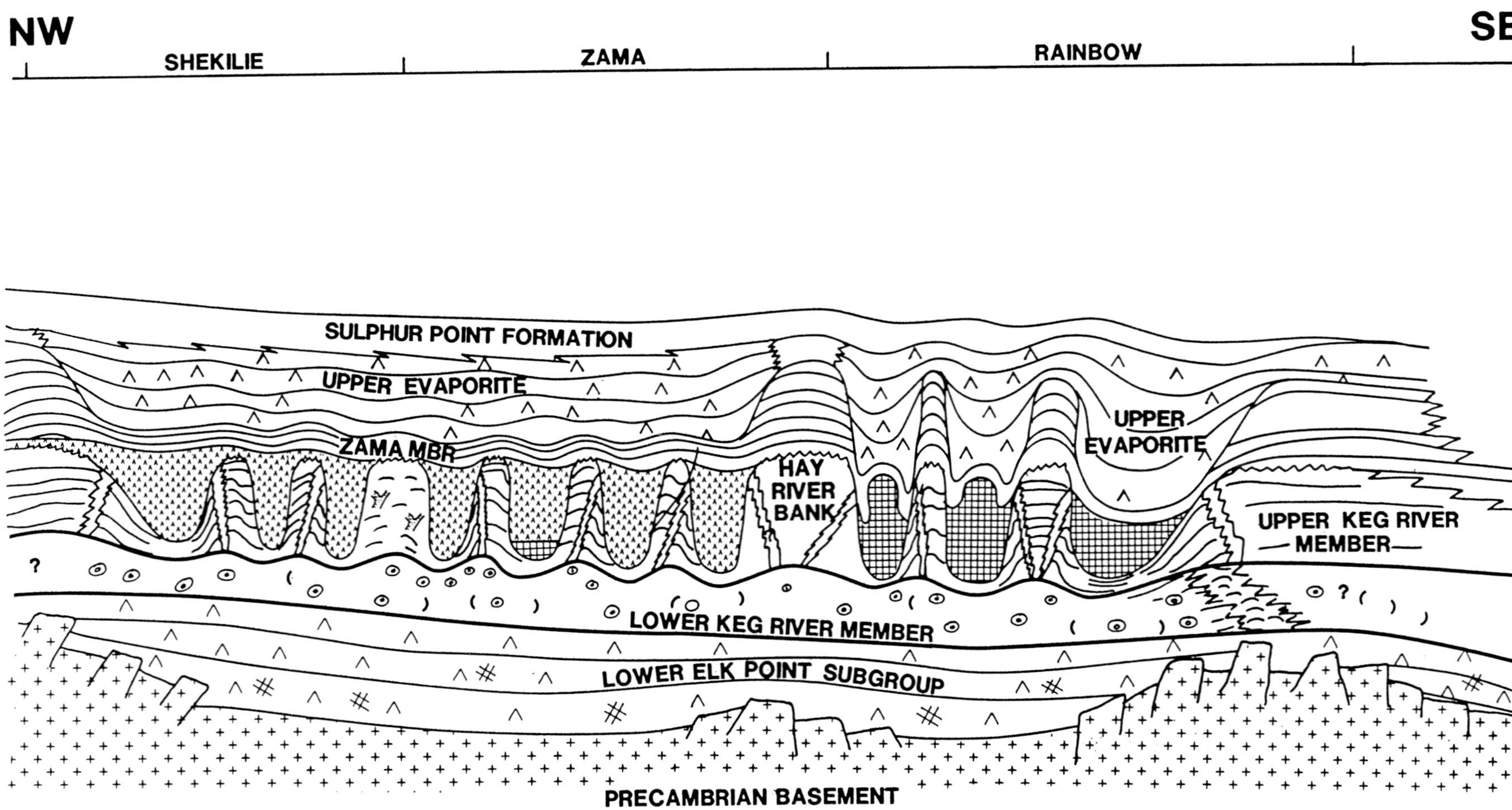


FIGURE 5. Schematic Cross Section from Shekilie Sub-basin to Rainbow Sub-basin

230 m) and have more variable areal dimensions and internal facies compositions than their Zama counterparts. The lithologic aspects of these reefs were initially described by Langton and Chin (1968). A more recent description of one of these reefs is provided by Frydl (1989). Unlike the reefs in the Zama sub-basin, they exhibit tremendous variations in thickness reflecting more than one episode of reef demise. They range from "pinnacle" reefs, similar in areal extent and facies make-up to those in the Zama sub-basin, to "atoll" complexes up to 5 km across. The foreslopes of Rainbow reefs, as those in the Zama sub-basin, are steep ($30^0$-$40^0$). Isopachous marine cements, described by Schmidt and others (1977), occur in an abundance within these foreslope deposits.

The greater thickness of Rainbow reefs reflects a much longer phase of reef growth as well as greater subsidence (or accomodation) in the Rainbow Sub-basin. Rainbow reefs have stages of growth that correspond to the reef stage at Zama, to the Zama carbonate, and to some portion of the succeeding Upper Anhydrite Member. The evolutionary development of one of the larger atoll reefs is shown in Figure 6. The initial stage of the reef corresponds to the Zama reef phase. The only major difference is the occurrence of a much larger lagoon in this complex. The upper stage of growth corresponds to both Zama carbonate and post-Zama carbonate phases in the Zama sub-basin. Commonly this stage lacks true reef margins and has a more restricted fauna. The distribution of facies corresponds to more of a "shoal" complex than to an atoll reef. The nature of this upper stage of growth reflects somewhat elevated salinities in the sub-basin at this time. Further and progressive basin restriction led to phases of reef demise.

## PEACE RIVER ARCH AREA

Although similar depositional units to those found in the Rainbow-Zama area may be recognized surrounding the Peace River Arch, the mode of deposition is somewhat different. The Peace River Arch was a positive feature all through deposition of the Elk Point Group. As a result, the marine incursion that initiated open-marine carbonate accumulation to the north caused the development of a widespread platform system surrounding the Arch. The initial phase of accumulation, the Lower Keg River, is similar to that in the northern sub-basins. The Lower Keg River has a ramp style profile with a deeper water aspect in a basinward direction away from the exposed Precambrian basement. As sea level continued to rise a broad platform system developed that fringed the Arch. Seaward of the platform margin a series of pinnacle reefs formed in the adjoining La Crete Basin. Figure 7 shows a generalized cross section depicting the evolution and distribution of the major domains in the Senex area. The shelf in this area is divided into three domains:

1) the inner shelf
2) the middle shelf
3) the shelf margin

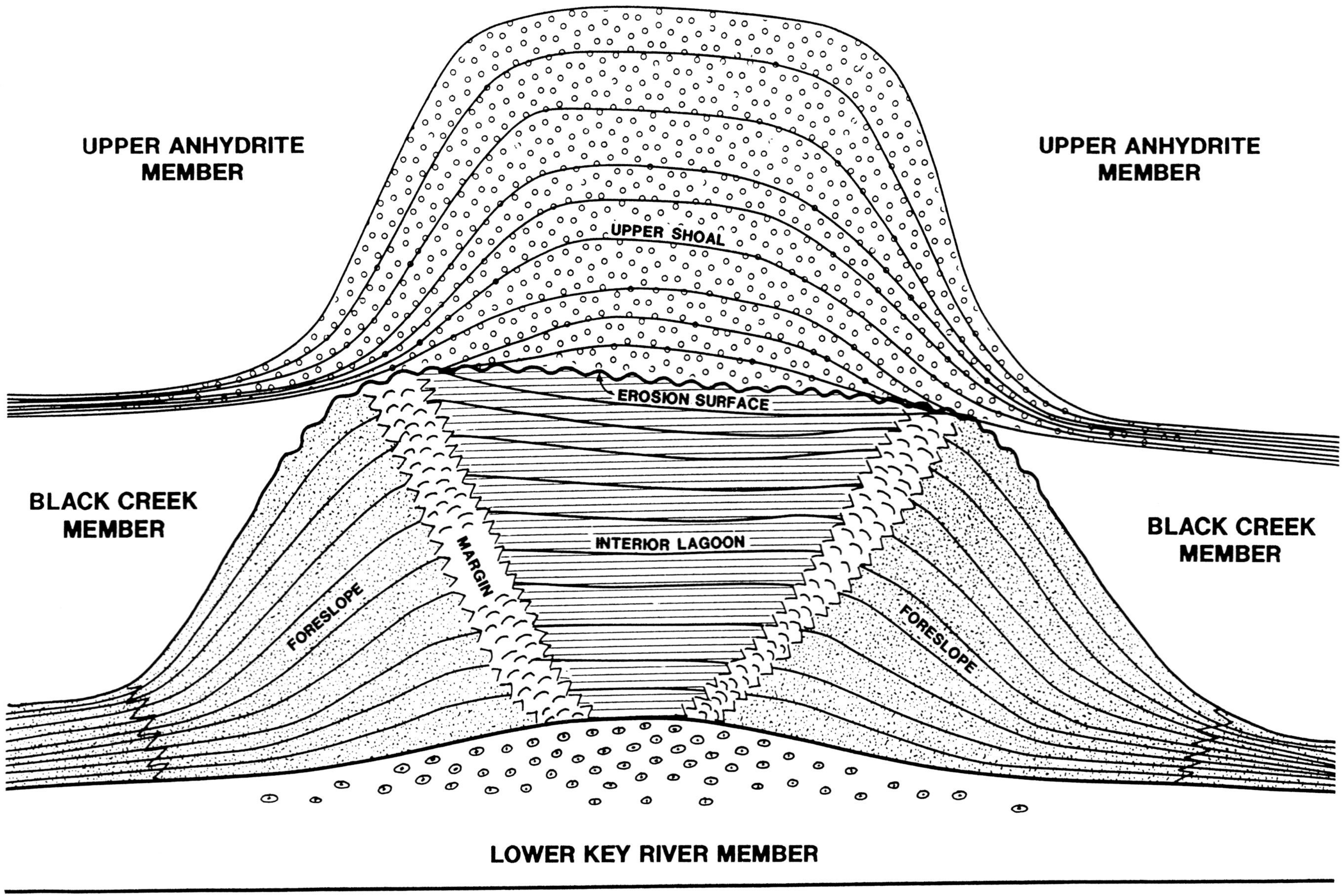


FIGURE 6. Model for Buildup Development in the Rainbow Basin

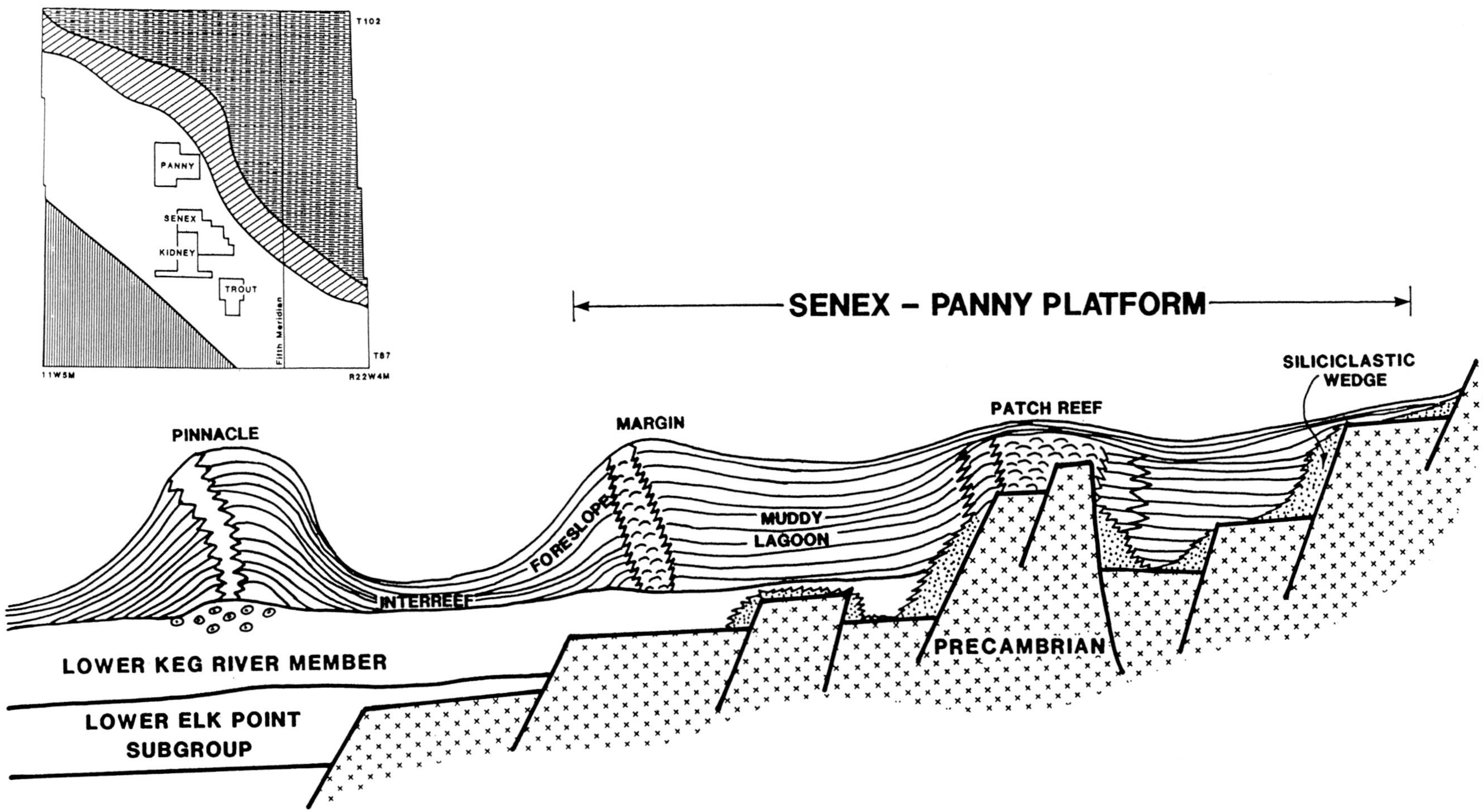


FIGURE 7. Schematic Cross Section Across Senex-Panny Area Shelf

The inner shelf is characterized by a high siliciclastic content and generally restricted carbonate deposits. The siliciclastics are derived from the exposed landmass of the Peace River Arch and overall fine upwards. The siliciclastics also change facies in a vertical manner becoming progressively more marine towards the top of the section. This change reflects a retreating style of deposition. Carbonates are more commonly found at the top of the section and generally consist of bioturbated and laminated micritic deposits.

The middle shelf belt contains the majority of hydrocarbon production from the Keg River carbonates around the Peace River Arch. It is characterized by repeated shale and carbonate cycles that can be correlated regionally. Each complete cycle appears to have been deposited during a pulse of sea-level rise and subsequent still stand. The carbonate portion of each cycle appears to have been deposited during times of rising sea level when the siliciclastics were trapped within valley systems futher up on the Arch. During stillstands the siliciclastics prograded out across the shelf.

In this realm, carbonate sedimentation was especially focussed around basement highs. Carbonate buildups fringing and overlying Precambrian highs provide the major hydrocarbon reservoirs in the Senex-Panny area. The nature of these bioherms vary, with more stromatoporoids and less interbedded shales occurring in a more basinward direction. Intervening deeper water areas consist predominantly of micritic sediments with scattered fossils.

The margin of the Senex-Panny shelf consists of a grainstone belt with only scattered coral and stromatoporoid bioclasts. This belt may correspond to a slightly deeper water deposit that formed on a distally deepened ramp. Isolate reefs generally comparable to those in the Zama area occur in the adjoining parts of the La Crete Basin. These reefs, however, are wet due to the lack of a thermally mature source (Hassler and others, 1987).

## SOUTHERN REGION

Upper Elk Point strata in the southern region of the Elk Point Basin in southern Saskatchewan have many features in common with equivalent age strata along the Peace River Arch and in the Zama sub-basin. The southerly portion of the basin is fringed by an extensive carbonate platform of the Winnipegosis Formation on its eastern, southern and western sides. This carbonate platform is contemporaneous with the Upper Keg River platform in the Senex-Panny area along the Peace River Arch. Isolate pinnacle reefs and banks in the basin are coeval with pinnacle reefs in the La Crete Basin and the Zama sub-basin of northern Alberta. However, until the Home Tablelands oil discovery in 1986, they were long thought to be devoid of hydrocarbons. This discovery has sparked new interest in the play (Martindale and MacDonald, 1989).

Figure 8 shows a generalized cross of Upper Elk Point strata in southern Saskatchewan. The basal Upper Elk Point strata are open-marine ramp carbonates of the Lower Winnipegosis Formation. The bathymetry on the top of this unit played an important role in the localization and distribution of the Upper Winnipegosis and

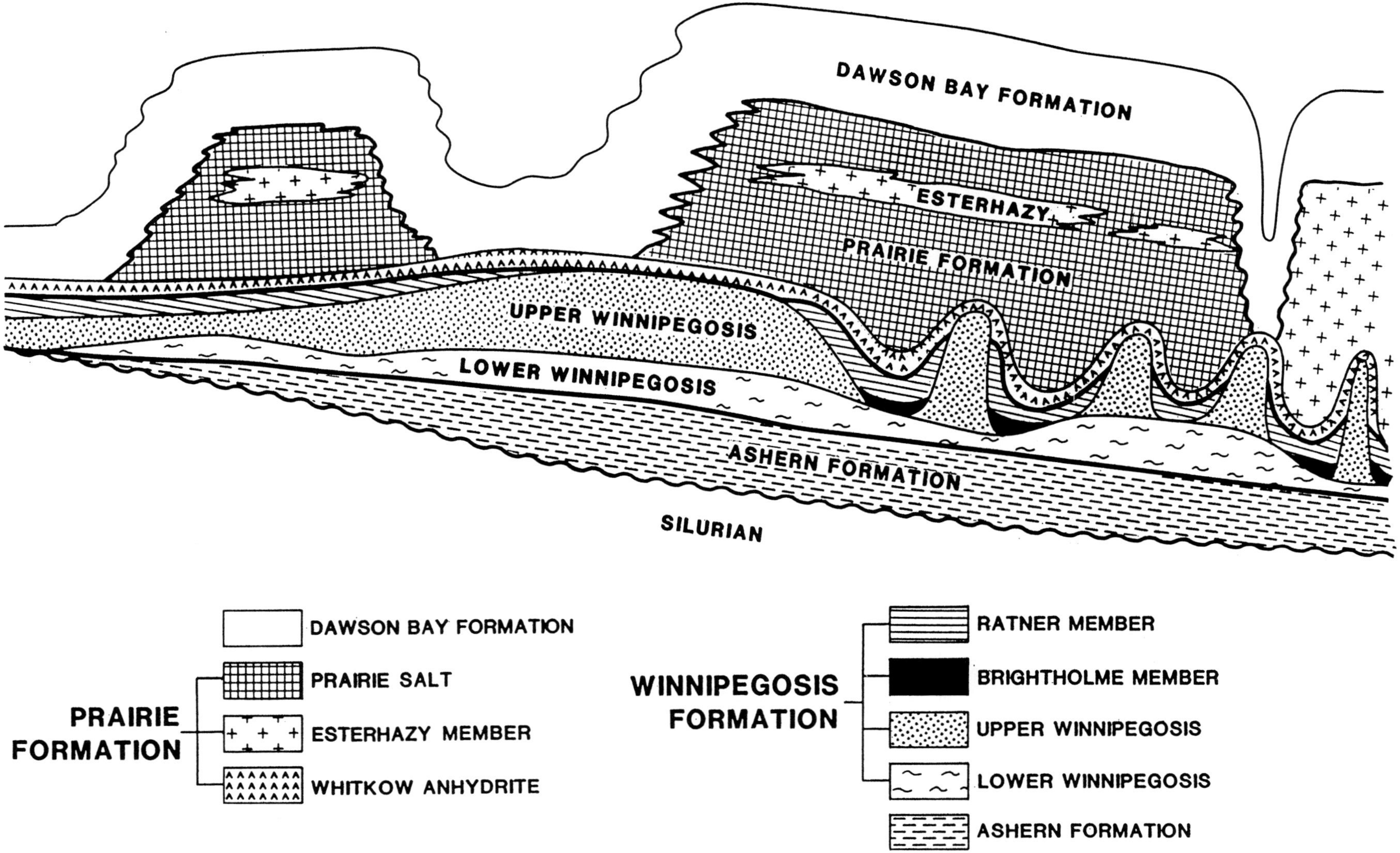


FIGURE 8. Schematic Cross Section Across Southern Saskatchewan Winnipegosis Shelf

Brightholme members.

The Upper Winnipegosis and Brightholme members were deposited simultaneously. The Upper Winnipegosis forms the main carbonate features in the area, consisting of a regionally extensive shelf attaining a maximum thickness of 38 m and isolate pinnacles up to 60 m thick. The location of the shelf margin appears to be controlled by a thickening of the Lower Winnipegosis ramp.

Basinward of the shelf edge a number of carbonate buildups occur. These buildups are composed of a variety of lithofacies including pelloidal and skeletal grainstones and algal wackestones. Only minor coral-stromatoporoid boundstones and rudstones are present (Martindale and MacDonald, 1989). The coeval Brightholme Member consists of dark laminated, bituminous deposits, reflecting euxinic conditions in parts of the basin. The presence of this unit appears to be controlled by the bathymetry on the underlying Lower Winnipegosis ramp. Brightholme laminites are present only in areas with thin Lower Winnipegosis accumulations. These deposits, where thermally mature, provide the source of oils found in the isolate Winnipegosis reefs (Osadetz and others, 1989).

Overlying the Brightholme laminites and downslope portions of the Upper Winnipegosis banks and isolate reefs is a distinct, organic-lean, light coloured laminated mudstone referred to as the Ratner Member (Reinson and Wardlaw, 1972). It is present throughout the area except where accumulation of the Upper Winnipegosis exceeds approximately 45 m. This succession is interpreted to be a mesosaline, possibly evaporitic deposit forming the basal unit of the brining-upward sequence of the Prairie Formation (Kendall, 1975).

Capping the Winnipegosis Formation and providing a permeability seal are the evaporites of the Prairie Formation. The base of the Prairie Formation is marked by a regionally extensive unit called the Whitkow anhydrite. Overlying the Whitkow anhydrite are a series of salt units of the Upper Prairie Formation. Encased within these salts are a series of potash beds called the Esterhazy Member. Locally salts of the Prairie Formation have undergone dissolution leading to collapse of the overlying Dawson Bay onto the top of the Whitkow anhydrite.

## SUMMARY

The evolution of the Upper Elk Point megacycle reflects the interplay between sea level, tectonics and climate.

The onset of Keg River and Winnipegosis deposition marks the initial inundation of open-marine seas into the Alberta Basin. Deposition of the initial phases of the Keg River and Winnipegosis formations occurred during times of rapid sea-level rise. Following the deposition of widespread ramp deposits of the Lower Keg River and Lower Winnipegosis members, shallow-water shelfal deposits became confined to slowly subsiding areas. More rapidly subsiding lower-lying areas became parts of either a widespread basin in Saskatchewan or in northeastern Alberta (the La Crete Basin) or embayments referred to as sub-basins. Isolate reefs formed on shoals within the basins or sub-basins. Both shelfs and isolate reefs continued to grow until killed by hypersalinity caused by at least a basin-wide lowering of sea level. Evaporites formed during this "low-stand" of the sea and partially to almost completely filled in basin and sub-basin topography.

As sea level rose again shallow-water carbonate deposition became confined to areas with normal-marine salinities close to the barrier margin. These carbonates include both areally widespread platforms and ramps as well as isolate reefs in the Rainbow sub-basin. Gradually, the area behind the barrier margin became hypersaline during a sustained phase of an overall slow rise in sea level. As a consequence, carbonate growth in northern Alberta continued up to varying levels and was replaced by evaporite formation. Evaporite deposition continued almost to the close of the megacycle, interrupted only by the transgressive Sulphur Point phase in northwestern Alberta and northeastern British Columbia.

## REFERENCES

CAMPBELL, C.V., 1987, Stratigraphy and Facies of the Upper Elk Point Subgroup, northern Alberta., *in* Krause, F.F., and Burrows, O.G., eds., Devonian lithofacies and reservoir styles in Alberta: 13th CSPG Core Conference and Second International Symposium on the Devonian System, Calgary, Alberta, p. 243-286.

FRYDL, P., 1989, Reservoir heterogeneity in the Keg River `AA' Pool Rainbow field, Alberta: Canadian Society of Petroleum Geologists, 1989 Core Conference, Geology and Reservoir Heterogeneity, Calgary, Alberta, p. 1.1-1.14.

GRAYSTON, L.D., SHERWIN, D.R., AND ALLAN, J.F., 1964, Middle Devonian, *in* McCrossan, R.G., and Glaister, R.P., eds., Geological History of Western Canada: Alberta Society of Petroleum Geologists, p. 49-59.

HASSLER, G.T., CAMPBELL, C.V., STOAKES, F.A., AND FOELLMER, K.E.H., 1987, Keg River stratigraphy and depositional facies, Senex area, Peace River Arch, West Central Alberta, *in* International Symposium on the Devonian System: Canadian Society of Petroleum Geologists, Program and Abstracts, p. 110.

HRISKEVICH, M.E., 1966, Stratigraphy of Middle Devonian and older rocks of Banff Aquitaine Rainbow West 7-32 discovery well, Alberta: Bulletin of Canadian Petroleum Geology, v. 14, p. 241-255.

KENDALL, A.C., 1975, The Ashern, Winnipegosis and Lower Prairie Evaporite formations of the commercial potash areas, *in* Christopher, J.E., and MacDonald, R., eds., Summary of Investigations 1975 by the Saskatchewan Geological Survey, Department of Mineral Resources, p.61-65.

LANGTON, R.J., AND CHIN, G.C., 1968, Rainbow Member facies and related reservoir properties, Rainbow Lake, Alberta: Bulletin of Canadian Petroleum Geology, v. 16, p. 104-143.

LAW, J., 1955, Geology of northwestern Alberta and adjacent areas: American Association of Petroleum Geologists Bulletin, v. 39, p. 1927-1978.

MARTINDALE, W. AND MacDONALD, R.W., 1989, Sedimentology and diagenesis of the Winnipegosis Formation (Middle Devonian) Tableland area, southeastern Saskatchewan: Canadian Society of Petroleum Geologists, 1989 Core Conference, Geology and Reservoir Heterogeneity, Calgary, Alberta, p. 2.1-2.52.

McCAMIS, J.G., AND GRIFFITH, L.S., 1967, Middle Devonian facies relationships, Zama area, Alberta: Bulletin of Canadian Petroleum Geology, v. 15, p. 434-467.

OSADETZ, K.G., SNOWDON, L.R., STOAKES, F.A., AND HASSLER, G.T., 1989, Geochemistry of Winnipegosis source rocks of the Williston Basin: Criteria for Exploration Success: CSPG - CSEG Exploration Update `89, Programs and Abstracts, p. 38.

REINSON, G.E., AND WARDLAW, N.C., 1972, Nomenclature and stratigraphic relationships, Winnipegosis and Prairie Evaporite formations, central Saskatchewan: Bulletin of Canadian Petroleum Geology, v. 20, p. 301-320.

SCHMIDT, V., McDONALD, D.A., AND McILREATH, I.A., 1977, Growth and diagenesis of Middle Devonian Keg River cementation reefs, Rainbow Field, Alberta, *in* McIlreath, I.A., and Harrison, R.D., eds., Supplement to the Geology of Selected Carbonate Oil, Gas and Lead-Zinc Reservoirs in Western Canada: Canadian Society of Petroleum Geology, Calgary, Alberta, p. 1-21.

WILLIAMS, G.K., 1984, Some musings on the Devonian Elk Point Basin, Western Canada: Bulletin of Canadian Petroleum Geology, v. 32, p. 216-232.

CHAPTER 7

# BEAVERHILL LAKE MEGASEQUENCE

C.V. CAMPBELL

## INTRODUCTION

This succession incorporates the entire Beaverhill Lake Group and the underlying Watt Mountain Formation. These units are interpreted to comprise a first-order cycle. Collectively, the progression of strata in this megasequence corresponds to an overall transgressive-regressive succession. Thicknesses of these strata range from nil along the Peace River Arch where these units onlap out, to over 200 m in east-central and northern Alberta. This pattern reflects greater rates of subsidence away from the Peace River and Western Alberta arches.

Figure 1 depicts the general paleogeography of the Beaverhill Lake megasequence in the WCSB. This map illustrates the distribution of both shallow-water and basinal facies. In Alberta the dominant element is a deep-water basin that extends as far south as south-central Alberta. Because of the truncation of Beaverhill Lake strata along the pre-Jurassic and pre-Cretaceous unconformities, the configuration of the connection of this basin to the deep-water seaway in the Northwest Territories is unknown.

The basin is fringed by shallow-water platform carbonates on both its east and west sides. The platform carbonates along the southeast side of the basin grade into the more restictive carbonate-evaporite successions of the Souris River Formation. Shallow-water carbonates along the west side of the basin and south of the Peace River Arch include backstepping platform and bank successions. These units onlap and ultimately cover the Western Alberta Arch. Of particular note in this area is the occurrence of the Swan Hills platform margin at Caroline. A giant gas-condensate pool was discovered here by Shell in 1985 which ranks as the most significant hydrocarbon discovery in Devonian rocks in the WCSB in the last quarter century. Further north, isolate Swan Hills reef complexes occur on a more areally extensive carbonate platform, basinward of time-synchronous bank development. Collectively, more oil is pooled in these reef complexes than any others in the WCSB.

Backstepping carbonates of the Slave Point Formation onlap against both the southeast and northern flanks of the Peace River Arch. The Peace River Arch remained emergent throughout the entire Beaverhill Lake interval and provided a source for terrigenous clastic sediment. A time-transgressive belt of these terrigenous sediments, generally called the Granite Wash, fringe the Arch and intertongues with the Slave Point carbonates.

Slave Point carbonates along the north side of the Peace River Arch extend to the barrier margin in northeast British Columbia and the southern Northwest Territories. Part of this reef-fringed margin formed on the Tathlina Arch. An open seaway occurs to the west and northwest of this barrier complex. An extension of this seaway, the Hotchkiss Channel, extended into Alberta along the north flank of the Peace River Arch. Significant gas and gas-condensate discoveries have been made in platform complexes fringing this embayment in the past 10 years.

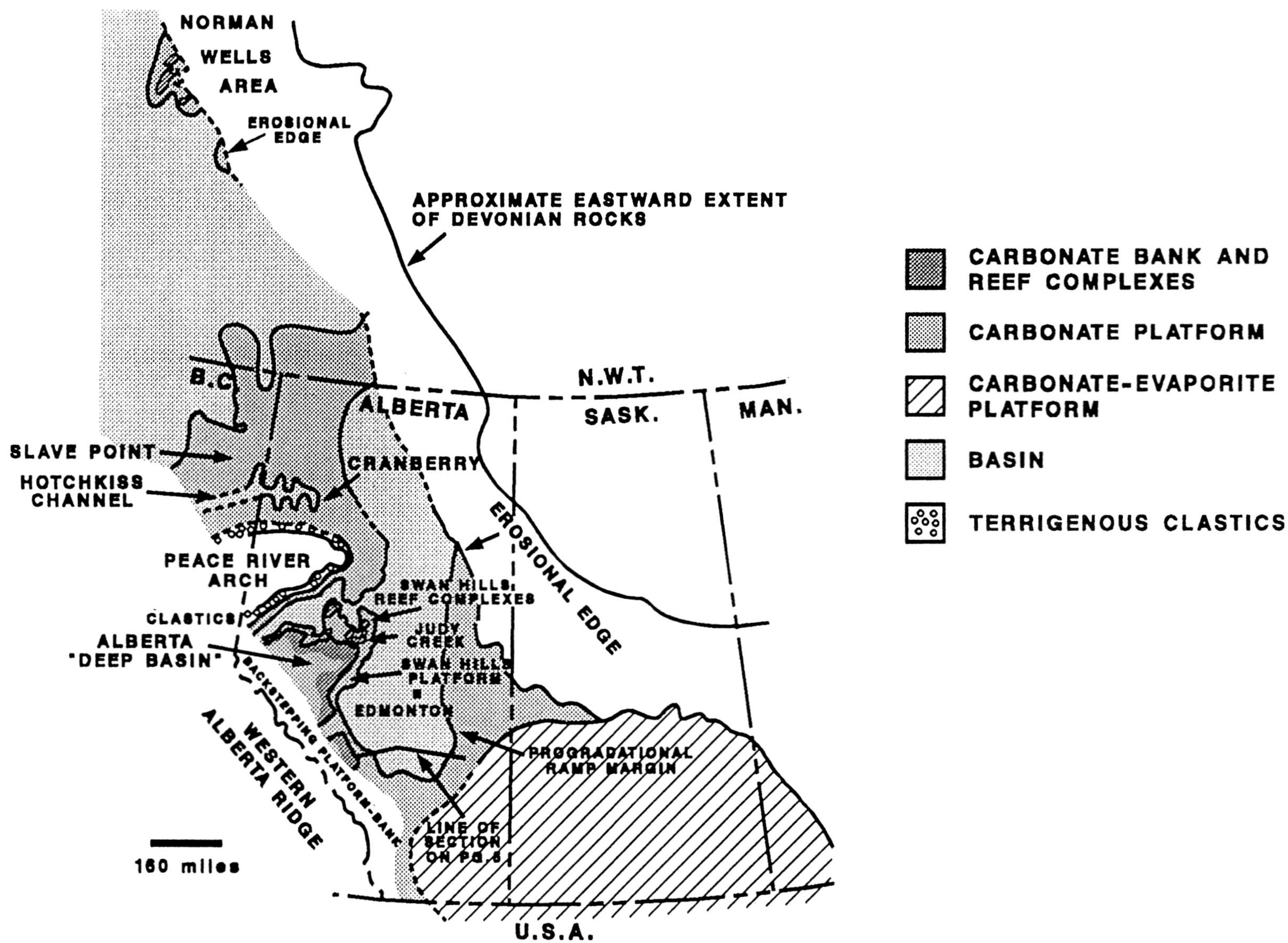


FIGURE 1. Paleogeography of the Beaverhill Lake Megasequence

## STRATIGRAPHIC NOMENCLATURE

Terminology used for Beaverhill Lake strata reflects the stratigraphic investigation in a number of areas. As a consequence, different names have been applied to similar lithologic units of the same age. The nomenclature used in this chapter corresponds to that of the Geological Staff, Imperial Oil Ltd. (1950), Law (1955), Fong (1959), Leavitt and Fischbuch (1968) and Keith (1990).

The basal two units of the Beaverhill Lake megasequence are the Watt Mountain and Fort Vermilion formations. Both of these units occur over a widespread area. The Watt Mountain Formation comprises an assemblage of dominantly siliciclastic strata. This unit is thickest along the flanks of the Peace River Arch and thins progressively away from this feature. The Fort Vermilion Formation is a restricted sequence of interbedded anhydrites and carbonates. North of the Peace River Arch, it changes facies in a generally northwestward direction into more normal-marine carbonates.

The table in Figure 2 reviews the stratigraphic nomenclature of Beaverhill Lake strata in Alberta and northeast British Columbia. Along the west side of the Beaverhill Lake basin, shallow-water lithofacies are included in both the Swan Hills and Slave Point formations. The distinction in usage corresponds purely to geographic area. Shallow-water carbonates flanking and north of the Peace River Arch comprise the Slave Point Formation. Shallow-water facies in the Swan Hills area and to the south make up the Swan Hills Formation.

A succession of deeper water, basinal limestones and calcareous shales in this same area comprises the Waterways Formation. This usage also applies to a similar facies in the Beaverhill Lake Basin to the east.

A somewhat different nomenclature is used to subdivide Beaverhill Lake strata in east-central Alberta, along the east side of the basin. Following a proposed recent revision by Keith (1990), the Beaverhill Lake megasequence is split into eight formations. The basal two formations, the Watt Mountain and Fort Vermilion units, are used in a manner consistent with that to the west. The overlying Slave Point Formation corresponds to only a very thin, relatively clean carbonate. The overlying Beaverhill Lake strata include both shallow-water and basinal carbonate lithofacies. This succession is divided into, in ascending order, the Firebag, Calumet, Christina, Moberly, and Mildred formations. Keith (1990) correlated these units to the west into the Waterways succession infilling the Beaverhill Lake Basin.

## MODEL FOR BEAVERHILL LAKE MEGASEQUENCE

A general model elucidating the evolution of the Beaverhill Lake megasequence is shown in Figure 3. The transect illustrated in this model extends from the area near Edmonton, on the east side of the basin, across the basin, through the Swan Hills area and onto the Swan Hills bank to the southwest.

The model depicts an overall asymmetric transgressive-regressive cycle of sedimentation. Initiation of siliciclastic deposition of the Watt Mountain Formation marks a period of relative lowering of sea level on the Arch. Terrigenous sediments, shed off the Arch, covered a wide area of Alberta during this period. Deposition of ensuing strata correspond to a transgressive phase of sedimentation. Inundation of the sea led initially to an episode of alternating meso- and hypersalinity. This resulted in the deposition of interbedded anhydrites and carbonates of the Fort Vermilion Formation throughout a wide area of Alberta. Subsequent inundation of open-marine waters into central Alberta triggered widespread carbonate deposition.

The initial stages of carbonate sedimentation show a backstepping style of

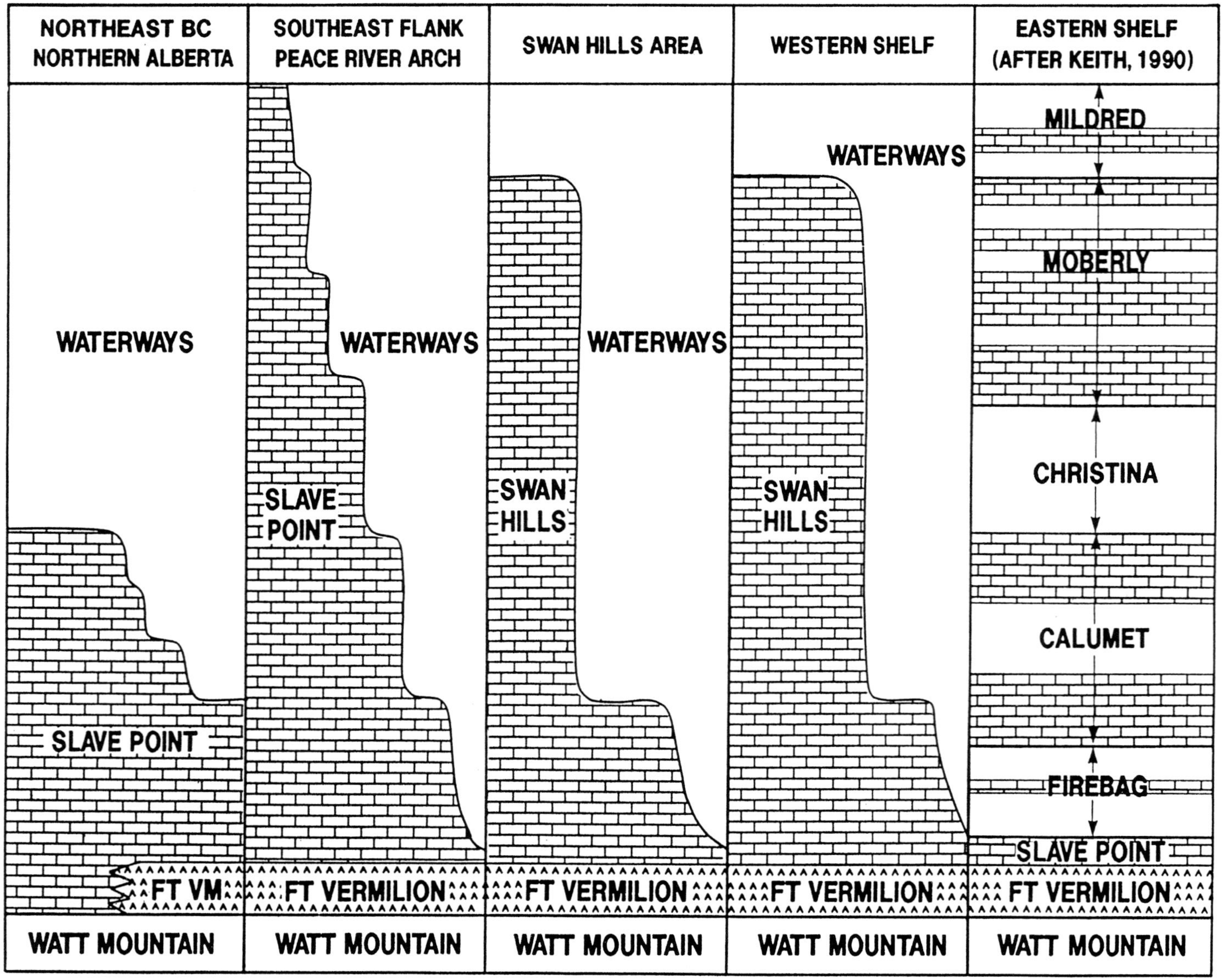


FIGURE 2. Stratigraphic Nomenclature for Beaverhill Lake Megasequence

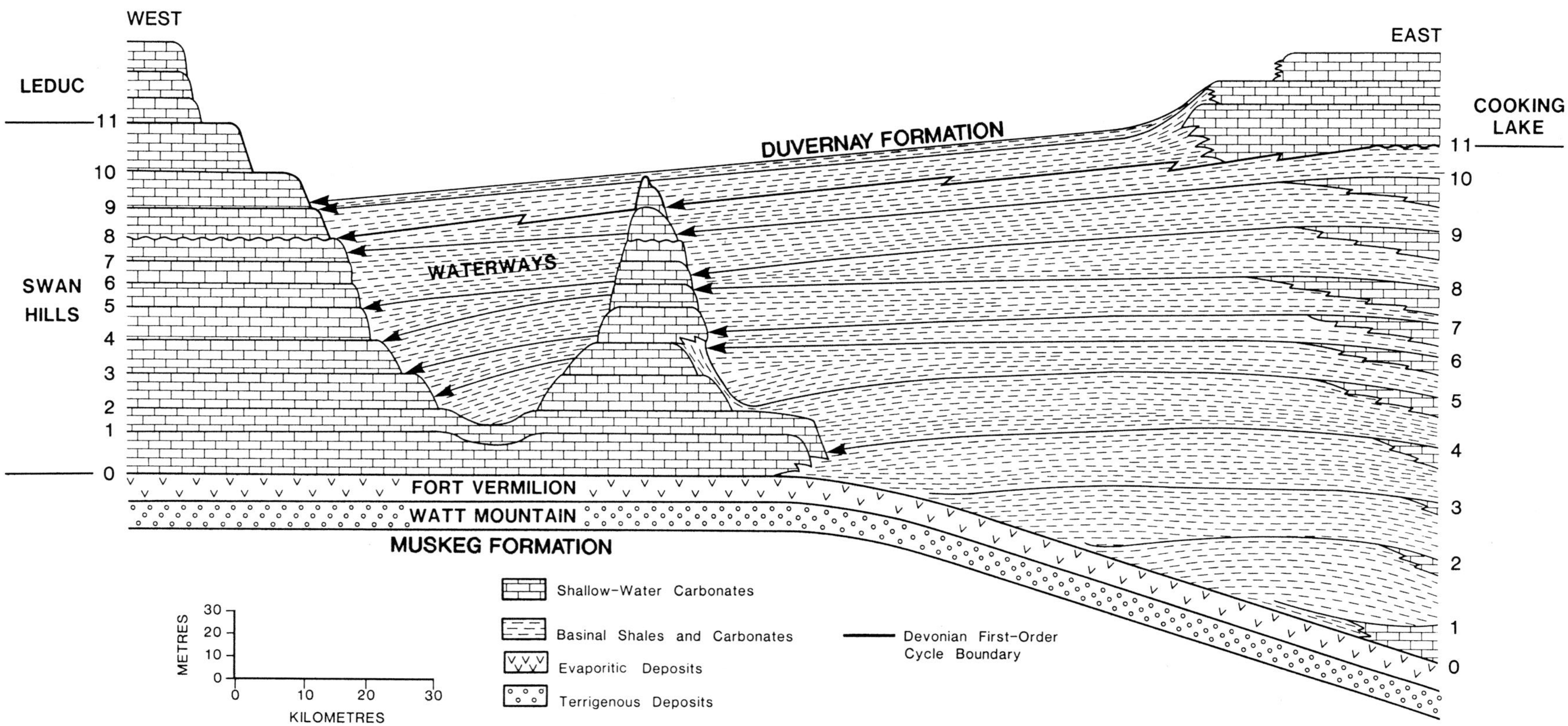


Figure 3. East to west cross section of Beaverhill Lake strata showing major styles of deposition.

evolution on both sides of the Beaverhill Lake basin. On the west side, backstepping resulted in a major shift of platform growth to the west. Isolate reefs grew on aborted platform highs, contemporaneous with more widespread carbonate deposition on the bank to the southwest. During this initial phase, the amount of growth on both the regional Swan Hills bank and the isolate reefs exceeded the increment of basin fill. As a result, the basinal area along the west side of the basin became deeper.

Ensuing growth on both the regional bank and isolate reef complexes was more aggradational. This growth occured in an incremental manner corresponding to second-order cycles of sedimentation. The top of a distinct cycle approximately 105 m above the top of the Fort Vermilion Formation at Judy Creek marks a period of emergence on both the bank and the reef complexes. Succeeding increments of bank and reef growth occur in an upstepping to backstepping manner.

The growth of the Beaverhill Lake platform along the east side of the basin records an overall forstepping pattern following the initial transgressive phase. Shallow-water carbonates prograded to the west or basinward in a ramp-like manner. This direction of progradation reflects the dispersal of sediment driven by the prevailing easterly trade wind belt. Lime mud, produced on the platform, was shed to the west into the basin. Successions of these basinal units offlap in a sigmoidal manner to the west and downlap against both the regional Swan Hills bank and isolate reef complexes. This forestepping pattern defines the regressive nature during most of the intermediate and later phases of Beaverhill Lake deposition.

The very final phase of Beaverhill Lake sedimentation records a transgressive episode followed by regression and ultimately subaerial exposure during a fall of relative sea level. On the west side of the basin, this is marked by drowning of the isolate reef complexes and the backstepping of the regional Swan Hills bank. On the east side of the basin, the same sea-level rise corresponds to the drowning of the Moberly platform and marks the onset of Mildred shale deposition. Lowering of sea level initially caused sediment progradation and eventually culminated in the emergence of the Mildred shale along shallow, upslope portions of the ramp to the east.

More in-depth summaries of Beaverhill Lake strata in the various paleogeographic domains in Alberta and northeast British Columbia are provided in the following sections.

## EASTERN SHELF AREA

The shelf fringing the east side of the Beaverhill Lake basin consists of an interbedded succession of shallow-water platformal carbonates and deeper water, more argillaceous basinal limestones. Figure 4 shows a reference section from Keith (1990) for these units. Shallow-water units correspond to those sections with a clean gamma-ray response; deeper water facies have a more radioactive log signature.

The interbedding of these facies record the cyclical intertonguing of platform carbonates and deeper water facies along a leeward platform margin. Times of rapid relative sea-level rise are marked by backstepping of carbonate platforms to the east. Platform successions subsequently aggraded and prograded basinward during times of slower relative sea-level rise. Lime mud derived from these prograding platforms was carried into the basin by waters driven by the easterly trade winds. The accumulation of these fine sediments provided a foundation that allowed carbonate platforms to build out in a ramp-like manner. The overall forestepping pattern of Beaverhill Lake strata in this area attests

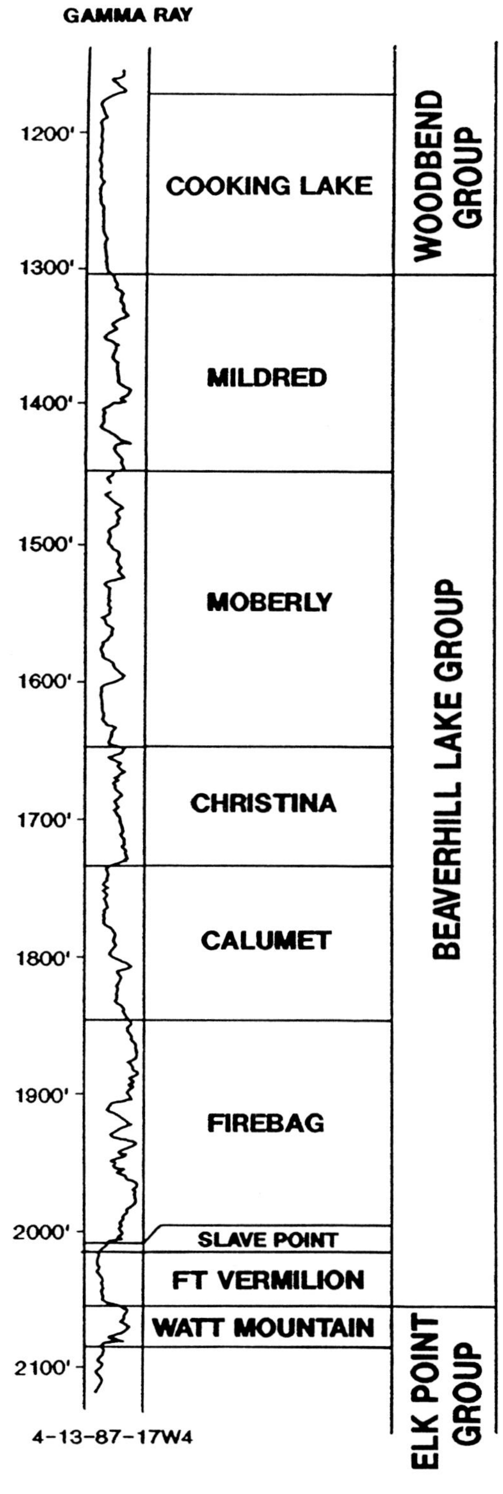


FIGURE 4. Reference Section for Beaverhill Lake Strata on Eastern Shelf. The Succession Consists of Interbedded Deeper Water, Shaley Sequences and Shallower Water, More Carbonate Rich Units (after Keith, 1990)

to conditions where sedimentation outpaced relative sea-level rise.

## WESTERN SHELF AREA

The western shelf is on the windward side of the Beaverhill Lake basin. Figure 5 illustrates the marked difference in stacking patterns and style of shelfal units on this side as compared to those on the eastern flank of the basin. As previously described, shelf successions on the east side generally forestep in a ramp-like manner into the basin. On the west side, they tend to backstep and upstep.

The backstepping and upstepping style on the west side of the Beaverhill Lake basin is attributed to the easterly trade wind. As a consequence of this circulation, fine sediments sourced on the platform tended to be pushed back into the interior portion of the shelf. Only minor amounts of fine sediment were deposited in the adjacent basin. As relative sea-level rise outpaced basin fill, the depth of water in the adjacent basin became progressively deeper. Without a fine sediment apron to build out on, platform growth was generaly limited to a backstepping or upstepping style.

Figure 6 shows a more detailed interpretation of platform growth along the west side of the basin at the giant Caroline gas-condensate pool. Stoakes and Dixon (1992) correlated seven depositional cycles throughout the field area. The cycles are separated by apparent deepening events which are indicated by the superposition of deeper water over shallower water facies. The initial unit (Cycle 1) consists of an areally extensive carbonate platform. Cycles 2 through 4 backstep progressively toward the west. Cycle 5 exhibits a forestepping style and may be equivalent to the initial cycle of reef growth at Judy Creek discussed in Chapter 4. The 6th and 7th cycles are comprised of a number of backstepping, gently basinward sloping rampal sequences. These latter sequences lack the distinct platform-margin facies that characterize the margins of the platform in cycles 2 through 5.

## SWAN HILLS AREA

The Swan Hills area includes a number of isolate carbonate reefs that occur on a more areally extensive carbonate platform. The isolate reefs are time equivalent to a regional Swan Hills carbonate bank complex further to the southwest. Kaufman and Meyers (1988) provide an account of the evolution of the Swan Hills bank at the Rosevear Field. The general evolution in these domains is depicted on Figure 3.

Swan Hills carbonates overlie the evaporitic Fort Vermilion Formation and terrigenous deposits of the Watt Mountain Formation. The Watt Mountain Formation includes deposits of an abandoned deltaic system. The slight relief created by these deltaic deposits led to localization of carbonate platform growth in this area (Jansa and Fischbuch, 1974).

These initial stages of carbonate growth were amalgamated to form part of a widespread platform. Figure 7 shows a transect along the edge of this platform. In this position, the platform is divided into two second-order cycles. The first cycle exhibits an aggradational manner of growth. Colonization and vertical aggradation by stromatoporoids along a break in slope created significant bank to basin relief. Shallow-water deposits of the upper cycle forestep a small distance basinward. A submarine-cemented hardground caps the entire succession.

Subsequent sea-level rise and backstepping led to localization of platform growth to discrete shoals. These shoals provided the nucleus for subsequent reef

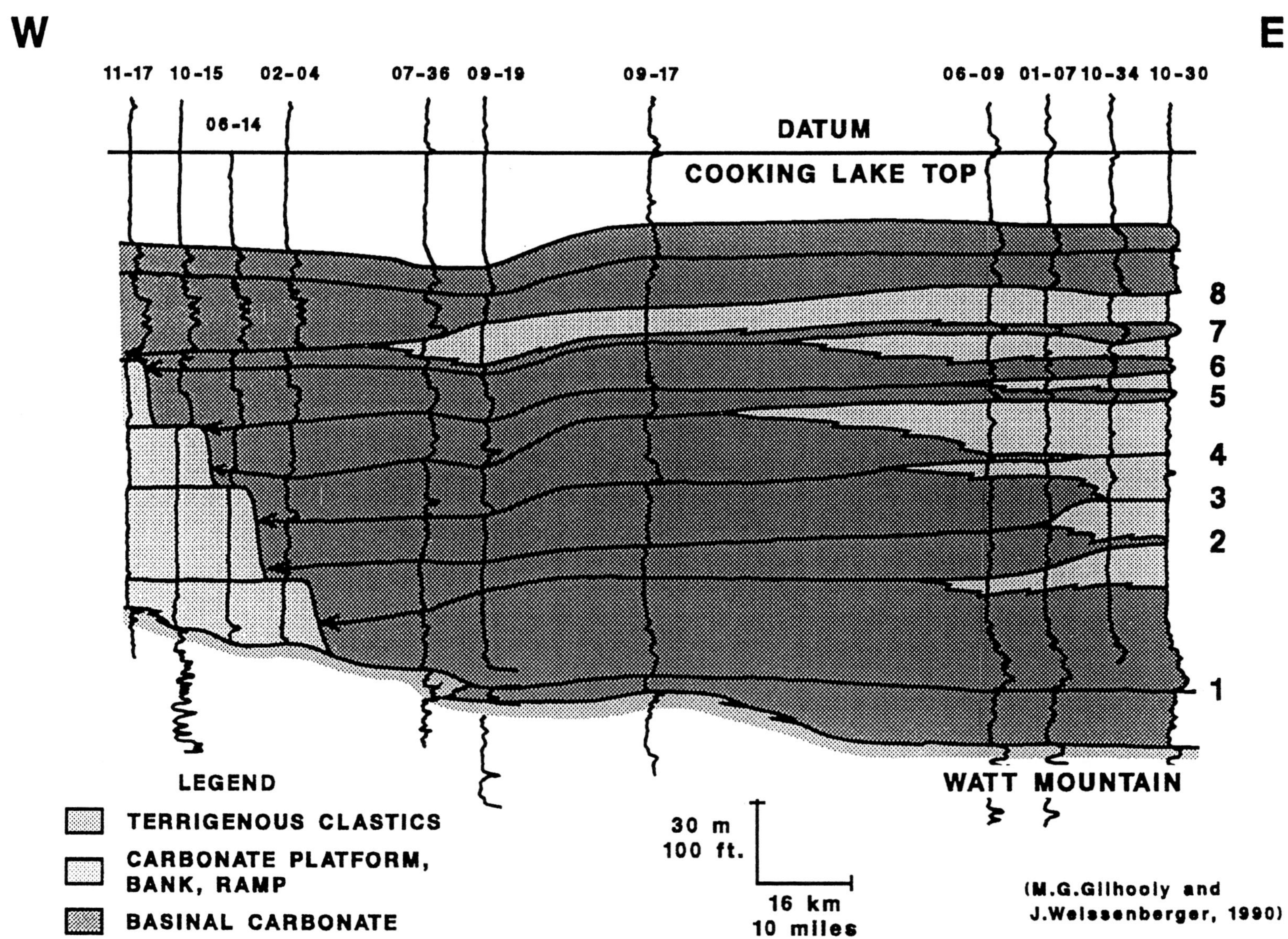


FIGURE 5. Stratigraphic Cross Section Across the Beaverhill Lake Basin in South Central Alberta

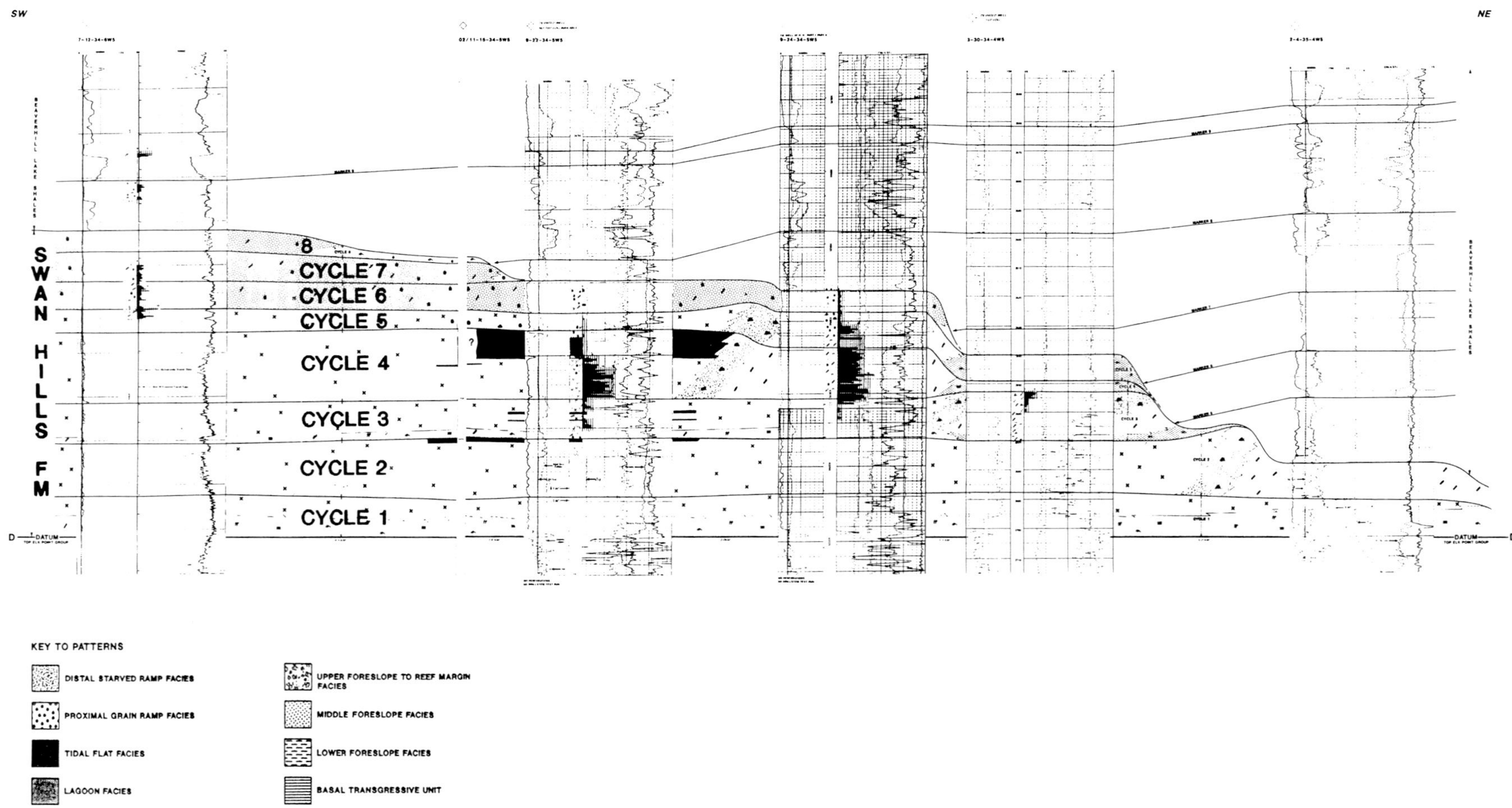


Figure 6. Stratigraphic facies cross section of the Swan Hills platform, Caroline gas field.

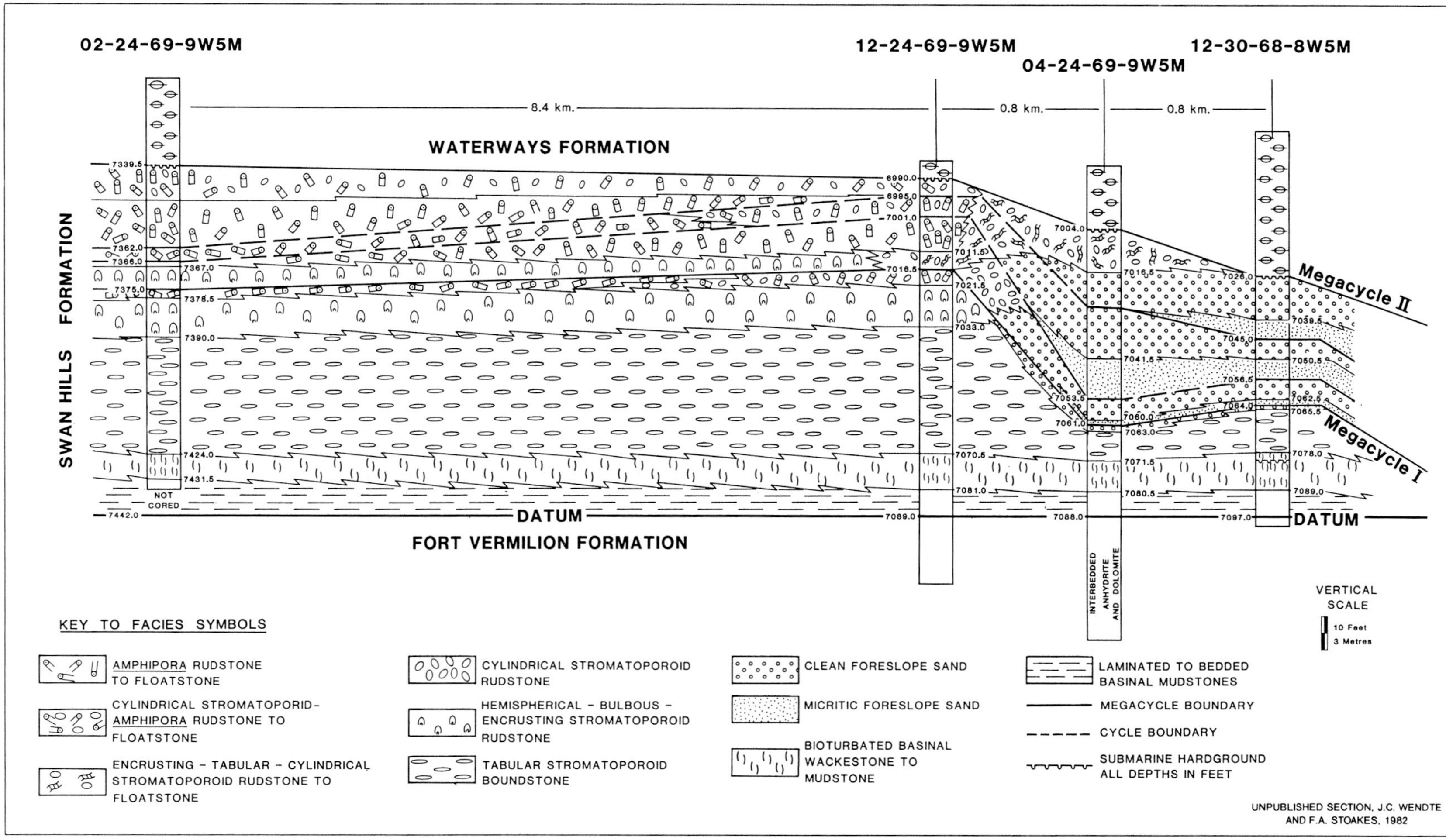


Figure 7. Stratigraphic facies cross section of the Swan Hills platform, House Mountain area.

growth, such as at Judy Creek (see Chapter 4 and Muir and others, 1990). The evolution of these reefs has been discussed in Chapter 4 and is only briefly summarized here. Summaries of these reefs are presented in Murray (1966), Fischbuch (1968), Leavitt (1968), Wendte and Stoakes (1982) and Viau (1983 and 1984). A transect across the Judy Creek reef complex is shown in Chapter 4-Figure 4. The evolution of the Judy Creek reef complex is summarized as follows:

(1) Localization of the reef on an underlying platform shoal.
(2) Progradation and subsequent aggradation during the first reef cycle with the area of the reef expanding during this phase.
(3) Overall backstepping during the next three reefal cycles.
(4) Emergence along the top of the reef complex.
(5) Reflooding and backstepping of shoal complexes on top of the pre-existing reef complex.
(6) Drowning and abandonment of the complex.

Following drowning, the reef complex was downlapped and covered by the westerly sloping clinothem successions of the Waterways Formation. Sheasby (1971) details the depositional patterns of these successions in the area near the Swan Hills reef complex.

## SOUTHEAST FLANK OF THE PEACE RIVER ARCH

The evolution of the Slave Point Formation along the southeast flank of the Peace River Arch records the interaction of relative changes in sea level with the influence of the Precambrian basement.

The structural history of the Arch is complex and not completely understood (see Chapter 1). What is known is that the major uplift of this feature preceeded the the deposition of Elk Point sediments that onlap its lower flanks. The uplift of the Arch during this phase involved block faulting. These faults occur in conjugate sets that strike northwest-southeast and northeast-southwest. Relief created by this faulting was strongly modified by fluvial incisement and erosion during emergence prior to Middle Devonian time. A pattern of radial drainage with deeply incised valleys was created. As a consequence of the faulting and erosion, the intermediate and upper flanks of the Peace River Arch had a very irregular surface. In areas of good well control, basement "bumps" can be seen to occur along northwest-southeast and northeast-southwest trends.

Figures 8 and 9 show schematic and well cross sections illustrating the onlap of Beaverhill Lake strata onto the southeast flank of the Arch. Deposits of this phase overlie evaporitic sediments of the Muskeg Formation. Succeeding units progressively onlap the Arch to the northwest. Papers by Dunham and others (1983), Tooth and Davies (1988) and Gosselin and others (1988) provide summaries of Slave Point complexes along the southeast flank of the Arch.

The initial deposits of the Beaverhill Lake cycle are terrigenous clastics of the Watt Mountain Formation. This Formation consists of an assemblage of fluvial, deltaic and marine facies that built out from the Arch. The onset of this unit is interpreted to correspond to a time of lowering of relative sea level in the vicinity of the Arch.

The overlying Fort Vermilion Formation was deposited during a period of rising sea level that followed the deposition of the Watt Mountain Formation. The unfossiliferous carbonates and anhydrites of the Fort Vermilion indicate mesosaline to hypersaline conditions during this episode. Four brining-upward carbonate to anhydrite cycles can readily be identified and correlated from density logs along the entire flank of the Arch. The divergence of overlying

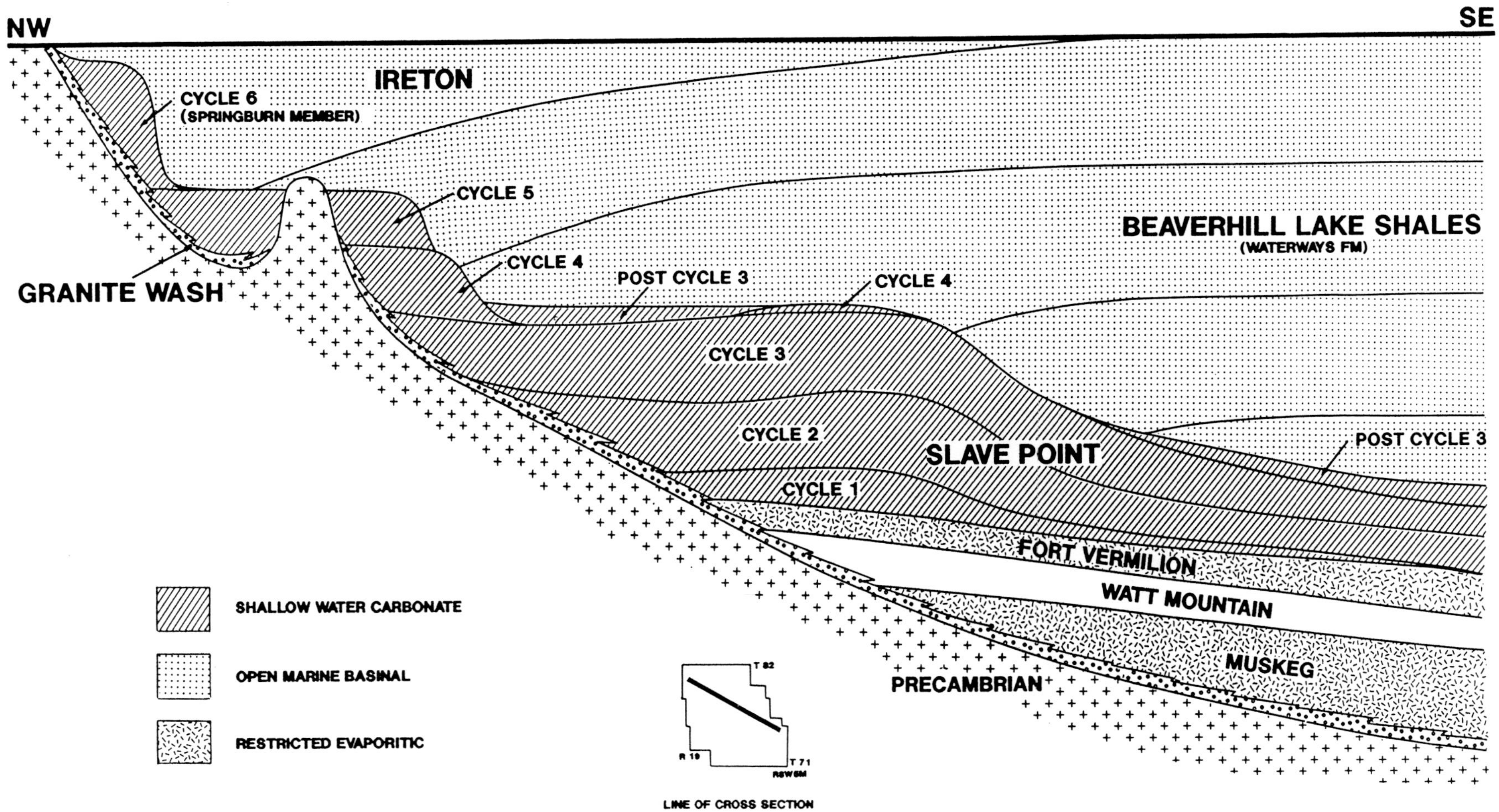


FIGURE 8. Schematic Cross Section of Beaverhill Lake Strata onto the Southmost Flank of the Peace River Arch

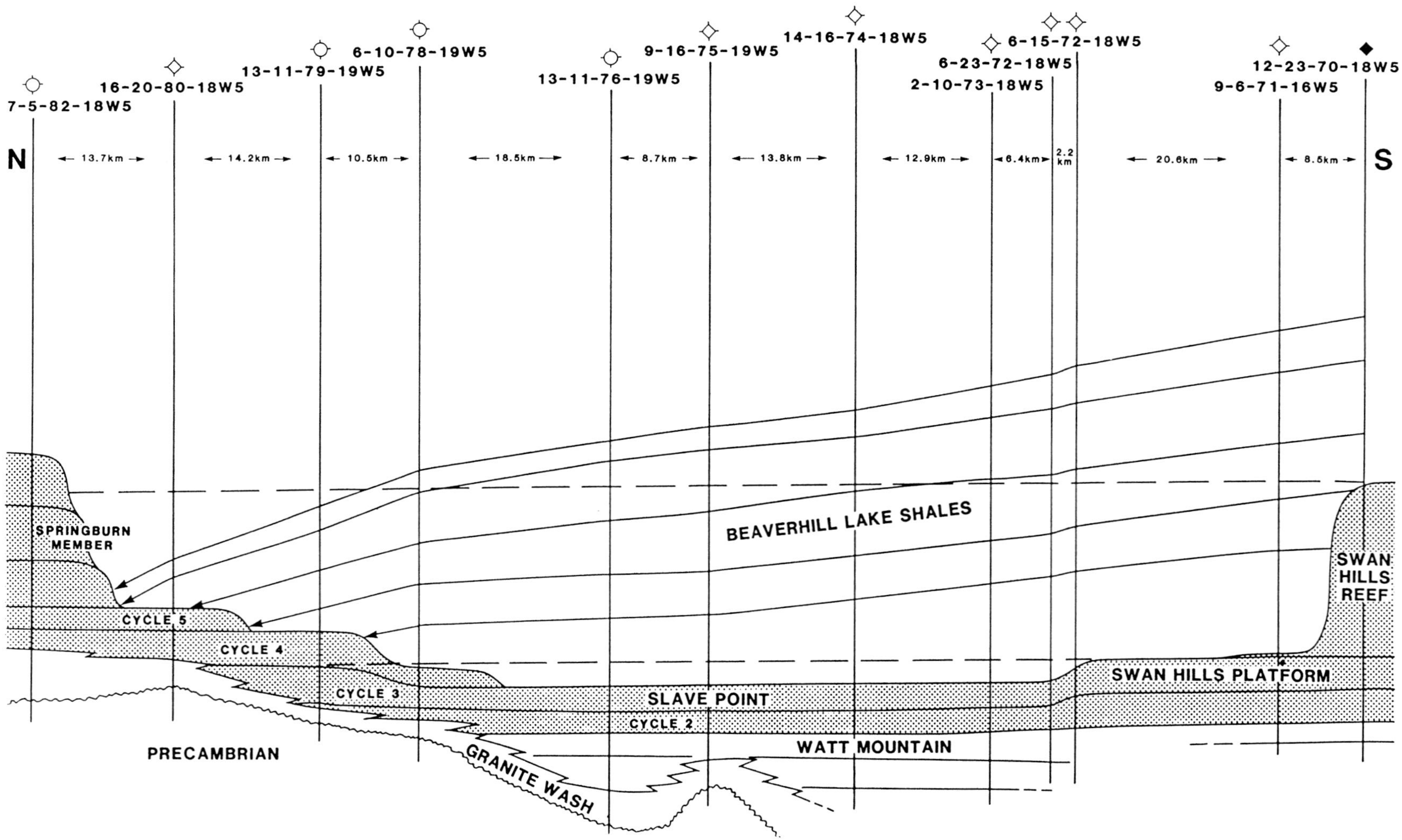


Figure 9. Well log cross section of Beaverhill strata from southeastern flank of the Peace River Arch to Swan Hills platform.

stratigraphic markers with the top of the Fort Vermilion points towards continued structural growth of the Arch during the deposition of this stage of the Beaverhill Lake megacycle.

The overlying Slave Point Formation consists of a series of backstepping shallow-water carbonate platforms and isolate reefs. Six major upward-shoaling sequences are recognized. Each sequence is capped by a submarine-cemented hardground in basinward positions.

The lowermost cycle is composed of a widespread succession of unfossiliferous carbonates up to a maximum thickness of approximately 10 m. This succession grades from laminated lime mudstones at its base into bioturbated lime mudstones at the top. The generally unfossiliferous nature of this deposit indicates a somewhat stressed environment. The subtle thinning of this sequence in a seaward direction indicates that this sequence was deposited on a ramp profile.

Cycles 2 and 3 make up a widespread platform that can be correlated around the Arch and is equivalemt to the initial Swan Hills platform described in the previous section (see Figure 9). These platforms show a distinct change in depositional style from the underlying sequence. Profiles across these platform successions show well-developed lagoon, platform-margin, foreslope and open-marine basinal facies with abundant macrofauna.

Cycles 4 and 5 mark a phase of major backstepping onto the previously exposed Precambrian surface. Deposits in these cycles vary from more widespread platform to isolate reef complexes. Dunham and others (1983) showed the influence of Precambrian knobs on the distribution of Slave Point carbonates of this phase. These workers documented the nucleation of a fringing reef around a bald Precambrian high at the Slave field. Wells drilled subsequent to that investigation document the presence of the least six other reef-fringed bald Precambrian highs at Slave. These reefs were initially isolate features, but later coalesced to make up a much larger complex.

Cycle 6 again backsteps and consists of platforms and isolate reefs onlapping the previously emergent Precambrian basement. The name Springburn Member has been given to these shallow-water carbonates (Mayer and others, 1964).

Each of these successive Slave Point stages is downlapped by limestones and calcareous shales of the Waterways and ultimately the Duvernay and Ireton formations.

## NORTHERN ALBERTA-NORTHEASTERN BRITISH COLUMBIA

A widespread Slave Point platform extended from the north flank of the Peace River Arch to the barrier complex in northeast British Columbia and the southern Northwest Territories. The Slave Point Formation in this area varies from as thin as 7 m along the eastern margin of the platform to approximately 150 m along the barrier-margin complex. This variation in thickness is indicative of both the backstepping of upper Slave Point units and the lateral facies transition of Fort Vermilion anhydrites into carbonates to the west of the 6th Meridian. The greater subsidence rates away from the Peace River Arch also plays a role in this thickening.

This platform is incised by a number of open-marine embayments. The most significant of these embayments is the Hotchkiss Channel just north of the Peace River Arch. A number of finger-like extensions, from the Slave Point platform fringing the Arch project to the north into the south side of the embayment. Transects across these projections show a profile of facies from open-marine basin, to foreslope and margin and into an interior lagoon. These platform extensions provide reservoir rocks which pool significant gas and gas-condensate reserves. The largest of the pools discovered to date is the Cranberry Field. Figure 10 shows a stratigraphic-facies cross section across this pool.

The Slave Point at Cranberry consists of two megasequences. The lower megasequence consists of an initial progradational stage in which shallow-water carbonates prograded basinward over deeper water lime mudstones. This degree of progradation is unique to this area and absent elsewhere in the Slave Point. This extensive degree of progradation appears to have occurred synchronously with the deposition of the Fort Vermilion anhydrite and possibly even with the Watt Mountain Formation further to the east. The rate of relative sea-level rise was probably not as rapid in this interval as during later Slave Point phases. Progradation, therefore, would be more likely. The upper phase of the first megacycle is more aggradational. A number of third-order sequences are recognized in the interior lagoon.

The upper megasequence exhibits a backstepping to aggradational style of growth. An elevated rim up to 7 m higher than the centre of the lagoon is evident along the periphery of the complex. Following deposition of these upper cycles the complex was drowned and abandoned by a subsequent rise in sea level. The top of the Cranberry complex equates to the cycle 3 top along the southeast flank of the Peace River Arch and to the top of the initial Swan Hills platform. The deeper water lagoon at Cranberry was later infilled by open-marine crinoidal lime mudstones and wackestones. This deposit appears to be synchronous in age to rhythmical banded lime mudstones that downlap onto the complex and fill the embayment. This initial basin-fill was then covered by a more argillaceous basin-fill during a later Beaverhill Lake stage.

## SUMMARY

Stacking patterns of sedimentary successions are known to be a consequence of the interaction of sea level (regional and eustatic), tectonic and environmental depositional controls. The Beaverhill Lake sequence in the WCSB provides a superb stratigraphic laboratory for investigating the significance of these controls.

One inherent aspect of the Beaverhill Lake in this regard is that basin-fill successions, unlike that of the later Woodbend and Winterburn megasequences, are predominately carbonate. This is especially true for the area south of both the Peace River Arch and the Swan Hills buildups. These basin-fill successions can be traced back to their shelfal equivalents on the east side of the basin.

The regional consistency in the pattern of cyclicity attests to basin-wide changes in relative sea level. A relative lowering of sea level, at least in the area of the Peace River Arch, triggered the initiation of the Beaverhill Lake megasequence. Throughout most of the Beaverhill Lake megasequence, relative sea level rose in a sawtooth, cuspate manner. Only a single episode of a relative lowering of sea level is evident within the megasequence. The megasequence is apparently terminated by another drop in relative sea level.

Almost all the basin-fill successions correlate with shelf sequences and hence can be said to represent "highstand" conditions. Fine carbonate sediment, sourced on the platform on the east side of the basin, was shed into the basin and formed uniformly westerly dipping clinoforms. Successions of this fine sediment forestep into and fill most of the basin in south-central Alberta. This style of basin fill provides an excellent example of how depositional controls, in this case circulation direction, can significantly modify the impact of relative sea level on carbonate stacking patterns.

Another aspect relating to stacking patterns is that the forestepping style along the east side of the basin also corresponds to the area with the greatest

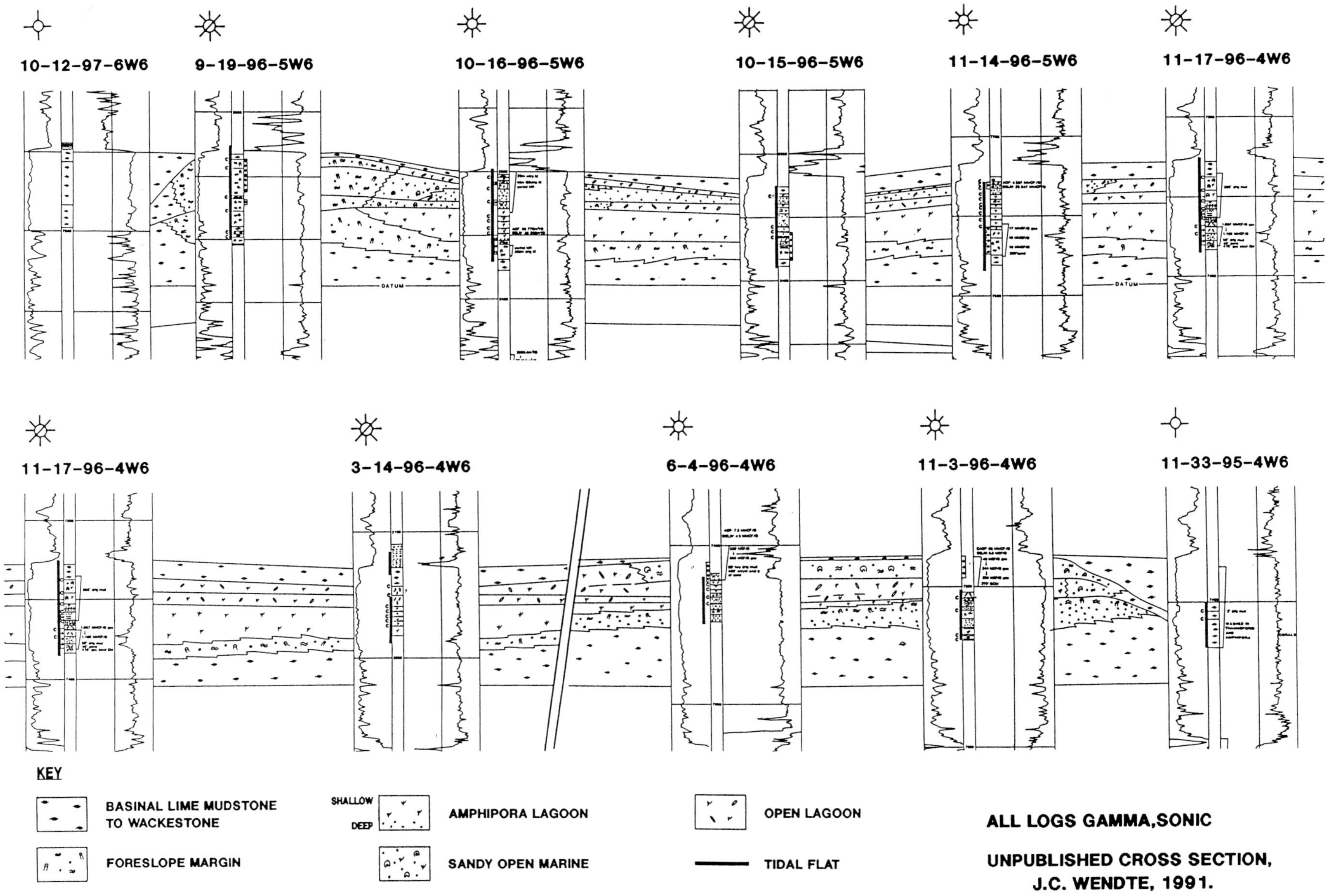


FIGURE 10. Cranberry Slave Point Stratigraphic Cross Section

rate of subsidence. Subsidence creates accomodation, which in itself favours a more transgressive style of sedimentation. In the WCSB high rates of carbonate productivity on the shelfs and the westerly dispersal system "overode" the effects of variation in subsidence.

Despite the relationship along the eastern side of the Beaverhill Lake basin, local tectonic controls exerted a strong influence and control on depositional patterns along the Peace River Arch. Erosionally modified fault scarps served to localize the nucleation of isolate reef complexes.

## REFERENCES


DUNHAM, J.B., CRAWFORD, G.A., AND PANASIUK, W., 1983, Sedimentology of the Slave Point Formation (Devonian) at Slave Field, Lubicon Lake, Alberta, *in* Harris, P.M., ed., Carbonate Buildups: Society of Economic Paleontologists and Mineralogists Core Workshop Number 4, p. 73-111.

FISCHBUCH, N.R., 1968, Stratigraphy, Devonian Swan Hills reef complexes of central Alberta: Bulletin of Canadian Petroleum Geology, v. 16, p. 444-556.

FONG, G., 1959, Type section Swan Hils Member of the Beaverhill Lake Formation: Journal of Alberta Society of Petroleum Geologists, v. 7, p. 95-108.

GEOLOGICAL STAFF, IMPERIAL OIL LTD WESTERN DIVISION, 1950, Devonian Nomenclature in Edmonton Area: American Association of Petroleum Geologists Bulletin, v. 34, p. 1807-1825.

GOSSELIN, E.G., SMITH, L,., AND MUNDY, D.J.C., 1988, The Golden and Evi reef complexes, Middle Devonian, Slave Point Formation, northwestern Alberta, *in* Geldsetzer, H.H.J., James, N.P., and Tebbutt, G.E., eds., Reefs Canada and Adjacent Areas: Canadian Society of Petroleum Geologists, Memoir 13, p. 440-447.

JANSA, E., AND FISCHBUCH, N.R., 1974, Evolution of Middle and Upper Devonian sequence from a clastic coastal plain-deltaic complex into overlying carbonate reef complexes and banks, Sturgeon-Mitsue area, Alberta: Geological Survey of Canada, Memoir 234, 105 p.

KAUFMAN, J., AND MEYERS, W.J., 1988, A backstepping platform reef, Swan Hills Formation, Rosevear Field, central Alberta, *in* Geldsetzer, H.H.J., James, N.P., and Tebbutt, G.E., eds., Reefs Canada and Adjacent Areas: Canadian Society of Petroleum Geologists, Memoir 13, p. 478-486.

KEITH, J.W., 1990, The influence of the Peace River Arch on Beaverhill Lake sedimentation: Bulletin of Canadian Petroleum Geology, v.38A, p.55-65.

LAW, J., 1955, Geology of northwestern Alberta and adjacent areas: American Association of Petroleum Geologists Bulletin, v. 39, p. 1927-1975.

LEAVITT, E.M., 1968, Petrology, paleontology, Carson Creek North reef complex, Alberta: Bulletin of Canadian Petroleum Geology, v. 16, p. 298-413.

LEAVITT, E.M., AND FISCHBUCH, N.R., 1968, Devonian nomenclature changes, Swan Hills area, Alberta: Bulletin of Canadian Petroleum Geology, v. 16, no. 3, p. 288-297.

MOYER, G.L., GREY, F.F., GRAYSTON, L.D., AND BELYEA, H.R., 1964, Upper Devonian, *in* McCrossan, R.G., and Glaister, R.P., eds., Geological History of Western Canada: Alberta Society of Petroleum Geologists, p. 60-88.

MUIR, I.D., SPRINGATE, G.L., AND MAWDSLEY, J.R., 1990, Geological model of the Middle-Upper Devonian Snipe Lake BHL "A" pool platform-reef complex, Western Canada: American Association of Petroleum Geologists Bulletin, v. 74, p. 726-727.

MURRAY, J.W., 1966, An oil-producing reef-fringed carbonate bank in the Upper Devonian Swan Hills Member, Judy Creek, Alberta: Bulletin of Canadian Petroleum Geology, v. 14, p. 1-103.

SHEASBY, N.M., 1971, Depositional patterns of the Upper Devonian Waterways Formation, Swan Hills area, Alberta: Bulletin of Canadian Petroleum Geology, v. 19, p.377-404.

STOAKES, F.A., AND DIXON, R.J., 1992, The Devonian Swan Hills Formation of the Caroline Field, Alberta: American Association of Petroleum Geologists Bulletin, abstracts for annual AAPG Convention (in press.).

TOOTH, J.W., AND DAVIES, G.R., 1988, Gift Lake Slave Point reef, Middle Devonian Alberta, *in* Geldsetzer, H.H.J., James, N.P., and Tebbutt, G.E., eds., Reefs Canada and Adjacent Areas: Canadian Society of Petroleum Geologists, Memoir 13, p. 440-447.

VIAU, C., 1983, Depositional sequences, facies and evolution of the Upper Devonian Swan Hills buildup, central Alberta, Canada, *in* Harris, P.M., ed., Carbonate Buildups: Society of Economic Paleontologists and Mineralogists Core Workshop Number 4, p. 73-111.

VIAU, C.A., AND OLDERSHAW, A.E., 1984, Structural controls on sedimentation in the Swan Hills Formation, Swan Hills Field, central Alberta, *in* Eliuk, L., ed., Carbonates in subsurface and outcrop : Canadian Society of Petroleum Geologists, 1984 CSPG Core Conference, p. 103-131.

WENDTE, J.C., AND STOAKES, F.A., 1982, Evolution and corresponding porosity distribution of the Judy Creek reef complex, central Alberta, *in* Cutter, W.G., ed., Canada's Giant Hydrocarbon Reservoirs: Canadian Society of Petroleum Geologists, Core Conference Manual at CSPG - AAPG Convention, Calgary, Alberta, p. 63-81.

WENDTE, J.C., WONG, P.K., MUIR, I.D., WEISSENBERGER, J., GILHOOLY, M.G., AND STOAKES, F.A., 1990, Devonian carbonate sequence of Western Canada: American Association of Petroleum Geologists Bulletin, v. 74, no. 5, p. 788.

CHAPTER 8

# WOODBEND MEGASEQUENCE

F.A. STOAKES

## INTRODUCTION

This sequence comprises the strata between the Beaverhill Lake Group (Chapter 7) and the Winterburn Group (Chapter 9). The stratigraphic section includes shallow-water carbonates of the Cooking Lake, Leduc and Grosmont formations and deep-water carbonates and shales of the Majeau Lake Member, Duvernay and Ireton formations. The isolate reef complexes of the Leduc Formation are especially well known because they contain some of the largest oil reserves in the WCSB. In central Alberta, the basin-fill succession of the Woodbend Group differs markedly from that of the underlying Beaverhill Lake Group. Rather than being comprised dominantly of carbonate muds derived from coeval carbonate banks to the east, Woodbend basin-fill units have a significant clay shale content. The occurrence of these shales marks the initial delivery of terrigenous deposits into the central Alberta Basin from an extrabasinal northerly source. The configuration of these shale "banks" controls the distribution of some overlying carbonate platform deposits.

Figure 1 depicts the configuration of the Alberta Basin during the Woodbend which marks the furthest incursion of the open-marine Devonian seaway onto the craton. Shallow-water platform carbonates occur along the periphery of this basin in east-central Alberta, in the Devonian "Deep Basin" area of west-central Alberta and around the Peace River Arch landmass.

In east-central Alberta, isolate reef complexes of the Leduc Formation overlie the areally widespread and drowned platform deposits of the Cooking Lake Formation. A distinct alignment of these reefs (the Rimbey-Meadowbrook trend) marks the western edge of the underlying platform. As illustrated by Ross and Stephenson (1989) this trend partially overlies and parallels the Precambrian basement grain of what these authors refer to as the "Rimbey Arc". The Rimbey-Meadowbrook trend of reefs is subparallel to the Bashaw-Duhamel trend as well as the margin of the Southern Alberta Shelf further to the southeast. It should be noted however that there are other Leduc reefs (such as Redwater, Golden Spike, Stettler-Fenn-Big Valley and Willingdon-Pinedale) which exhibit occurrences that do not correspond to these gross orientations and may require a different explanation. Jones (1980) attempted to explain some of these trends in terms of deep seated vertical faulting but this mechanism has yet to be substantiated.

Cratonward of the Southern Alberta Woodbend Shelf Complex, deposits become progressively more restricted in lithologic aspect and consist of shallow-water carbonates and anhydrites.

Widespread platform deposition of the Grosmont Formation post-

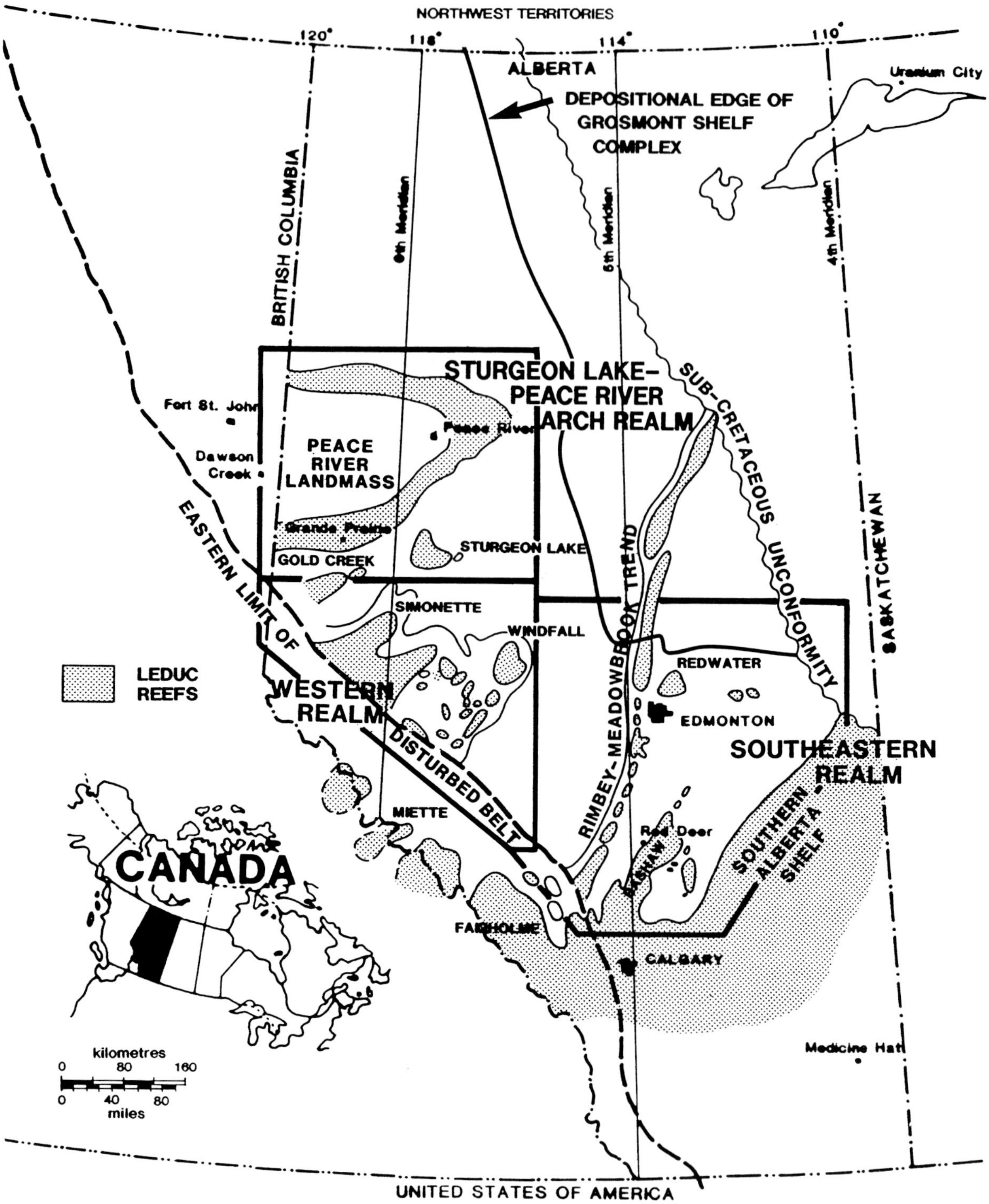


FIGURE 1. Paleogeography of the Woodbend Group

dates that of the Cooking Lake Formation. This platform complex is limited in occurrence to northeast Alberta and contiguous portions of the Northwest Territories (Figure 1). It overlies a succession of basinal deposits of the Ireton and Duvernay formations.

The carbonate complexes in the Devonian "Deep Basin" area of west-central Alberta occur along the northeast flank of, and out from, the Western Alberta Ridge. The areal distribution of these shallow-water complexes indicate that by this time the crest of this feature had already been submerged. The carbonate complexes in this area consist of both large carbonate banks and smaller, less areally widespread isolate reefs of the Leduc Formation. Most of the bank complexes and isolate reefs overlie a somewhat more areally widespread low Leduc platform. However some small isolate reefs may sit directly on top of the underlying Swan Hills bank. Leduc reef complexes of the Devonian "Deep Basin" show a relatively irregular distribution in contrast to the aligned trends of the eastern part and encircle a small sub-basin often referred to as the Wild Hay Basin. This basin had connections to the open seaway to the west (called the Jasper Basin by Mountjoy, 1978) as well as to the north and east.

Leduc carbonates around the Peace River Arch form a narrow carbonate fringe along the emergent landmass. To the south, between the Arch and the "Deep Basin" area large isolate Leduc complexes such as Sturgeon Lake and Gold Creek occur.

## STRATIGRAPHIC NOMENCLATURE

Figure 2 shows the stratigraphic nomenclature of the Woodbend succession in the various portions of the WCSB. In both east-central and northeast Alberta, areally widespread platform carbonates of the Cooking Lake Formation form the basal part of the Woodbend succession. These are overlain by isolate reefs of the Leduc Formation. Basinal successions in both areas consist of a lower more carbonate-rich unit, the Duvernay Formation, and an upper more argillaceous unit, the Ireton Formation. The Duvernay Formation passes upwards from micritic, downslope-ramp lithologies that directly overlie the Cooking Lake Formation into dark-coloured, organic-rich laminites. The organic-rich laminites form the bottom-set or fondothem beds of large clinoforming shale "banks" of the Ireton Formation (Oliver and Cowper, 1963). Carbonate platform deposits overlie the uppermost, shallow-water undathem portions of these shale "banks". A large carbonate platform complex in northeast Alberta, termed the Grosmont Formation and a thinner unit in east-central Alberta referred to as the Camrose Member of the Ireton Formation are two examples of this depositional style.

In central Alberta, basinward of the Rimbey-Meadowbrook reef trend, the entire Woodbend succession consists of deep-water facies of, in ascending order, the Majeau Lake Member of the Cooking Lake Formation, and the Duvernay and Ireton formations. The upper boundary of this sedimentary package is difficult to pick precisely but occurs somewhat below a prominent bed which we refer to as the Z Marker (see Stoakes *in* Lexicon of Canadian Stratigraphy, 1990 and further discussion in Chapter 9). Recognition of the occurrence

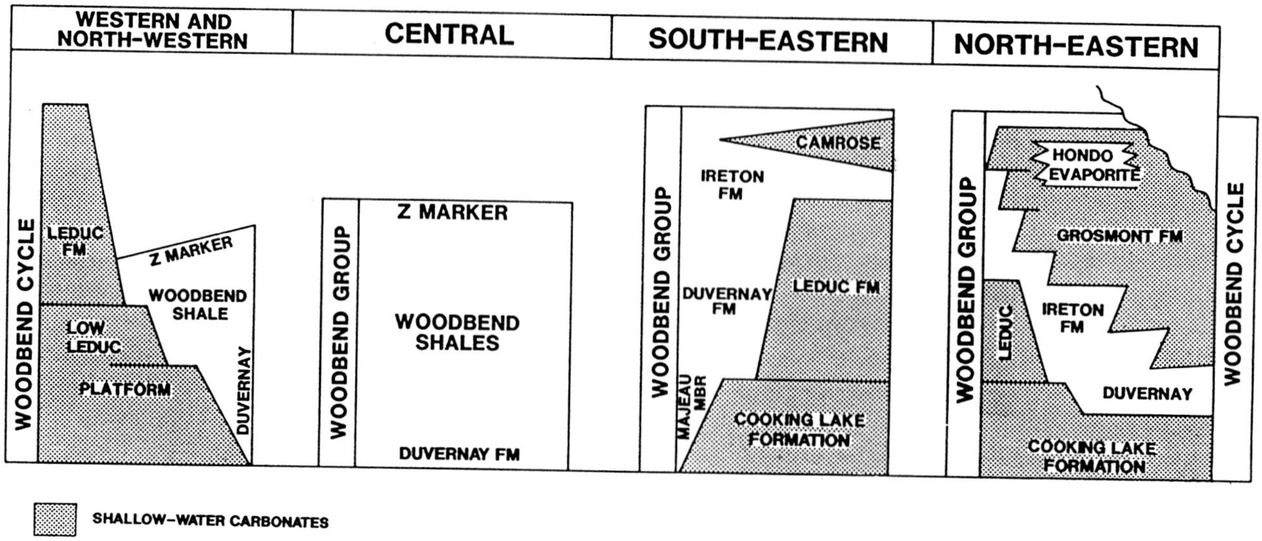


FIGURE 2. Stratigraphic Nomenclature of the Woodbend Group

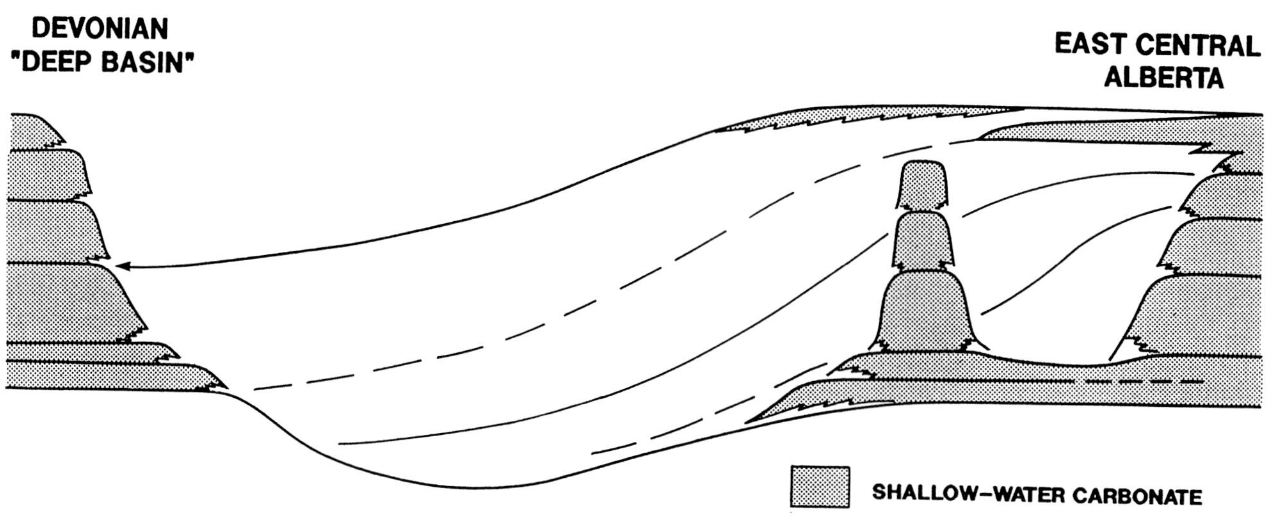


FIGURE 3. Conceptual Model of Woodbend Group Sedimentation

and significance of this marker is paramount in understanding the relative timing and stratal relations of reef growth and subsequent basin fill on opposite sides of the Alberta Basin. The stratigraphic terminology on the west side of the Alberta Basin and around the Peace River Arch generally conforms to usage in east-central Alberta differing only in two respects: 1) Widespread shallow-water carbonate deposition similar to that of the Cooking Lake is often referred to simply as the low Leduc platform; 2) The upper part of the off-reef shale succession overlies the Z Marker and is therefore considered in this account as part of the overlying Winterburn sequence. The shale beneath the Z Marker corresponds to that of the Ireton Formation of the type section, (refered to as Woodbend shale) but is a significantly thinner development than in east-central Alberta.

## MODEL FOR WOODBEND MEGASEQUENCE

Sediments of the Woodbend Group comprise a single major cycle of deposition consisting of a lower transgressive and upper regressive portion as depicted in Figure 3. The transect shown in this figure extends from the shelf area in east-central Alberta west into the Devonian "Deep Basin" area. The transgressive phase is dominated by shallow-water platform and reefal carbonates and the regressive phase by prograding, fine-grained calcareous and argillaceous basin-fill units. This megacycle of sedimentation was initiated by an overall major transgressive episode over the Beaverhill Lake topography, shifting facies belts away from the basin centre. This transgression marks the furthest incursion of the late Devonian open-marine seaway into the WCSB and the establishment of the greatest amount of bathymetric relief in central Alberta between the shelf and the basin floor.

### *Transgressive Phase*

Despite the overall transgressive character of this phase, the Woodbend megasequence appears to have been initiated by a very gradual lowering of sea level. As discussed in Chapter 3, this relative lowering led initially to the basinward progradation of lowermost Cooking Lake deposits over the Beaverhill Lake, Mildred Member shales. This occurred on a westward sloping ramp resulting in emergence in updip locations to the east. After this brief regressive episode the carbonate platform and isolate reef complexes of the transgressive phase grew during an incrementally rising sea level that expanded the basin margins and accentuated bathymetric relief between reef and basin. This time of maximum marine inundation corresponds to the deposition of the middle Leduc carbonates (see later discussion).

The initially rapid rates of relative sea-level rise caused backstepping of carbonate platforms on both the east and west sides of the Alberta Basin. In east-central Alberta, the Cooking Lake platform evolved through three major phases or cycles of sedimentation (see Chapter 3). During this phase of backstepping, an initially broad, lower Cooking Lake high became progressively more pronounced and localized. By the end of Cooking Lake deposition, large discrete accretionary highs had formed along the western margin of the platform and in certain selected positions behind this edge. Likewise, the same phase of rapid sea-level rises led to backstepping of lower Leduc carbonate platforms over an underlying Swan Hills bank on the west side of the basin.

Increased rate of sea-level rise resulted in drowning of the Cooking Lake platform on the east side of the basin. This drowning is marked by the widespread occurrence of a submarine-cemented hardground on top of the shallow-water platform. Following this drowning, shallow-water carbonate sedimentation resumed in two preferred sites: 1) On aborted discrete platform highs; 2) On a regional platform in a backstepped position further cratonward. Localization of shallow-water deposition on aborted platform highs led to the establishment of isolate reefs of the Leduc Formation, and regionally, deposition of the backstepped Southern Alberta Shelf. Ensuing episodes of sea-level rise led to more local backstepping and upstepping in both the isolate reef complexes and coeval shelf further to the southeast.

Many authors have noted the apparent alignment of reef trends suggesting an underlying tectonic control (Martin, 1967; Jones, 1980; Mountjoy, 1980) and this is especially true for the strikingly linear Rimbey-Meadowbrook trend. More recently Ross and Stephenson (1989) have demonstrated a partial coincidence between the Rimbey-Meadowbrook trend and the structural grain of the Precambrian "Rimbey Arc" (compare Figure 5 in Chapter 1 with Figure 1 in this chapter). However the existence of structural offset of Devonian units below the reef trend has never been demonstrated. What is known is that Leduc reefing is localized by depositional highs on the underlying Cooking Lake platform, many of which, but by no means all, occur along the west margin of the platform beneath the Rimbey-Meadowbrook trend. This Cooking Lake platform margin appears to be controlled primarily by the configuration of the underlying Beaverhill Lake ramp or clinoform profile. Clearly what remains unknown is how the Precambrian basement grain controls Beaverhill Lake or even earlier sedimentation and how this initial paleorelief is transferred to the Cooking Lake level.

On the west side of the basin, during the same time interval, rapid sea-level rises led to the initiation and localization of isolate reefs and banks of the Leduc Formation. The same transgressive phase led to backstepping and onlap of other large Devonian bank complexes onto the Western Alberta Ridge.

During the transgressive phase only a minor increment of sediment was added to basinal areas over most of the western Alberta Basin. A possible exception to this occurs locally in reentrants and embayments of the Leduc bank complexes where a significant accumulation of deep-water limestones constitutes the lower and middle portions of the basin fill. In particular this occurs in the southwest portion of the Alberta Basin in the subsurface Harmatton-Elkton area and in nearby outcrops, where it is referred to as the Westerdale Member of the Ireton Formation. The proximity of large Leduc reef complexes and the enclosed nature of the Westerdale-bearing basins indicates the probable origin of the Westerdale facies from hemipelagic sedimentation of shelf-derived carbonate (Eliuk in Lexicon of Canadian Stratigraphy,

1990). It is uncertain, at least to the present author, whether this basinal sedimentation equates to the transgressive or regressive phase of Woodbend sedimentation, or both. The significant thickness and basin occurrence tends to favour the last alternative.

### *Regressive Phase*

The regressive phase of the Woodbend megacycle corresponds to the partial infilling of the basin by prograding shales and thin carbonate platform units of the Duvernay and Ireton formations. Although reef growth and basin filling took place in an overall reciprocal fashion (Stoakes, 1980 and Chapter 5) the major portion of basin fill took place following deposition of the uppermost middle Leduc. Filling of the basin was asymmetric, largely from east to west and continued to the onset of deposition of the Nisku Formation of Winterburn age. Basin filling was succeeded on the east side of the basin by forestepping carbonate platform units extending progressively over shale clinoforms in a "piggyback" fashion. Only a very thin increment of basinal sedimentation occurs around the base of Leduc reefs on the west side of the basin and around the Peace River Arch to the north.

These regressive units were deposited during an intermittently rising sea-level history. However, an overall slower rate of sea-level rise combined with the delivery of fine terrigenous sediment to the basin permitted material to progressively infill deeper-water areas (see Chapter 5), gradually encroaching upon and finally terminating some areas of reef growth.

On the east side of the Alberta Basin the growth of individual stages on isolate reefs and on the Southern Alberta Woodbend Shelf correspond to increments of Ireton basin fill. The source of the shale is thought to be the Ellesmerian Fold Belt in the Canadian Arctic Archipelago (Stoakes, 1979, p.250) This shale gained access to the Alberta Basin via a seaway that existed between the Peace River Arch landmass and the Grosmont Complex (Stoakes, 1980) and was fed laterally to the basin by circulating currents. The asymmetric pattern of Ireton basin fill indicates that fine argillaceous input was largely along the eastern edge of the basin with the easterly trade wind belt imparting a westerly progradational aspect. As a result increments of Ireton shale deposition progressively forestep basinwards as a series of sigmoidal clinothem successions. The tops or undaform portions of individual shale banks built up to fairweather wave base which allowed carbonate platforms to become established. As additional increments of Ireton shale were added to the basinward portion of the profile, these carbonate platforms forestepped in an overall "piggyback" fashion. Basin-filling Ireton clinoforms gradually and progressively encroached upon areas of already established Leduc reef growth. Ultimately the level of turbidity began to adversely affect reef growth, especially it's ability to recover following sea-level rises. In response to this encroachment, upper stages of reef growth backstepped and eventually drowned. Further aggradation and progradation resulted in downlap and overlap of Ireton shales onto and over these complexes.

The west side of the Alberta Basin was far removed from the undaform portions of the clinoforming Ireton shales. As a consequence only a small increment of basinal Ireton deposits was added to this part of the basin during this phase. The lack of shale influx and hence better water quality allowed many isolate reefs and banks and fringing banks to sustain growth for the full duration of the regressive phase. These complexes grew in a predominantly upbuilding or upstepping fashion and at the end of the regressive phase of Woodbend deposition stood up to 100 m above the adjacent sea floor. This figure increases to almost 160 m northwards around the Sturgeon Lake reef and to 200 m along the flank of the Peace River Arch Leduc fringing reef (Stoakes, 1987).

Ensuing sections provide more detail on the growth of the isolate reef complexes, the stratigraphic and temporal relationships between the reefs and adjacent basin fill and further discussion of the nature and occurrence of "piggyback" platforms in the basinal succession. The Woodbend basin can be divided into three separate areas or geographic realms following Stoakes and Wendte (1987):

1) Southeastern Realm
2) Western Realm
3) Sturgeon Lake-Peace River Arch Realm

(see Figure 1).

## SOUTHEASTERN REALM

*Carbonate Platform and Isolate Reef Succession* - Isolate Leduc reef complexes directly overlie the drowned Cooking Lake platform. The growth of these complexes reflects the interplay between relative sea-level rise and environmental controls. In the same way as Swan Hills reefs evolved, Leduc complexes grew in a number of stages that correspond to the second-order cyclicity described in Chapter 2. Furthermore many of the cycle levels can be correlated from one complex to another. This correlation is best displayed in the limestone buildups (such as Redwater and Golden Spike) and is less distinct in the massively dolomitized equivalents. Figures 4 and 5 show wireline log-based cross sections across two of these complexes: the eastern margin of the Redwater reef and Golden Spike (see Figure 1 for location). A schematic cross section across the Golden Spike reef portraying the evolution of the entire complex, is presented as Figure 6.

The second-order cycles are characterized by lateral shifts in facies that are a response to relative sea-level rises. Furthermore, second-order cycles may be grouped into larger composite growth stages corresponding, in this case, to lower, middle and upper Leduc stages. Individual cycle levels and composite growth stages are labelled on the cross sections in Figures 4, 5 and 6.

At Redwater (Figure 4) the line of cross section passes through an arcuate shaped marine embayment that developed along the east side of the reef complex. The recognition of the cyclicity and the depositional style of cycles and growth stages provides a basis for understanding reef evolution. The lower Leduc at

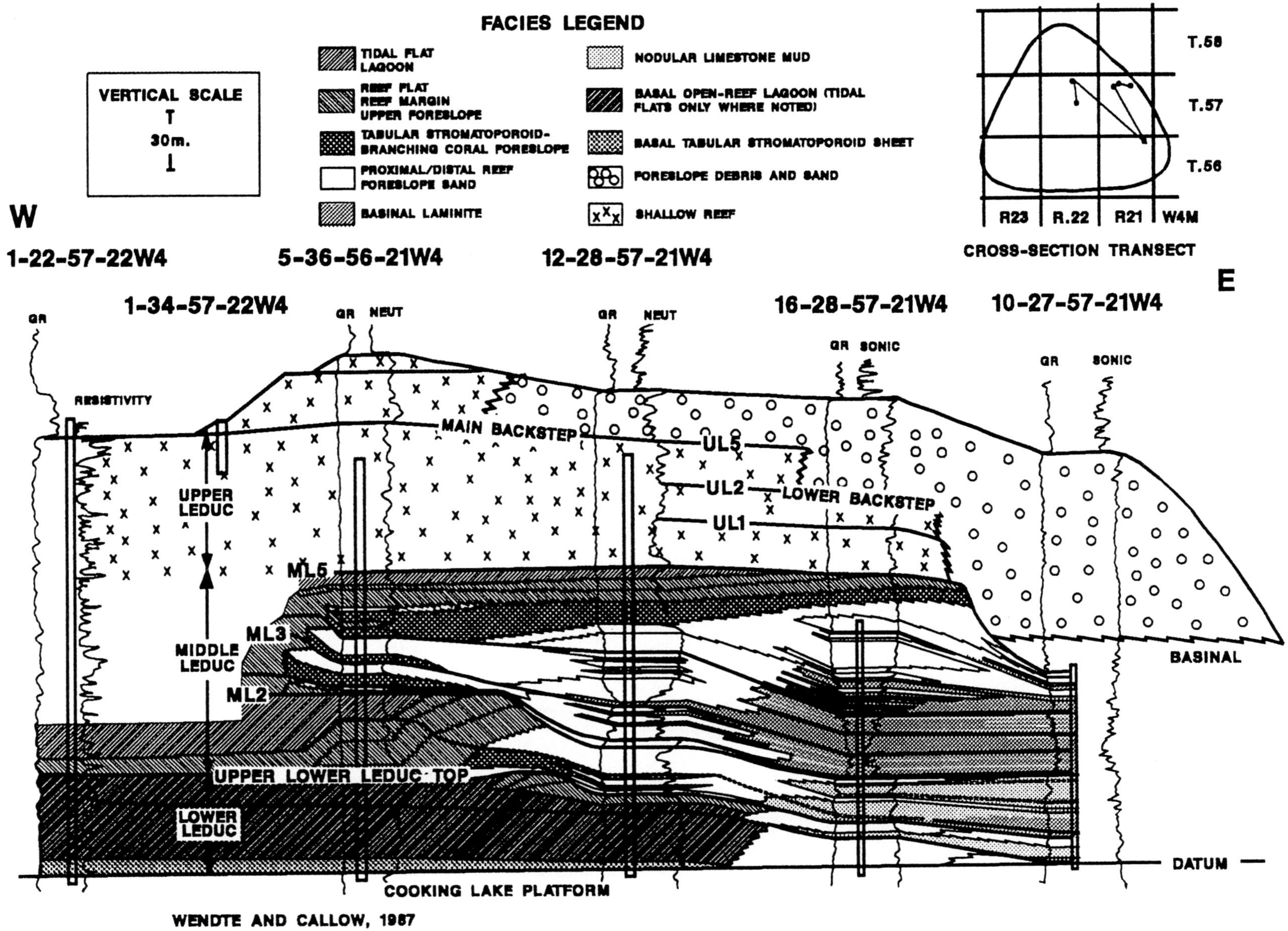


FIGURE 4. Facies Cross Section of the Eastern Margin of the Redwater Reef

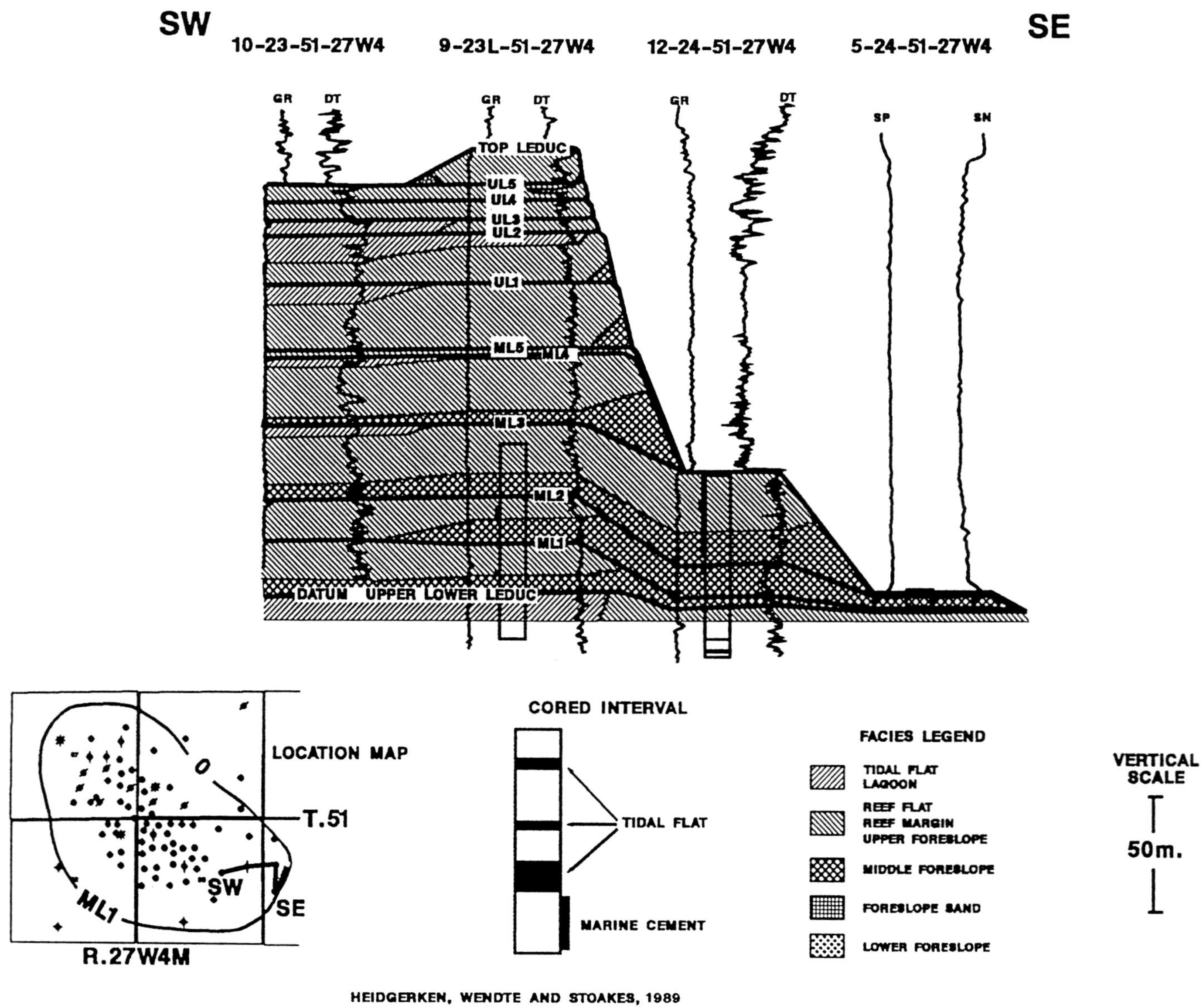


FIGURE 5. Facies Cross Section of the Eastern Margin of the Golden Spike Reef

Redwater consists of a shoal complex lacking well-developed reef margins sitting directly on a large depositional high of the drowned Cooking Lake platform (see Chapter 3). The corresponding growth stage at Golden Spike is readily recognizable as it comprises a more areally widespread platform or bank complex that the more localized middle and upper Leduc complex sits on. The lower Leduc at Golden Spike consists of many upward-shoaling cycles. The uppermost sequence on the cross section (Figure 5) is referred to as the "upper" lower Leduc. Thus the top of the lower Leduc of McGillivray and Mountjoy (1975) and Walls (1983) corresponds to the top of the "upper" lower Leduc of Heidgerken and others (1989) (see Figures 5 and 6). It should be noted that McGillivray and Mountjoy (1975) consider the lower Leduc at Golden Spike to equate to the Cooking Lake below Redwater but our work shows this clearly not to be the case. At Golden Spike, the Cooking Lake and lower Leduc stages comprise distinct successions that are up to 79 m and 58 m thick, respectively. The subsequent middle and upper Leduc stages comprise what is generally referred to as the "reef". At both Redwater and Golden Spike, the cycles comprising the middle Leduc (ML1 to ML5) show a classic pattern of transgressive retreat (ML1 to ML3 on Figure 4) and regressive forestepping (ML3 to ML5 on Figure 4). Backstepping and upstepping of the reef during the transgressive episode resulted in progressively greater reef to basin relief. At Redwater this interval marks the furthest encroachment towards the reef interior of deep-water, organic-rich basinal laminites (Figure 4). In off-reef areas these deep-water laminites constitute a widely correlatable unit, the upper part of the Duvernay Formation, which marks maximum marine inundation of the basin. The upper part of the middle Leduc at both Redwater and Golden Spike (ML4 and ML5 on Figures 4,5 and 6) reflect regressive progradation of shallow-water reefal facies over deeper-water foreslope deposits. The ML4 level at Golden Spike and some small increment just below ML5 at Redwater (ML4 lacks expression in available cored sections and could not be precisely located at Redwater) mark the most progradational levels in both reef complexes. This level can be correlated east of Redwater to a clinoforming shale bank that reflects infill of the adjacent basin.

The ML4 level corresponds to the top of the most significant progradational cycle within the Woodbend succession. Even though the succession of strata immediately following this level indicate only an increase in the degree of aggradation and <u>not</u> a lowering of sea level, this surface conforms to the definition of a *Depositional Sequence* boundary by Jervey (1988). In significance and order of magnitude, the level of this cycle boundary may correspond to the level of a cycle alluded to in Chapter 2 as being intermediate between our "normal" second-order and our first-order cycles.

The uppermost increment of middle Leduc (ML5) and the upper Leduc (UL1 to UL6) at both Redwater and Golden Spike display initially renewed upstepping (UL1 and UL2) and then predominantly backstepping (UL3 to UL6). The most pronounced backstep at Redwater occurs at the UL5 level (Figure 4). This level equates to the major drowning seen at Golden Spike (Figure 6) such that only

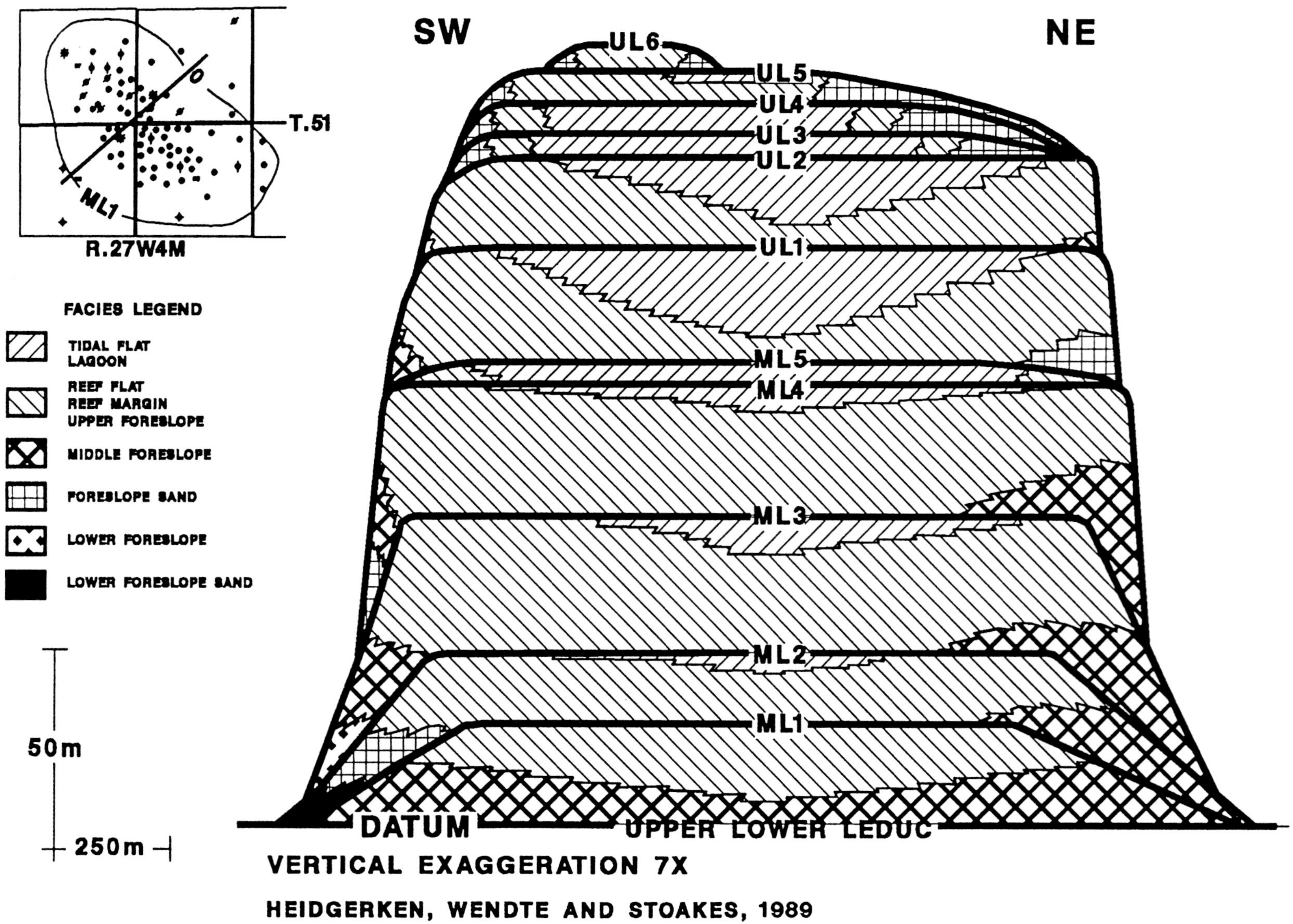


FIGURE 6. Schematic Cross Section of the Golden Spike Reef

isolate small shoal complexes survive. Thus the termination of reefing at Golden Spike appears to precede that at Redwater. This earlier demise is possibly attributable to the smaller areal extent of the complex which affected its potential to re-establish itself following intermittent drownings. The termination of reef growth at Redwater corresponds to a later phase when shale banks finally encroached into close proximity to the complex. Storms often reworked shale from the approaching banks onto the reef complex. Abundant green shales, analogous to those described in Chapter 4 at Judy Creek, occur in the reef complex at Redwater (Klovan, 1964). These become increasingly more common upsection in the upper Leduc attesting to the progressive approach of the basinal shale banks. This basin filling had two effects on reef growth: 1) Restriction of water circulation, driven by the northeasterly trade winds, reflected in the increasingly lower energy micritic nature of the uppermost Leduc reef stages; 2) High levels of turbidity which adversely affected the growth of "frame-building" organisms.

*Basin-fill Succession* - During this phase large amounts of fine terrigenous sediment derived from an extrabasinal source to the north was delivered to the basin. Progradation of shale banks and consequent basin filling permitted the establishment of widespread carbonate platforms that extend progressively basinward in a "piggyback" fashion. These shale banks initially prograded in a north to south direction but gradually turned to the west. Because of the progressive nature of this basin-fill process, carbonate platforms that grew on these shale banks became established at different times in various parts of the basin. The largest of these platform complexes comprises the Grosmont Formation of the Woodbend Group (Figure 2). It is made up of a thick succession (up to 170 m) of shallow-water carbonates limited to the northeastern portion of the basin (Figure 1) and covering some 100,000 $km^2$. This thick carbonate package progrades south and west and occupies the same stratigraphic interval as the Leduc and Ireton formations of south-central Alberta. Cutler (1983) subdivided this succession into four broadly sigmoidal depositional units that individually consist of a number of shoaling-upward cycles. The initial three units show platform aggradation with little associated basin filling and probably equate to the transgressive phase of the Woodbend cycle. The uppermost unit has more of an aggradational aspect. This aggradation of carbonates along the basinward edge of the complex isolated an interior lagoon which became filled with evaporites of the Hondo Member. This is followed by progradation and basin filling of the Ireton Formation. As such the Grosmont represents the landward stacked shallow-water portions of individual Ireton shale clinoforms. The forestepping of the Grosmont Complex southwards progressively terminates reef growth to the south along the northern part of the Rimbey-Meadowbrook trend (Figure 8).

Deposits of another one of these "piggyback" carbonate platforms comprise the Camrose Member of the Ireton Formation (Stoakes, 1977 and 1980). This platform is appreciably thinner than the Grosmont Platform Complex and is confined to a position south of the Grosmont depositional limit. The initiation of this

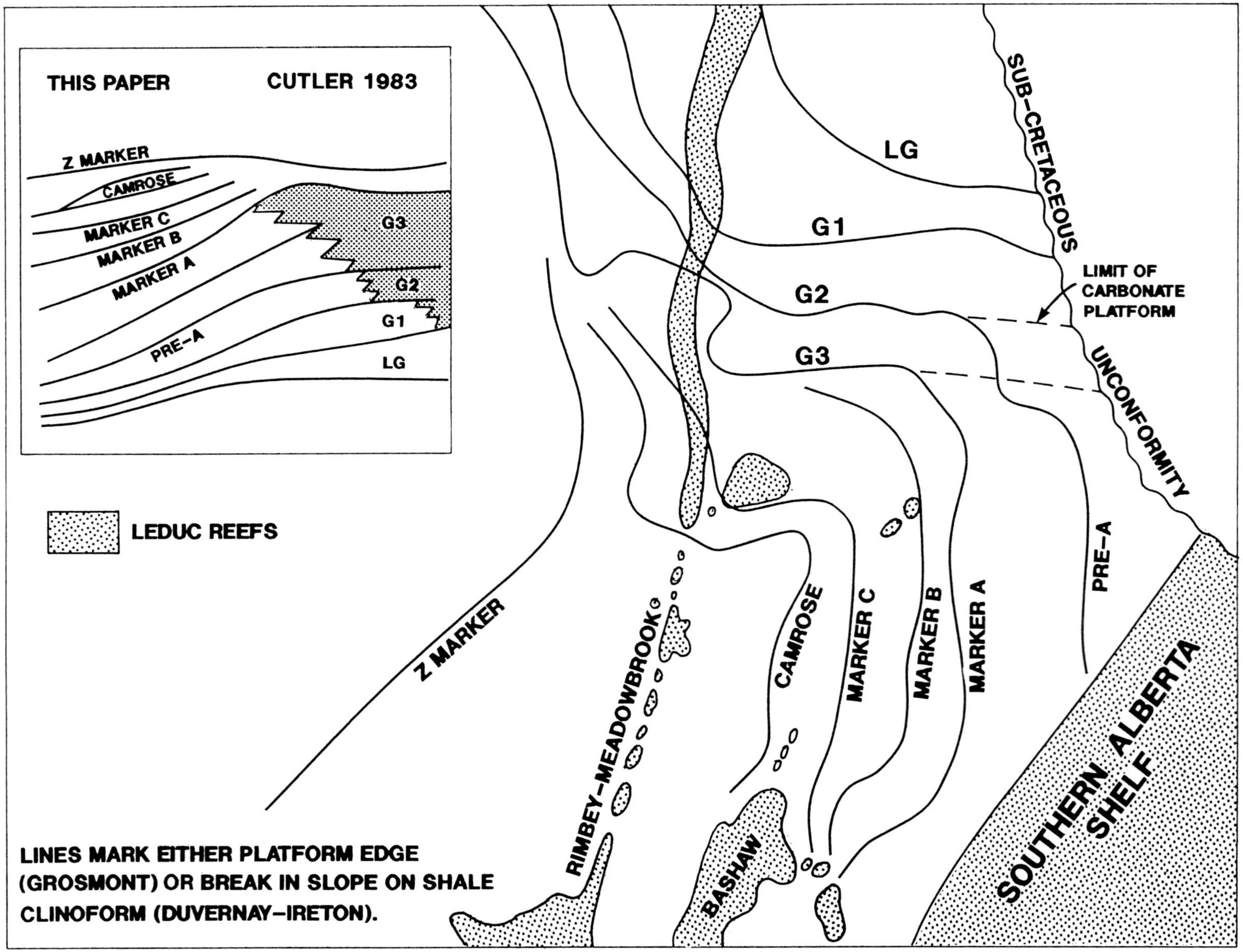


FIGURE 7. Successive Depositional "Edges" of the Grosmont-Ireton

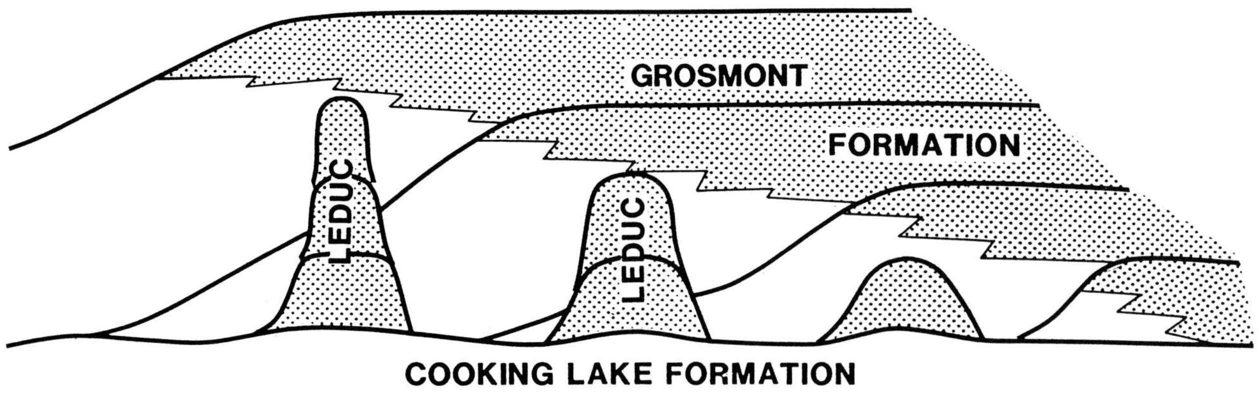


FIGURE 8. Leduc-Grosmont Relations, Northern Part of the Rimbey-Meadowbrook Trend

carbonate platform postdates that of the Grosmont to the north.

## WESTERN REALM

*Carbonate Platform, Bank and Reef Succession* - Consists of backstepping and upstepping shallow-water carbonates of the Leduc Formation. These rest on a backstepped, drowned platform equivalent to the Cooking Lake Formation of the Southeastern realm but is referred to informally as the low Leduc platform. Figure 9 shows a cross section across a portion of the Devonian "Deep Basin" area of the Western realm. The transgressive portion of the Woodbend succession in this area corresponds to backstepping carbonate platforms and small isolate reefs that are developed on carbonates of the underlying Swan Hills bank. Each of the lower Leduc platform stages shows a progressively more areally restricted occurrence and two distinct backstep levels or terraces may be distinguished (Stoakes, 1987). Thereafter carbonate growth becomes more isolate giving rise to individual reefs and banks with a rather irregular geographic distribution. Coinciding with this is backstepping of carbonate platforms towards the crest of the Western Alberta Ridge. These large platform complexes outcrop in the Alberta Front Ranges.

Individual isolate reefs and banks grew in a number of distinct correlatable stages. However lack of significant penetration and the coarsely crystalline dolomite fabric of the rock have discouraged construction of detailed models similar to those illustrated from Redwater and Golden Spike.

*Basin-Fill Succession* - The thickness of the basin-fill succession on the west side of the Alberta Basin is significantly reduced as compared to that of the eastern side. This reflects incomplete filling of the basin during Woodbend "time". The basinal succession corresponds to deeper-water carbonate laminites of the Duvernay Formation and shales of the Ireton Formation which represent the "toes" of clinoforming assemblages to the east and southeast. This succession is capped by a widespread gamma-log unit referred to as the Z Marker. This marker approximates the top of the Woodbend cycle and can be traced upslope where it is equated to the top of the lower Lobstick in the West Pembina area. In the Western depositional realm and northwards towards the Peace River Arch it downlaps onto the drowned flanks of Leduc buildups (Figure 10). It is directly overlain by deep-water shales of the succeeding Winterburn megacycle that downlap onto this unit (see Chapter 9). The pattern of progradation of the shale clinoforms above the Z Marker level is distinctly different from those below. Historically these shales above the Z Marker have been included as part of the Ireton Formation. In this account these shales are referred to as the Winterburn shales in contrast to the underlying Woodbend shale section (Stoakes, 1980 and 1987). This divergence from the previously accepted stratigraphy represents a major step forward in understanding the stratal relations and sequence stratigraphy of the Upper Devonian across central Alberta. At the termination of the Woodbend cycle, reefs of the Southeastern realm

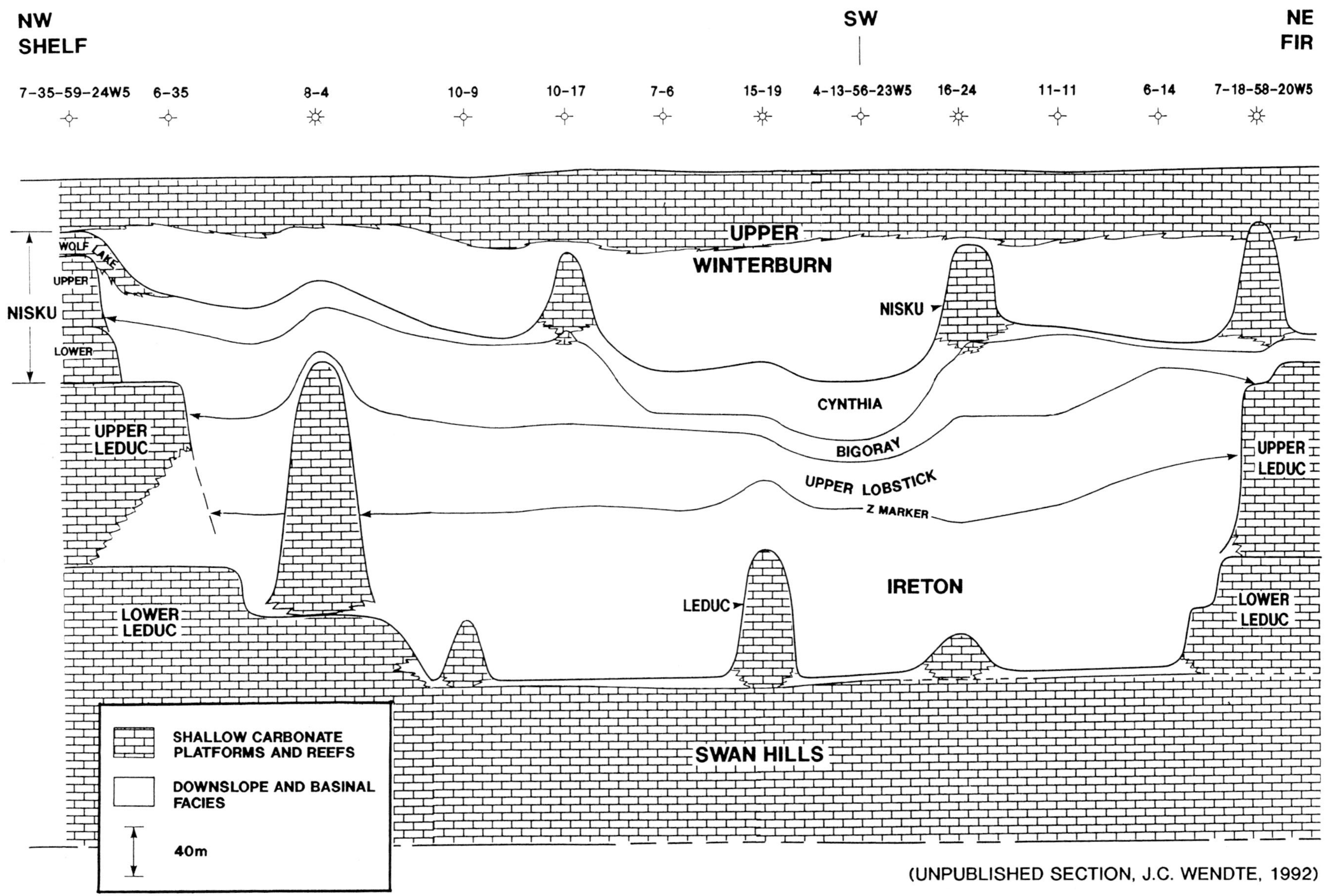


FIGURE 9. Stratigraphic Cross Section Across the Devonian "Deep Basin"

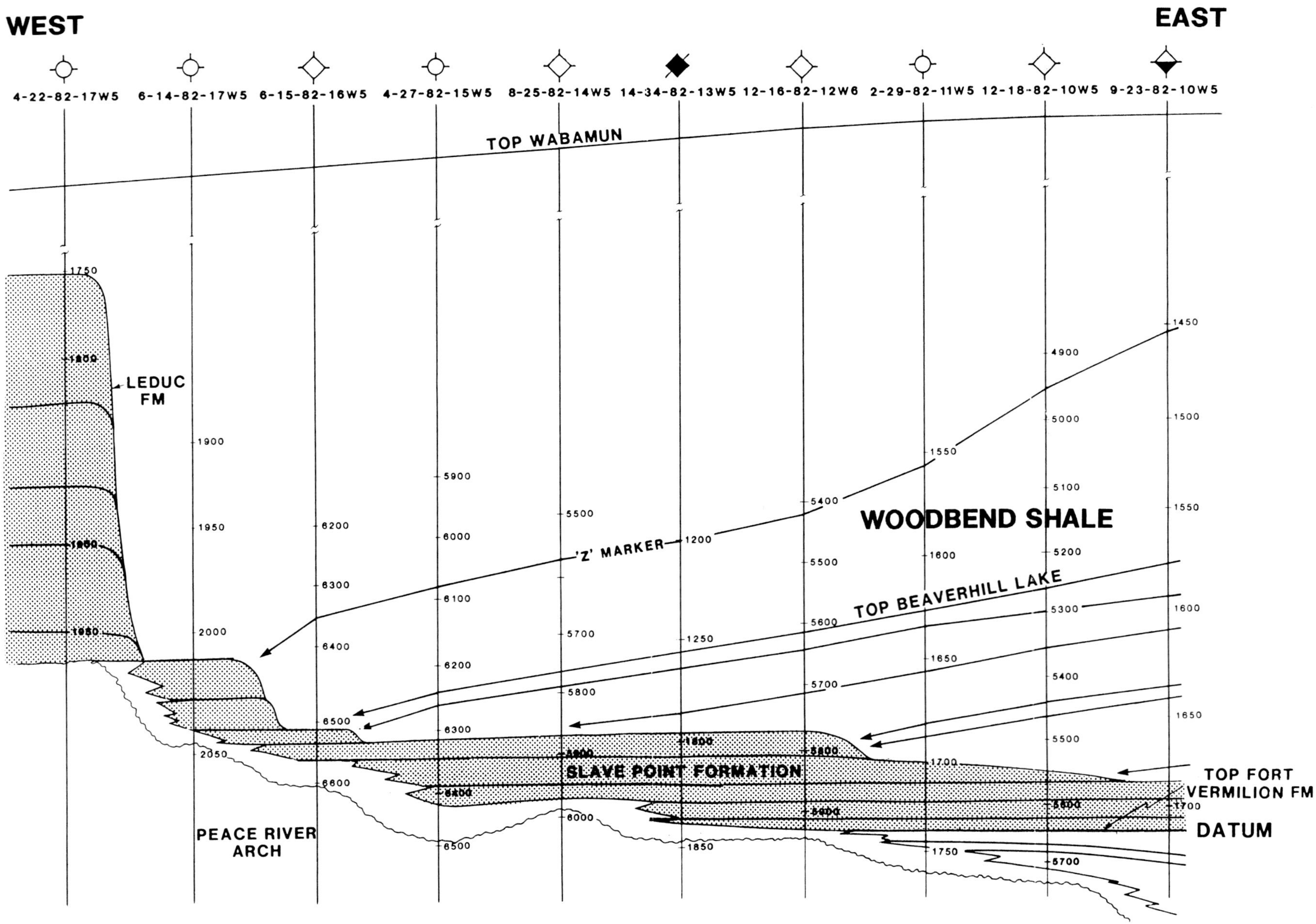


FIGURE 10. Stratigraphic Cross Section East of the Peace River Arch

had been engulfed by prograding Ireton shale banks but those of the Western realm still thrived in shallow water. The transgression that initiated the Winterburn megacycle could have permitted, in some locations, continued aggradation of some of these isolate reefs such that a portion of the uppermost "Leduc" is in fact equivalent to Lobstick deposition of the Winterburn (Stoakes and Wendte, 1987).

## STURGEON LAKE - PEACE RIVER ARCH REALM

*Carbonate Bank and Reef Succession* - Consists of backstepping and upstepping shallow-water carbonates of the Leduc Formation developed on top of the Swan Hills bank or in some cases siliciclastic sediments of the Granite Wash. The Leduc Formation occurs both as a fringing reef to the Peace River Arch landmass and as large isolate buildups (Gold Creek and Sturgeon Lake, Figure 1).

The Sturgeon Lake Leduc reef complex reaches over 300 m in total thickness. It sits directly on a shallow-water more areally extensive platform of the Swan Hills level with no stratigraphic subdivision of the Cooking Lake Formation. Cross sections reveal however that lower levels of the complex exhibit a backstepping geometry (Figure 11). At positions in front of individual shallow-water platform backsteps, isolate pinnacle reefs occur that form a narrow fairway around the main Sturgeon Lake reef (Aukes and others, 1987). These pinnacles are up to 90 m in thickness and are largely comprised of detrital foreslope material. They appear to have been "drowned" early in the Woodbend transgressive phase while the main complex continued to grow. The rest of the Sturgeon Lake complex shows a predominantly aggradational geometry as it kept pace with relative sea-level rise. The final stages of reef growth are marked by abandonment of the atoll configuration and replacement by an overall retreating grain-ramp profile that may equate to lower Lobstick deposition. Successive grain ramps backstep to the southwest where they are directly overlain by shallow-water carbonates of the Nisku Formation of the succeeding Winterburn cycle.

Around the Peace River Arch the Leduc comprises a fringing reef system up to 250 m in thickness. Along more basinward portions the Leduc sits on shallow-water carbonates of the Slave Point Formation (Beaverhill Lake megacycle, see Figure 10). In positions higher on the Arch Leduc carbonates onlap onto siliciclastic sediments of the Granite Wash. Dix (1990) has divided the Leduc in this area into three sequences comprising, in ascending order: a basal transgressive unit; a rimmed shelf up to 150 m in thickness and an uppermost carbonate ramp unit which marks the demise of Leduc deposition. In many respects this pattern mirrors that of the nearby Sturgeon Lake complex and differs only in the reduced space available for accommodation.

*Basin-Fill Succession* - Sediments of this succession comprise the basinal Duvernay and Ireton formations that onlap the Leduc complexes. As in the area to the south (Western realm) a division may be made in the basin-fill succession based on the definition and correlation of the Z Marker. In the Sturgeon Lake - Peace

## STURGEON LAKE AREA

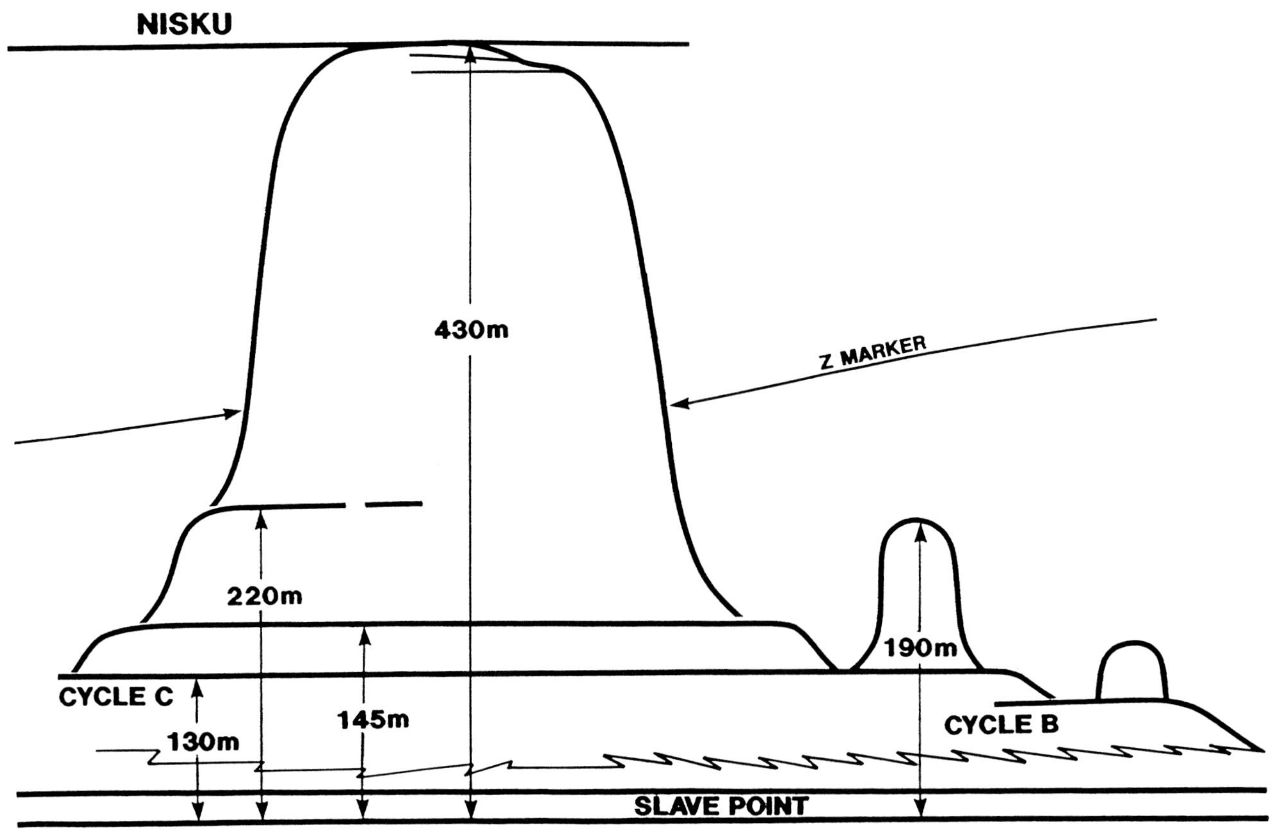


FIGURE 11. Schematic Cross Section of the Sturgeon Lake Leduc Reef

River Arch area the Z Marker clinoforms and downlaps onto the lower parts of the Leduc buildups (Stoakes, 1987; Dix, 1990) revealing that significant bathymetry still existed at the end of the Woodbend cycle. Mapping of the Z Marker surface locally (Dix, 1990) shows that it downlaps to the northwest and this is consistent with the pattern of basin-filling exhibited by the earlier Ireton clinoforms. Consequently the balance of the "Ireton Formation" basin fill in this area is part of the Winterburn cycle of deposition.

## SUMMARY

The Woodbend megacycle marks the point of maximum transgression of deep marine waters of the Upper Devonian sea into the WCSB. As such it represents the "hinge-point" between longer term depositional phases that are dominantly transgressive in aspect (Beaverhill Lake and early Woodbend) and those that are dominantly regressive (late Woodbend and Winterburn). It also corresponds to the first significant delivery of fine argillaceous-terrigenous sediments, derived from a northerly source, into the central portion of the Alberta Basin.

The pre-existing basin geometry gave rise to three depositional realms that shared a common sea-level history. Sediments of the Western realm and Sturgeon Lake area initiated on shallow-water sediments of the Beaverhill Lake megacycle (Swan Hills bank). Those of the Southeastern realm initiated on the shallow-water portion of the Beaverhill Lake (Waterways) clinoforms. Finally, the Peace River Arch fringing system initiated on shallow-water carbonates of the Slave Point Formation (Beaverhill Lake Group) and onlapped onto progressively higher portions of the Peace River Arch, a feature largely mantled with siliciclastics of the Granite Wash. Consequently this transgressive phase produced a generally similar pattern basinwide of backstepping and upstepping carbonate platforms, and isolate reef and bank complexes. This differs markedly from the pattern of the regressive phase of the megacycle which shows a distinct assymetry east to west. On the east side of the basin significant argillaceous influx early in Woodbend "time" led to basin filling that took place in a reciprocal relationship to reef growth (Stoakes, 1980). As individual Ireton clinoforms built seawards, carbonate platforms developed on their uppermost part in a "piggyback" fashion (Grosmont Formation). These clinoforms progressively encroached upon and finally engulfed areas of Leduc reef growth. On the west side of the basin carbonate banks and isolate reefs continued to grow in an overall upbuilding manner. Only a portion of the basin in this area was infilled by prograding Ireton clinoforms. The top of this phase of basin fill is discernible as a correlatable unit referred to as the Z Marker which downlaps out against the lower flanks of these Leduc complexes. Recognition and correlation of this marker is paramount in understanding the stratal relations and sequence stratigraphy of this portion of the Upper Devonian. This correlation indicates that the majority of the basin fill in the Western realm and around the Peace River Arch belongs to the succeeding Winterburn megacycle and that the duration of reef growth in this area could have been longer in this area than in the Southeastern realm.

The evolution of this megacycle reflects the interplay of relative sea level and depositional controls, such as wind direction, delivery of extrabasinal argillaceous sediment and antecedent topography. The overall transgressive-regressive nature of sedimentation corresponds to an early phase of relatively rapid sea-level rise followed by reduced rates of rise. Carbonate platforms, reefs and banks on both sides of the basin were initiated on bathymetrically high areas and grew through an incrementally rising sea level. Fine argillaceous sediment, derived from a northerly source, was fed laterally into eastern parts of the basin by a clockwise gyre, resulting in largely westerly directed progradation. This initiated filling of the basin largely from east to west. This pattern of infill was arrested by renewed relative sea-level rise that initiated the transgressive phase of the ensuing Winterburn megacycle.

## REFERENCES

AUKES, P.G., STACEY, E.C., AND WEBB, T.K., 1987, Early Leduc reef growth: an example at Sturgeon Lake South, *in* Second International Symposium on the Devonian System, Program and Abstracts, p. 268.

CUTLER, W.G., 1983, Stratigraphy and sedimentology of the Upper Devonian Grosmont Formation, northern Alberta: Bulletin of Canadian Petroleum Geology, v. 31, p. 282-325.

DIX, G.R., 1990, Stages of platform development in the Upper Devonian (Frasnian) Leduc Formation, Peace River Arch, Alberta: Bulletin of Canadian Petroleum Geology, v. 38A, p. 66-92.

HEIDGERKEN, L.D., WENDTE, J.C., and STOAKES, F.A., 1989, Stratigraphic evolution and its control on reservoir facies of a Leduc reef complex, Golden Spike, Alberta, *in* Exploration Update. Canadian Society of Petroleum Geologists - Canadian Society of Exploration Geophysicists Joint Conference, Programs and Abstracts, p. 67.

JERVEY, M.T., 1988, Quantitative geological modelling of siliciclastic rock sequences and their seismic expression, *in* Wilgus, C.K., Hastings, B.S., Kendall, C.G. St. C., Posamentier, H.W., Ross, C.A., and Van Wagoner, J.C., eds., Sea-Level Changes: An Integrated Approach: Society of Economic Paleontologists and Mineralogists, Special Publication no. 42, p.47-70.

JONES, R.M.P., 1980, Basinal isostatic adjustment faults and their petroleum significance: Bulletin of Canadian Petroleum Geology, v.28, p.211-251.

KLOVAN, J.E., 1964, Facies analysis of the Redwater reef complex,

Alberta, Canada: Bulletin of Canadian Petroleum Geology, v. 12, p.1-100.

LEXICON OF CANADIAN STRATIGRAPHY, 1990, Volume 4 Western Canada, including Eastern British Columbia, Alberta, Saskatchewan and Southern Manitoba, *in* Glass, D.J., ed., Canadian Society of Petroleum Geologists, 772 p.

MARTIN. R., 1967, Morphology of some Devonian reefs in Alberta, a paleogeomorphological study, *in* Oswald, D.H., ed., International Symposium on the Devonian System: Canadian Society of Petroleum Geologists, v. 2, p.365-385.

McGILLIVRAY, J.G., and MOUNTJOY. E.W., 1975, Facies and related reservoir characteristics Golden Spike Reef Complex, Alberta: Bulletin of Canadian Petroleum Geology, v. 23, p.753-809.

MOUNTJOY, E.W., 1978. Upper Devonian reef trends and configuration of the western portion of the Alberta Basin, *in* McIlreath, I.A. and Jackson, P.C., eds., The Fairholme Carbonate Complex at Hummingbird and Cripple Creek: Canadian Society of Petroleum Geologists, p. 1-30.

MOUNTJOY, E.W., 1980, Some questions about the development of Upper Devonian carbonate buildups (Reefs), Western Canada: Bulletin of Canadian Petroleum Geology, v. 28, p.315-344.

OLIVER, T.A., AND COWPER, N.W., 1963, Depositional environments of the Ireton Formation, central Alberta: Bulletin of Canadian Petroleum Geology, v. 11, p. 183-202.

ROSS, G., and STEPHENSON, R.A., 1989, Crystalline basement: the Foundation of Western Canada Sedimentary Basin, *in* Ricketts, B.D., ed., Western Canada Sedimentary Basin - A Case History: Canadian Society of Petroleum Geologists, Special Publication no. 30, p.33-46.

STOAKES, F.A., 1977, The Camrose Member of the Ireton Formation in the Stettler area of central Alberta, *in* McIlreath, I.A. and Harrison, R.S., eds., The Geology of Selected Oil, Gas and Lead-Zinc Reservoirs in Western Canada: 5th Canadian Society of Petroleum Geologists Core Conference, Calgary, Alberta, p. 89-106.

STOAKES, F.A., 1979, Sea level control of carbonate deposition during progradational basin-filling: The Upper Devonian Duvernay and Ireton formations of Alberta, Canada: Unpublished Ph.D. thesis, University of Calgary, Calgary, Alberta, 346 p.

STOAKES, F.A., 1980, Nature and control of shale basin fill and its effect on reef growth and termination: Upper Devonian Duvernay and Ireton formations of Alberta, Canada: Bulletin of Canadian Petroleum Geology, v.28, p.234-410.

STOAKES, F.A., 1987, Evolution of the Upper Devonian of Western Canada, *in* Principles and concepts for exploration and exploitation of reefs in the Western Canada Basin: Short Course Notes, Canadian Reef Inventory Project.

STOAKES, F.A. AND WENDTE, J.C., 1987, The Woodbend Group, *in* Krause, F.F. and Burrowes, O.G., eds., Devonian lithofacies and reservoir styles in Alberta: 13th CSPG Core Conference and Second International Symposium on the Devonian System, p.153-170.

WALLS, R.A., 1983, Golden Spike Reef Complex, Alberta, *in* Scholle, P.A., Bebout, D.G. and Moore., C.H., eds., Carbonate Depositional Environments: American Association of Petroleum Geologists, Memoir 33, p.445-453.

WENDTE, J.C., and CALLOW, G.O., 1987, Upper Devonian Woodbend platform and reef evolution; Redwater, Alberta, *in* Canadian Reef Inventory Project Symposium, Program and Abstracts booklet, Banff, Alberta, p.110.

CHAPTER 9

# WINTERBURN MEGASEQUENCE

F.A. STOAKES

## INTRODUCTION

This sequence comprises the strata between the Woodbend Group (Chapter 8) and the Wabamun Group (Chapter 10). Stratigraphically it includes carbonates of the Nisku Formation and a succession of strata that comprise the "upper" Winterburn (Figure 2). This succession includes carbonates of the Blue Ridge Member of the Graminia Formation and fine siliciclastic sediments of the Calmar and Graminia "silt" intervals.

Figure 1 depicts the basin configuration towards the end of Winterburn "time". It marks a major departure from the Woodbend configuration when a northeasterly opening to the basin existed between the Peace River Arch landmass and the Grosmont Platform Complex. Towards the end of Upper Nisku "time" a lobe of shale, along with coarser detrital sediment had blocked this entryway. This divided the Winterburn Basin into two geographic domains: firstly, a basin in central Alberta bordered by the "Meekwap" shelf to the north and northeast, the West Pembina shelf to the southeast and a less continuous shelf system in the Devonian "Deep Basin" to the west that presumably connected to an open seaway; and secondly, north and northwest of the Peace River Arch landmass, an early Winterburn basin that became infilled by shales of the upper part of the Fort Simpson Formation as far to the northwest as northeast British Columbia and contiguous areas of the Northwest Territories. Limestones of the Jean Marie Member of the Redknife Formation form a shelf bordering the open seaway to the west (Figure 1).

The Winterburn Group consists of a lower transgressive phase and an upper regressive phase of deposition. In the lower transgressive phase shelfal carbonates of the Nisku Formation backstep away from the basin in central Alberta. During the upper regressive phase, carbonate platforms with varying amounts of evaporitic and terrigenous components, forestep into the basin.

## STRATIGRAPHIC NOMENCLATURE

Figure 2 shows the stratigraphic nomenclature for the Winterburn succession in the various paleogeographic domains of the WCSB. The succession of strata in the Bashaw-Joffre area reflects the stratigraphy of the Nisku type section and is typical of the southern shelf. The Nisku Formation in this area consists of a lower carbonate interval and an upper evaporitic unit. The upper evaporitic unit grades vertically into siliciclastic deposits of the Calmar Formation. Further basinwards in the West Pembina area of central Alberta, the stratigraphic terminology is that proposed by the Exploration Staff, Chevron Standard Limited (1979). The development of this terminology followed the discovery of hydrocarbon-bearing pinnacle reefs along the basin margin. In this area the off-reef succession of the Nisku Formation is divided into three units - the Lobstick, Bigoray and Cynthia members. The upper part of the Lobstick, Bigoray and Cynthia members correspond to phases of isolate reef growth collectively termed the Zeta Lake Member. The uppermost unit of the Nisku Formation is the

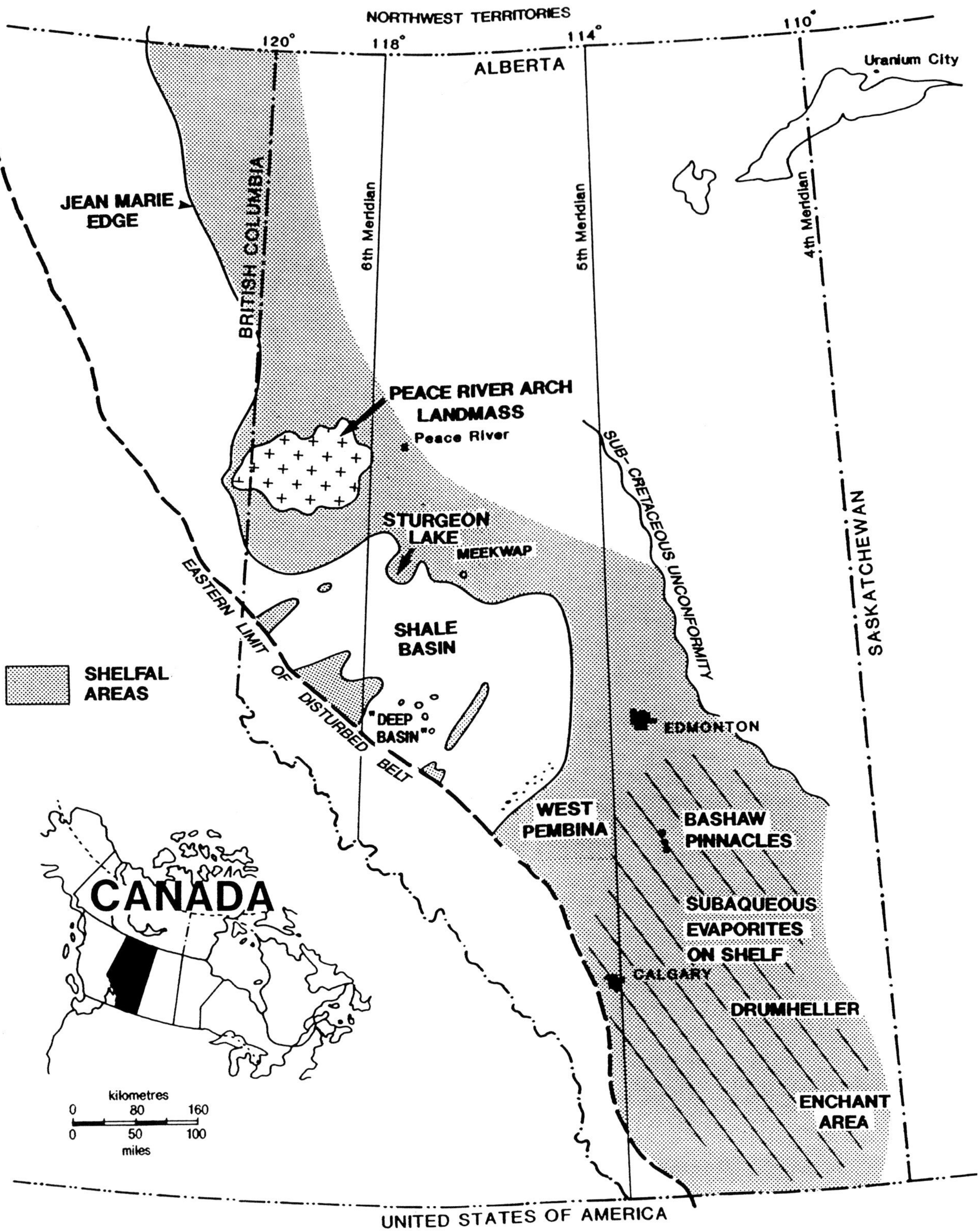


FIGURE 1. Paleogeography During "Late" Winterburn Time

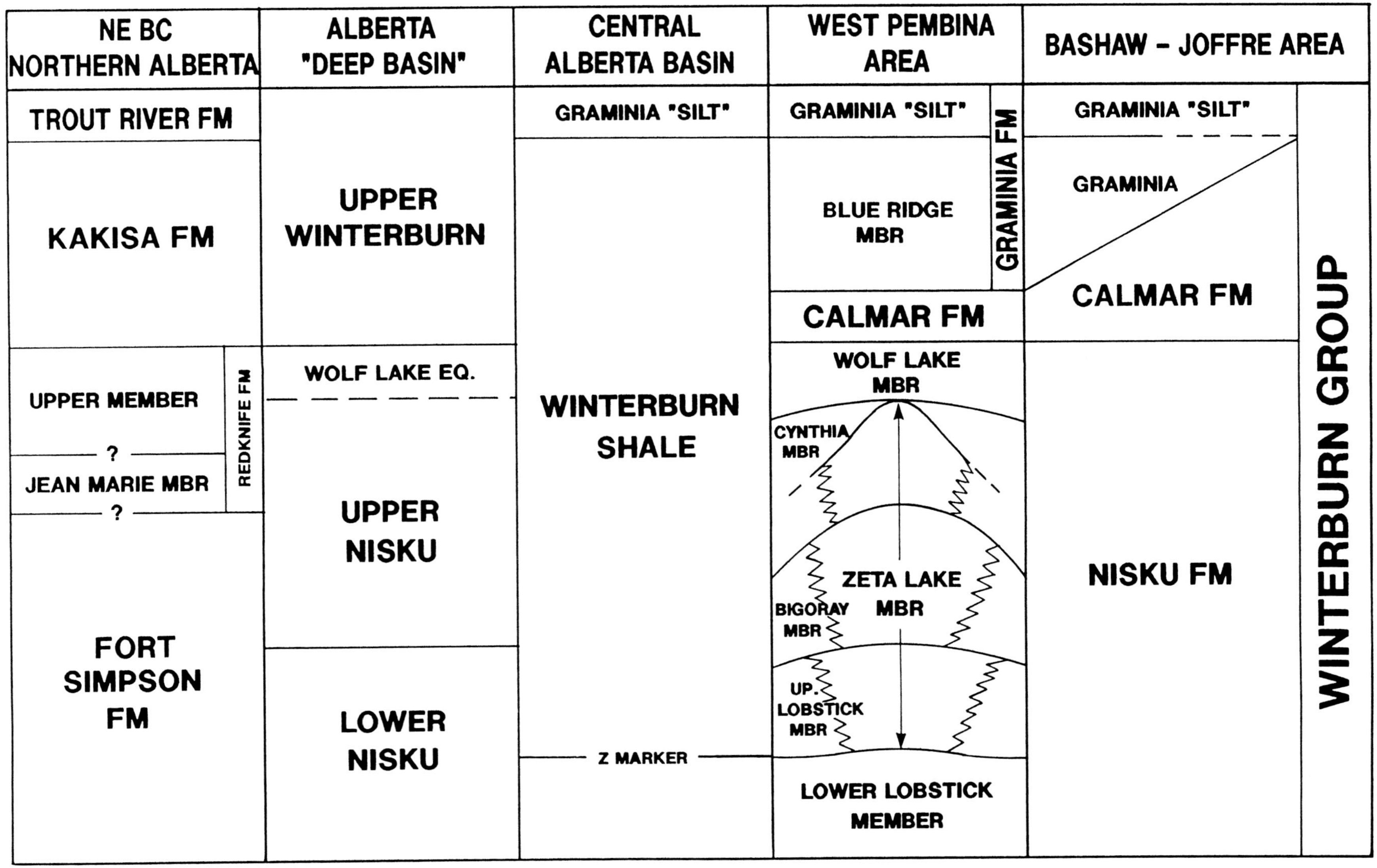


FIGURE 2. Stratigraphic Nomenclature of the Winterburn Group

Wolf Lake Member which occurs in basinward positions, and apparently correlates to a stage of growth on the isolate reefs themselves (Anderson and Machel, 1988). The "upper" Winterburn succession consists of silty carbonates of the Calmar Formation and carbonates and silty carbonates of the Blue Ridge Member and the Graminia "silt" Member respectively of the Graminia Formation. Bankwards from West Pembina, off-reef ramp units of the Bigoray and Cynthia members grade into shallow-water carbonates deposited along, and just back from, the Nisku platform edge. This succession corresponds to the "outer shelf" of the Exploration Staff Chevron Standard Limited (1979), the Nisku "bank" of Watts (1987) and the Dismal Creek Member (informal) of Machel (1983).

Basinward from the shelf margin, the Nisku, Calmar and Graminia successions clinoform into a thick shaly Winterburn sequence. This Winterburn shale section is separated from the underlying Woodbend shales by a widespread correlatable surface termed the Z Marker. Stoakes and Wendte (1987) clarified the stratigraphic position and significance of this marker and referred to it as the Z Marker, which was following internal Imperial Oil usage. This marker correlates to a mid-Lobstick cycle level at West Pembina (Figure 2). As such the lower part of the Lobstick Member has more depositional affinities to the Woodbend sequence than to the Winterburn sequence into which it is historically included, based on lithostratigraphic criteria.

Less is known in detail about the Winterburn succession on the shelfal areas towards the western side of the basin (the so-called Devonian "Deep Basin" of Figures 1 and 2). The Nisku Formation is often divisible into upper and lower units, with a portion of the upper unit considered to be the time-equivalent of the Wolf Lake Member. The overlying "upper" Winterburn succession consists of shallow-water carbonates, anhydrites and some silty shales. It is rarely systematically sub-divided and is treated collectively in this account.

In northern Alberta and northeast British Columbia, basin-filling shales of the Fort Simpson Formation comprise the lower part of the Winterburn succession. This basinal unit is overlain by a widespread sheet-like limestone, assigned to the Jean Marie Member of the Redknife Formation and interbedded deposits of ramp carbonates and more argillaceous basinal units of the succeeding Upper Member of the Redknife Formation and the Kakisa Formation. The Winterburn Group is capped by shallow-water and restricted, argillaceous and silty limestones and calcareous siltstones of the Trout River Formation.

## MODEL FOR THE WINTERBURN MEGASEQUENCE

A generalized model depicting the evolution of the Winterburn megasequence is shown in Figure 3. The transect extends from the West Pembina area northwestwards to the Nisku shelf in the Devonian "Deep Basin" area. The model portrays an overall asymmetrical transgressive-regressive cycle of sedimentation. This pattern is perhaps best exemplified at the basin margin in West Pembina, where the patterns are first described and then extended basinwards and bankwards. In the West Pembina area along the southeast margin of the Winterburn basin, ramp carbonates of the upper Lobstick Member backstep away from the basin. Zeta Lake Member reefs are initiated in deeper water on small accretionary highs on the underlying lower Lobstick Member. Subsequently, a major backstep initiates more of an aggradational style of growth along the shelf margin established to the southeast. This shelf margin complex and the isolate Zeta Lake Member reefs then grew in increments corresponding to the off-reef Bigoray and Cynthia members (Figure 2). During an ensuing relatively stable period of sea level, shales of the Cynthia Member prograded into the basin largely infilling the area between the isolate reefs and the shelf margin complex. So

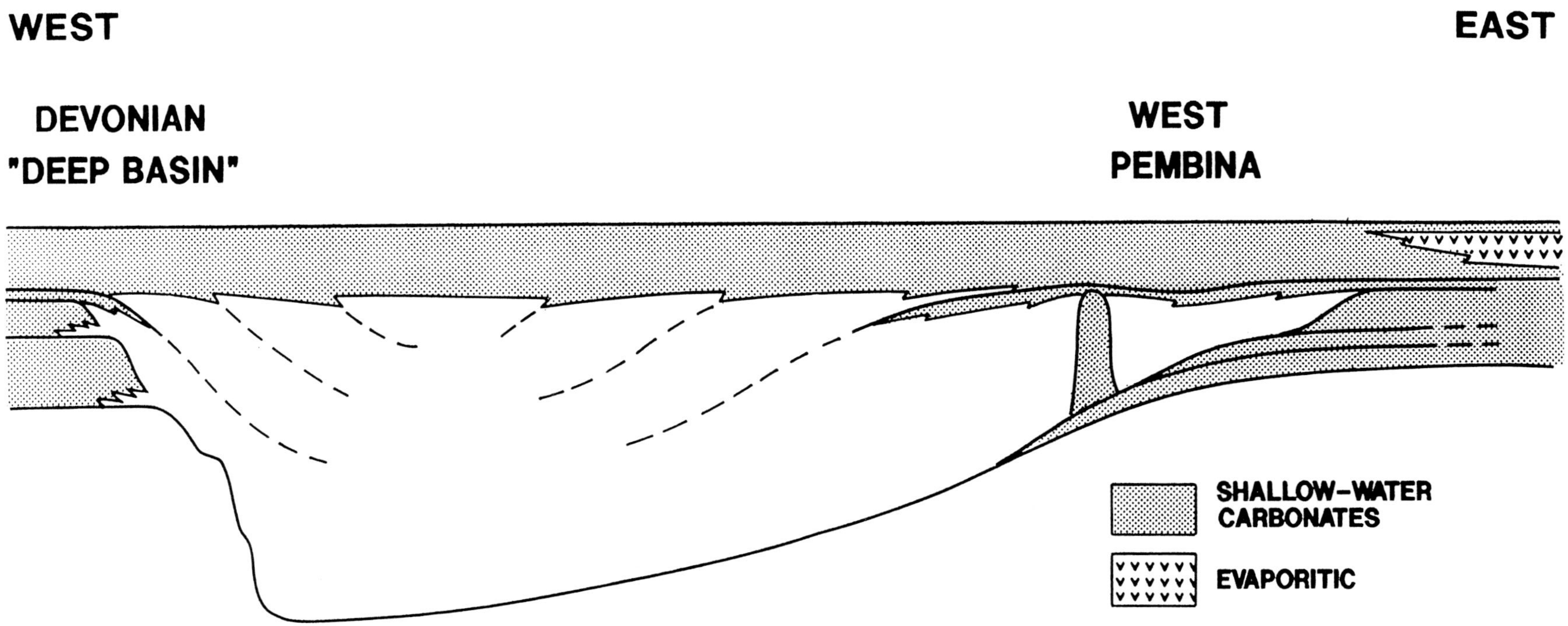


FIGURE 3. Conceptual Model of Winterburn Group Sedimentation

although basin-filling shales and isolate reefs occur within the same cycle of deposition, the shales apparently largely postdate the growth of the reefs and shelf margin. The upper part of the cycle is marked by progradation of tidal flat sediments of the Wolf Lake Member across the Cynthia basin fill and isolate reefs, and is succeeded by silty carbonates of the Calmar Formation. This marks infilling of the basin on the southeast and the termination of reef growth. Overlying units of the "upper" Winterburn, in particular the Blue Ridge Member, mark renewed rates of relative sea-level rise that resulted in reflooding of this part of the basin. As such, the Winterburn megasequence may be viewed as being comprised of at least two "subsequences". This transgressive event is apparent on regional sections from the renewed onlap of the Graminia Formation (Slingsby and Aukes, 1989, their Figure 2) onto the Calmar Formation top. This flooding may have implications for reef growth in other parts of the Winterburn basin. The final depositional unit of the Winterburn in the West Pembina area is the Graminia "silt" which marks final infilling of the basin.

The shelfal areas on the west side on the basin have depositional stages that correspond to both the transgressive and regressive phases observed in the West Pembina area. The transgressive phase consists of backstepping to upstepping carbonate platforms of the Nisku Formation. These deposits differ from synchronous deposits along the southeast margin of the basin in that they have more of an aggradational aspect. Instead of exhibiting a ramp profile, their geometry is more of a platform with distinct margins. This culminates in the regressive Wolf Lake equivalent which may mark a slight relative rise in sea level.

The following sections describe the stratigraphy, sedimentology and evolution of the Winterburn megacycle, in the various portions of the basin, in more detail.

### *West Pembina Area*

The overall evolution of the Winterburn succession in this area was discussed in the previous section. This section now provides more detail by incorporating the work of Anderson (1985), Machel (1983, 1984 and 1985) and Watts (1987). Figure 4 provides a well log cross section of this interval. Initial deposits of the Winterburn megacycle are open-marine downslope ramp carbonates of the lower part of the Lobstick Member. As mentioned previously, this deposit represents a continuation of progradation of the underlying Ireton Formation. Isolate Zeta Lake Member reefs initiated on small accretionary highs on top of this unit during a rapid rise in sea level. This timing of reef inception corresponds to formation of the Z Marker surface in off-reef and basinal positions. The reefs initiated in deeper water and grew in stages corresponding to equivalent deposition of the upper Lobstick, Bigoray and Cynthia members. During each of these stages the amount of reef to basin relief and the steepness of the foreslope increased. However, due to the fact that these reefs initiated and grew in relatively deep water, keeping pace with rising sea level, this cyclicity is not readily apparent in core. In off-reef positions this cyclicity is apparent with each successive unit showing a progressively deeper-water aspect. This can be seen in core but is also reflected in the incremental shifts of the gamma log to the right at the onset of each of these units (see 6-21 well on Figure 4). In each of the lower two units (upper Lobstick and Bigoray) there is a vertical gradation from calcareous and argillaceous siltstones into bioclastic limestones reflecting an upward-shoaling nature.

Within the reefs the succession of two basic facies indicates an overall upward-shoaling sequence. Figure 5 shows a schematic representation of these

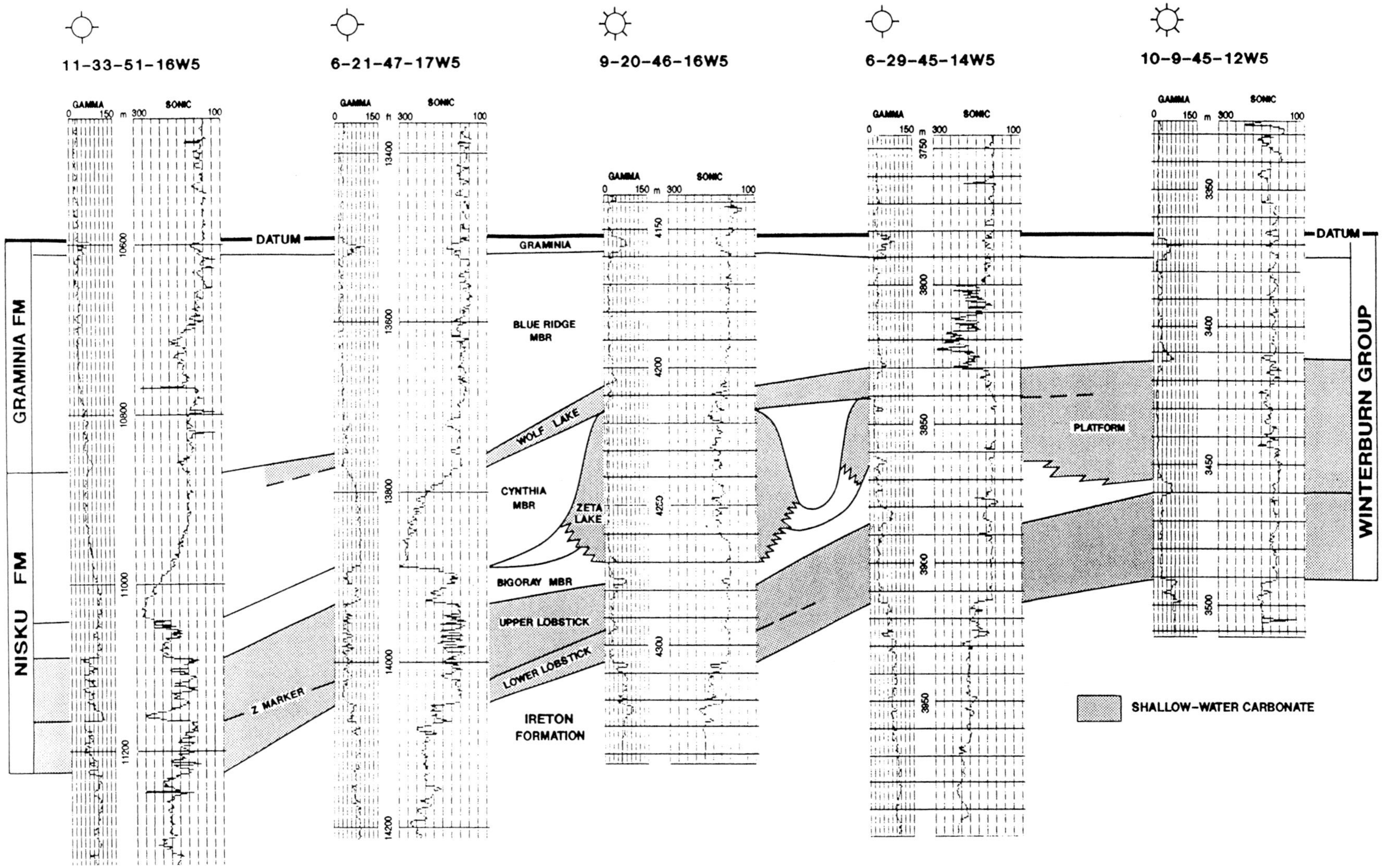


(MODIFIED AFTER UNPUBLISHED CROSS SECTION OF MYLES BOSMAN, 1991)

FIGURE 4. Well Log Cross Section, West Pembina Area

# NISKU REEFS AND RELATED LITHOFACIES

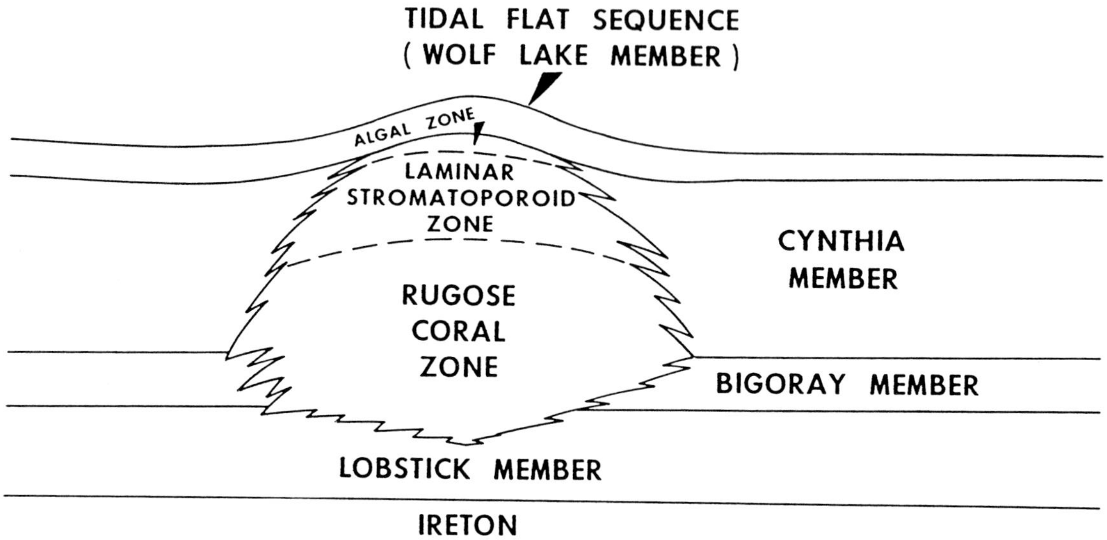


FIGURE 5. Simplified diagram showing the vertical zonation within a Nisku reef. Although only one vertical sequence is shown here, in most Nisku limestone reefs several sequences (often incomplete) can be recognized indicating that the reefs grew incrementally, possibly in response to relative sea level rises. (Watts, 1987)

reef facies from Watts (1987). The organic framework of these reefs in their lower and intermediate parts consists predominantly of spectacular *in situ* dendroid and fasciculate rugose corals. Thin to thick laminar anastomosing stromatoporoids are the dominant framebuilder in the upper part of the reefs, and some may be capped by a poorly developed algal crest. Where present, this facies consists of *in situ* dense growths of Solenoporid red algae. Isopachous marine cements infill in many of the larger interparticle voids and cavities in the reefs. Otherwise the matrix is generally comprised of green argillaceous mudstones.

The Bigoray and Cynthia increments of isolate reef growth are synchronous with the development of a backstepped platform sequence to the southeast (Figure 4). Along certain promontories the margin of this platform consists of an upward-shoaling sequence capped by thick encrusting laminar stromatoporoids.

A thick sequence of shales of the Cynthia Member infills in front of the platform and between the reefs. Deposition of these "shales" is interpreted to largely postdate the increment of "Cynthia" carbonate deposition on both the isolate reefs and platform. Between the reefs the Cynthia Member grades from a basal unit of dark calcareous shales up into finely and poorly laminated and bedded argillaceous siltstones. This overall shoaling sequence grades very gradually in more basinward positions, and relatively sharply in platformward positions, into shallow-water carbonates of the Wolf Lake Member.

The Wolf Lake Member records an increment of shallow-water deposition on both the shelf and isolate reefs. The shelfal unit progrades basinwards eventually engulfing the isolate Zeta Lake reefs. This is overlain by a silty sequence referred to the Calmar Formation. Following this, nodular bioturbated limestones of the Blue Ridge Member mark reflooding of the periphery of the Winterburn basin, initiating another sub-cycle of deposition that culminates with deposition of silty carbonates of the Graminia "silt" level. This Blue Ridge "cycle" becomes progressively anhydritic bankward towards its onlap edge as well as in northwestern portions of the Winterburn basin.

### *Bashaw - Joffre - Southern Alberta Shelf*

Deposition of the Winterburn succession in this area occurs south and east of the shelf margin complex in the West Pembina region (Figure 1). The evolution of these deposits records the interaction of relative sea level on a more restrictive setting with significant precursor topography.

The Winterburn interval conformably overlies carbonates and shales of the Woodbend Group. The dominant shallow-water elements of the Woodbend Group in this area are the reefal complexes (Rimbey-Meadowbrook and Bashaw-Duhamel trends and the Southern Alberta Shelf - see Chapter 8). The ensuing discussion will focus primarily on the Bashaw-Joffre area.

Figure 6 is a well log cross section across the Bashaw area illustrating patterns in the Nisku Formation. Depositional patterns in the latest Woodbend exerted a strong influence on Nisku sedimentation. "Piggyback" carbonates of the Camrose Member of the Ireton Formation occur to the northeast on top of basin-filling Ireton shale clinoforms (Figure 6). These clinoforms downlap onto the flank of the Leduc complex and equate to a stage of Camrose carbonate deposition on the uppermost part of the complex to the southwest. Uncovered portions of the Leduc surface exhibit small, isolate crinoid-bearing shoals equivalent to this Camrose carbonate ramp. Renewed basin filling of the post-Camrose Ireton shale changes this configuration only a little.

The Nisku succession comprises an initial open-marine carbonate and an

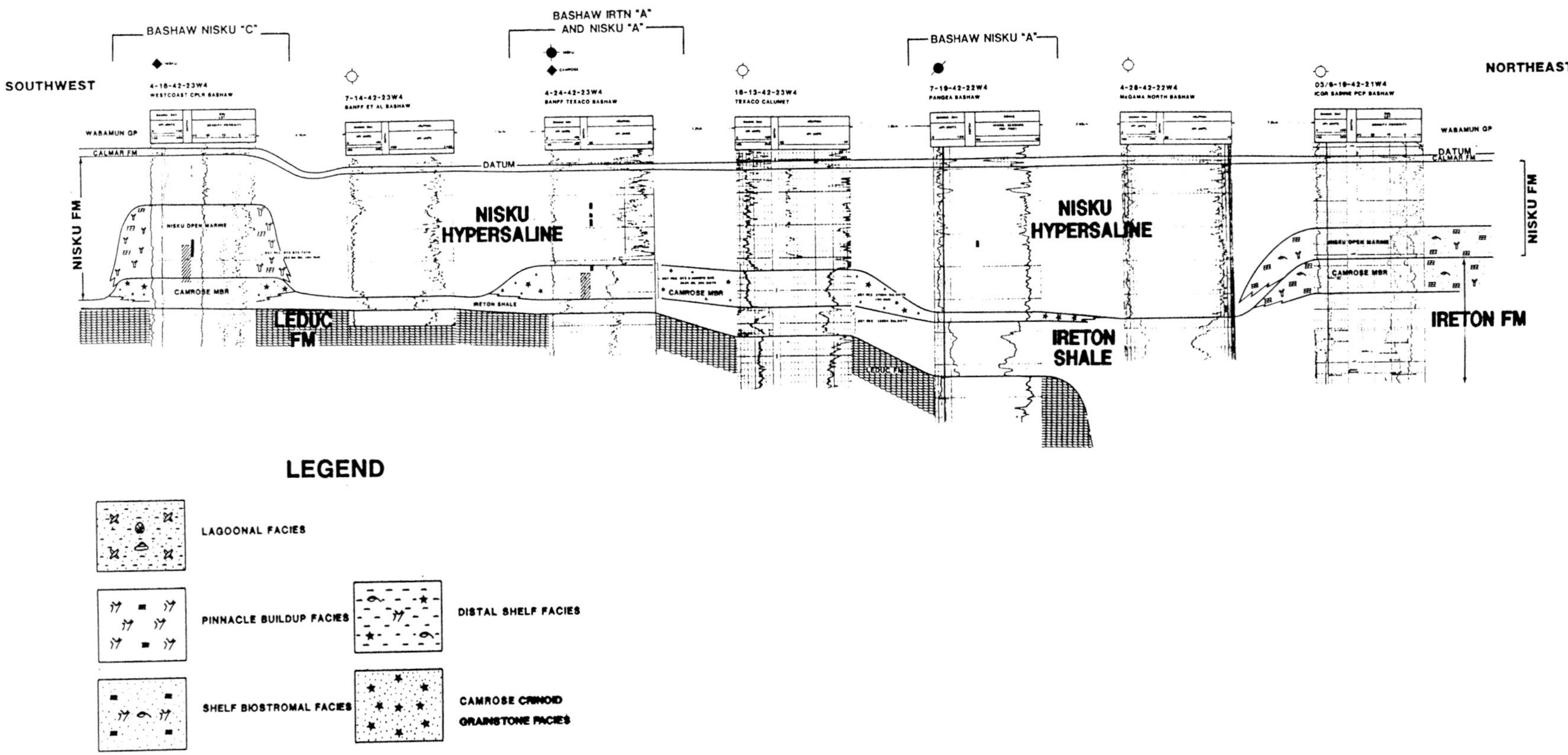


FIGURE 6. Well Log Cross Section of the Nisku Formation in the Bashaw Area

upper more restricted hypersaline unit (Andrichuk and Wonfer, 1954). The Nisku Formation along with the overlying Calmar averages 60+ m in total thickness and equates to the entire Winterburn succession further basinwards in the West Pembina area. Figure 6 illustrates the evolution of these phases.

The open-marine phase records an increase in rate of relative sea-level rise which initiated stromatoporoid-rich platform carbonate growth on pre-existing shallow-water areas on both the regional clinoform and on the Bashaw Leduc complex (Figure 6). Deeper-water areas were "left behind" with only a thin condensed unit being deposited (the so-called `moat' of Dixon and others, 1991). Within this deeper-water area pinnacle reefs comprised predominantly of thamnoporoid corals nucleated and grew on the underlying Camrose grain shoals (Figure 6). These pinnacles reach over 40 m in total thickness and grew during the transgressive event that initiated Nisku platform sedimentation elsewhere in the area. Other isolate carbonate features occur in this "moat" that appear to represent small outliers of the otherwise more extensive Nisku shelf.

The open-marine carbonate phase of the Nisku in this area equates to the upper Lobstick, Bigoray and early Cynthia sedimentation at West Pembina. This transgressive pulse reached far into southern Alberta with peritidal, sabkha conditions being the dominant component of the shelf to the south (Slingsby and Aukes, 1989).

As the Cynthia shelf margin grew up to sea level it reduced circulation from the open basin onto the inner platform and sedimentation became highly restricted. This gave rise to deposition of the upper more hypersaline phase of the Nisku which consists of repeated brining-upward cycles grading from laminated dolomudstones to laminated and bedded anhydrites. These deposits were formed during the progressive evaporation and subsequent desiccation of this sub-basin and onlap-fill the lows between the isolate reefs and platforms, eventually capping the entire succession (Figure 6). It exhibits a reciprocal thickness relationship to the preceding open-marine phase. Gilhooly (1987) recognized the hypersaline character of the upper Nisku deposits but interpreted at least the lower portion of the evaporitic assemblage as a lateral facies of the carbonate bodies in this area, instead of being a later infill. In this account the hypersaline phase is thought to represent largely subaqueous accumulation and entirely postdates the open-marine carbonate phase of the Nisku Formation. The hypersaline phase of the Nisku in this area equates to the latter part of Cynthia and Wolf Lake member deposition at West Pembina. It relates to a relative stillstand of sea level during which the shelf margin at West Pembina aggraded vertically and then commenced progradation seawards.

A succession of fine silts and shales comprises the overlying Calmar Formation in the Bashaw-Joffre area and marks final filling of the basin. Renewed marine transgression of the Blue Ridge Member of the Graminia Formation is not readily apparent across the majority of the inner shelf area. Increased accommodation towards the basin margin means that this unit onlaps and changes facies south and east, and the Graminia "silt" zone merges into the Calmar Formation.

### *Meekwap - Sturgeon Lake Area*

Winterburn sedimentation in the Meekwap-Sturgeon Lake area represents deposition along the north and north-northeast side of the basin (Figure 1). Well log cross sections reveal that the Z Marker drops stratigraphically from the West Pembina area north and west below the level of the Meekwap Member. This leads to the conclusion that the Winterburn basin, at the end of the transgressive phase had an opening to the northeast between the Peace River Arch and Grosmont Platform Complex (Stoakes, 1987, his Figure 19). It appears that this seaway was closed in mid-Winterburn "time" by prograding shale clinoforms moving from northeast to southwest (Figure 1). The upper part of this basin-fill unit often contains a thick siliciclastic interval suggesting proximal source conditions. This basin fill led to the establishment of a broad "piggyback" carbonate shelf of the so-called Meekwap Member (Cheshire and Keith, 1977). Production is obtained from small wafer stromatoporoid - *Renalcis* buildups at the shelf margin. Stabb (1990) equates this development to the Bigoray-Zeta-Dismal members at West Pembina, whereas in this account they would relate more to the latter part of the Cynthia and Wolf Lake. This shelf connects westwards from the Meekwap field area to the Sturgeon Lake Leduc reef complex where it sits directly on the Leduc with no intervening shale section (Chapter 8, Figure 11). Dix (1990, his Figure 6) maps out the extent of the Winterburn sub-basin west of the Sturgeon Lake reef. Presumably it's location reflects the sheltering effect of the Leduc complex on early Winterburn basin-filling coming from the east-northeast. As alluded to in the preceeding chapter (Chapter 8) the upper parts of the Sturgeon Lake Leduc reef comprise backstepping grain ramps that may equate to the transgressive phase of the Winterburn megacycle. This small sub-basin connects southwards to the main Devonian "Deep Basin" area. Further details on the pattern of basin fill and growth of pinnacle reefs is given in the next section.

### *Devonian "Deep Basin" Area*

The Devonian "Deep Basin" area lies on the west side of the Winterburn Basin in west-central Alberta, south of the Peace River Arch (Figure 1). This area includes large Leduc complexes that project to the northeast from the Leduc shelf on the west, and both small and large isolate Leduc reef complexes.

The Winterburn succession includes more open-marine carbonates of the Nisku Formation and an assemblage of carbonates, anhydrites and some argillaceous deposits of the "upper" Winterburn. Deeper-water carbonates and shales, equivalent to both the Nisku and "upper" Winterburn, infill the basin areas. Figure 7 shows a cross section depicting the evolution of this sequence. Nisku features in this area include: (1) shelfal complexes and their basinal extensions, (2) widespread banks over the Windfall Leduc complex, but separated from it by thin shales, (3) isolate reefs that overlie Leduc complexes, but separated from them by shales or deeper water limestones and (4) a belt of isolate pinnacle reefs overlying a thick succession of basin-filling shales and limestones. Elements 1, 3 and 4 are depicted on Figure 7. The transect on this section extends from the Fir Leduc complex, southwest into the adjacent basinal area and then to the northwest up onto the shelf. Nisku basin-fill units are correlated and named according to their West Pembina equivalents. Stratigraphic relations depicted on Figure 7 indicate water depths in the basin prior to Nisku deposition in excess of 100 m. This "inherited" accommodation reflects the incomplete nature of preceeding Woodbend basin fill.

The shelf contains increments of Lobstick, Bigoray and Cynthia equivalents, all backstepped away from the underlying Leduc edge, a pattern that occurs at the same stratigraphic level in other parts of the basin (Andrews, 1988). A Wolf Lake-equivalent carbonate progrades a small distance basinward.

The basinal succession records the progradation and clinoforming of large Nisku shale and carbonate, deeper-water "banks" into the basin. Progradation occurred both in a southwestward direction, extending out from the Fir Leduc

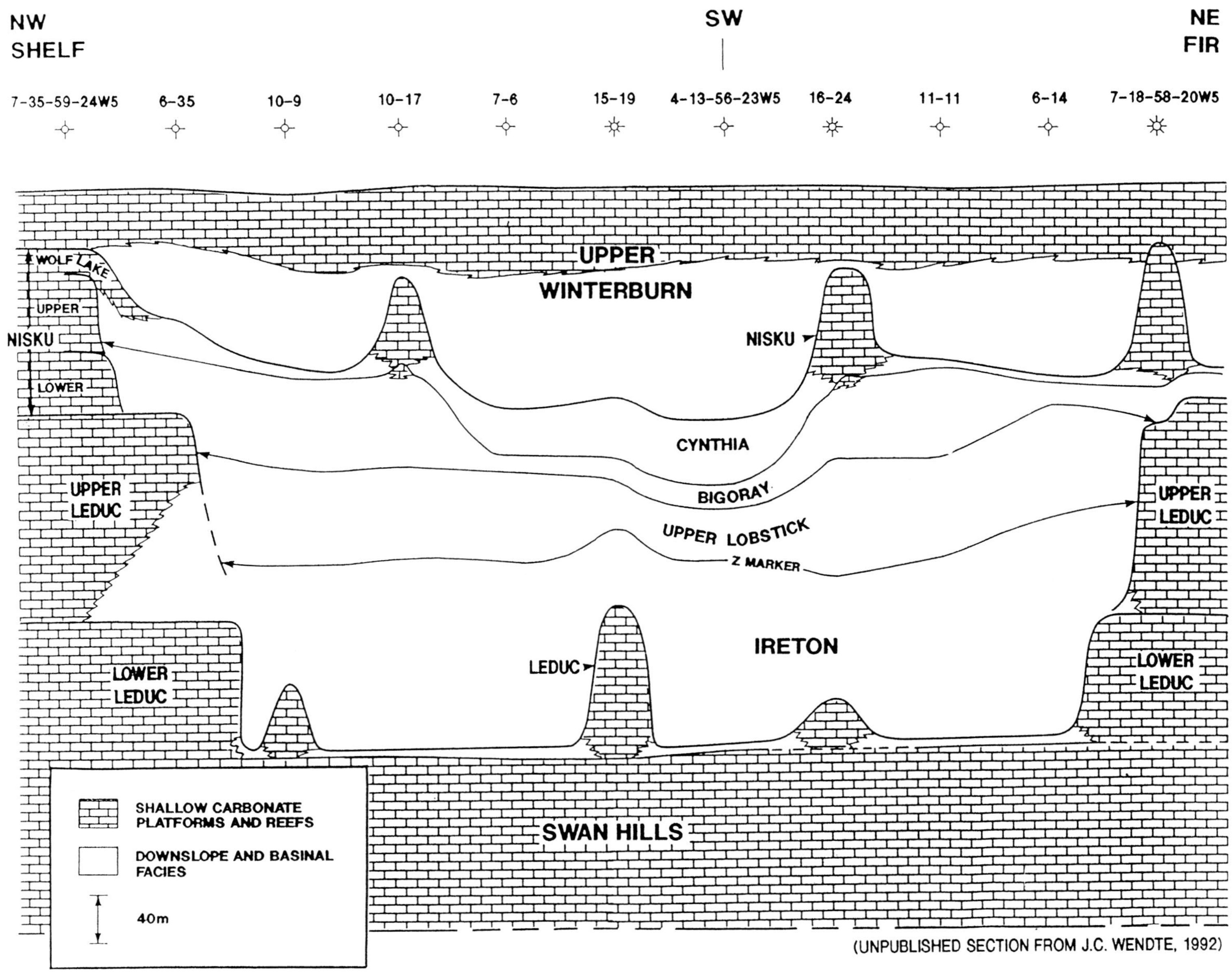


FIGURE 7. Stratigraphic Cross Section Across the Devonian "Deep Basin"

complex, and in a southeast direction, extending out from the shelf complex. A deep-water succession accumulated in the basin centre.

By the end of Lobstick deposition, appreciable bank to basin relief still existed. During the deposition of the Bigoray equivalent, an initial stage of isolate reefs grew along a belt paralleling the "margin" of the underlying Lobstick-equivalent clinoforms. Deposition of sediments also continued on intervening areas of the deep-water "banks". A thin succession of Bigoray shale also covered the Fir Leduc complex while only minor deposition occurred in the basin centre. Bank to basin relief remained much the same with the clinoform slope tending to steepen somewhat.

In contrast to the West Pembina area, it is Cynthia deposition that marked the phase of major growth on the isolate reefs along the belt paralleling the "margin" of the underlying clinoforms. Examination of the one available core indicates a deep-water thamnoporoid coral assemblage. An isolate reef also formed in a position overlying the Fir Leduc complex. A lobe of Cynthia shale prograded into the basin from the northwest. This shale succession filled in most of the pre-existing Bigoray basin, onlapping out against the basin flanks. Only a thin increment of shale, however, was deposited between the isolate reefs on the deep-water bank. At the end of this stage of growth, the isolate reefs attained a relief of approximately 60 m above the adjacent sea floor.

The belt of isolate reefs was drowned at the onset of Wolf Lake deposition, whereas the isolate reef overlying the Fir Leduc complex continued to grow until the end of this stage. A succession of mainly "upper" Winterburn calcareous muds infilled both the basin centre and the adjacent deep-water bank. By the end of Winterburn deposition, a widespread coastal plain existed over the entire area.

### *Northern Alberta - Northeast British Columbia*

A thick wedge of Fort Simpson shale infilled the pre-existing Woodbend Basin north of the Peace River Arch out to a longitude of approximately 122 degrees in northeast British Columbia and contiguous areas of the Northwest Territories. This shale bank is flat topped and exhibits evidence of having grown up to the zone of wave agitation.

Limestones of the Jean Marie Member formed a widespread sheet-like deposit covering this shale bank. The portion of the Jean Marie overlying the precursor Slave Point Petitot salient in northeast British Columbia contains significant gas reserves. Most of these reserves occur where the Jean Marie either overlies or flanks the margin of the Cordova Embayment, a Slave Point re-entrant extending into the centre of the Petitot Bank. Production from existing wells indicate pores in this limestone are gas bearing with no mobile water.

The depositional style in the Jean Marie over the Cordova Embayment indicates the lack of an effective depositional barrier to the north. As a consequence, the Jean Marie in this area exhibits more of a modified ramp profile. Figure 8 shows a north-south transect through the July Lake Field, a pool currently being developed. The pool corresponds to a shoal complex with small isolate patch reefs. On this shoal, the Jean Marie attains a maximum thickness of approximately 25 m on the reef crests and by the end of Jean Marie deposition, the crest of these reefs stood only about 8 m above the surrounding sea floor. On Figure 8, the d-37-I, a-41-A and a-25-E wells all occur on the shoal and encountered reef facies. Off the shoal, the Jean Marie thins and shows the development of deeper water facies. On Figure 8, the b-86-E well occurs off the shoal.

The Jean Marie consists of four upward-shoaling sequences. All cycles,

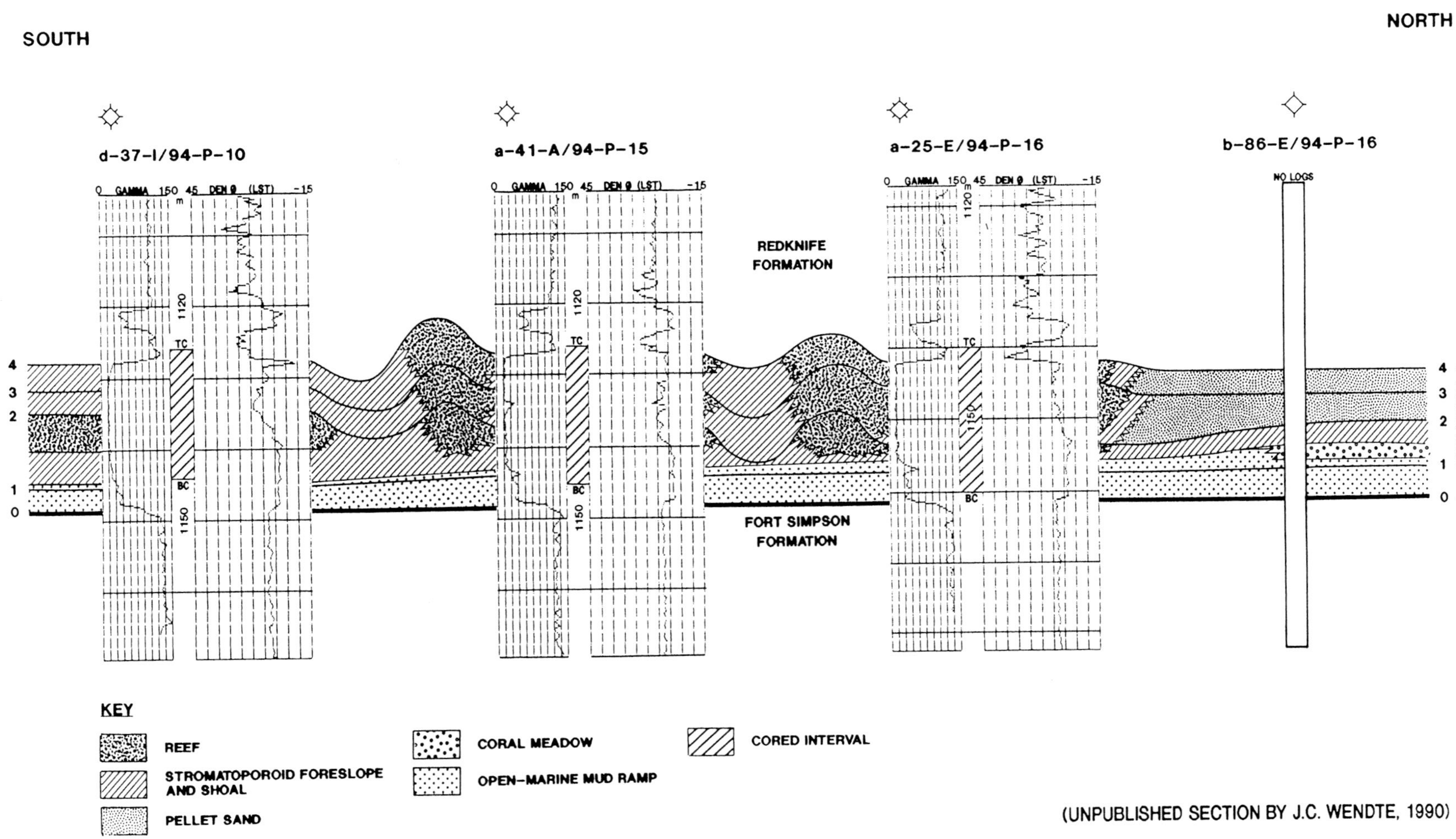


FIGURE 8. Well Log Cross Section of the Jean Marie Member, July Lake Field

however are not always expressed. At the a-25-E location, for example, the top of cycle 3 is not expressed because it occurs within the competent growing reef facies. The initial cycle corresponds to the widespread deposition of open-marine lime wackestones, following the marine incursion that initiated Jean Marie development. The second cycle correlates with the onset of patch reef growth. The third and fourth cycles exhibit a transgressive style, with backstepping and progressively greater reef to off-reef relief. At the end of cycle 4, the Jean Marie was drowned and covered by deeper-water shales of the Redknife Formation. Overlying Winterburn sequences, not shown on Figure 8, developed on ramps and shale out both to the north and west.

The nature of the isolate patch reefs deserves special mention because the internal composition of these buildups is so markedly different from earlier and coeval Devonian reefs. The core of these reefs consists of platy stromatoporoids that grew as inclined and upright forms. In growth form they more closely resemble Carboniferous and Permian aged Phylloid algae than typical Devonian stromatoporoids. The growth of these platy stromatoporoids created large shelter cavities. Within these cavities, the algae *Renalcis* encrusted the under surface of the stromatoporoids, along the more protected cavity tops. *Renalcis* then grew down into the cavities. Internal marine sediment filled in most of the remaining space.

## SUMMARY

Deposition of the Winterburn sequence records a mega transgressive - regressive cycle of sedimentation that in overall aspect is dominantly regressive. The initial transgressive phase consists of backstepping ramps and platforms, and isolate reefs. The upper more dominant regressive phase exhibits a forestepping style. During this phase, ramp carbonates built basinward in a "piggyback" manner, as the basin was being filled. The configuration of the basin which characterized preceeding Beaverhill Lake and Woodbend sedimentation changed markedly in mid-Winterburn time with the infilling of the open-marine seaway between the Peace River Arch landmass and the Grosmont Platform Complex. By the end of this cycle, all but an isolated depression south of the Peace River Arch had been filled. This small sub-basin, often referred to informally as the "Economy sub-basin" existed south of the Peace River Arch and west of the Sturgeon Lake Leduc reef (see Figure 6 of Dix, 1990). The final depression was subsequently filled by a basin centre evaporitic sequence that preceded deposition of the overlying Wabamun Group. The dominantly asymmetric east to west infill corresponds to the easterly trade wind belt.

On a grander scale, the Winterburn succession caps the top of an even larger scale depositional phase. In the WCSB, an overall transgressive and backstepping phase occurs from the onset of Upper Devonian deposition up until the deposition of the middle Leduc (see Chapter 8). During this phase, the areal extent of the Alberta Basin continually expanded. From the middle Leduc to the end of the Winterburn, sedimentation is marked by overall forestepping and basin filling and the Alberta Basin progressively shrank in areal extent. It is the Winterburn megasequence that completes this basin-filling episode.

## REFERENCES


ANDERSON, J.H., 1985, Depositional facies and carbonate diagenesis of the the downslope reefs in the Nisku Formation (Upper Devonian), central Alberta, Canada: Unpublished Ph.D. thesis, University of Texas, Austin, Tx. 393p.

ANDERSON, J.H., and MACHEL, H.G., 1988, The Upper Devonian Nisku reef trend in Central Alberta, *in* Geldsetzer, H.H.J., James, N.P., and Tebbutt, G.E., eds., Reefs Canada and adjacent areas: Canadian Society of Petroleum Geologists, Memoir 13, p. 391-398.

ANDREWS, G.D., 1988, Devonian Leduc outcrop reef-edge models and their potential seismic expression: Canadian Society of Petroleum Geologists Short Course Notes, March 1988.

ANDRICHUK, J.M., and WONFOR, J.S., 1954, Late Devonian geologic history in Stettler area, Alberta, Canada: Bulletin of American Association of Petroleum Geologists, v. 38, p. 2500-2536.

CHESHIRE, S.G., and KEITH, J.W., 1977, Meekwap Field - a Nisku (upper Devonian) shelf edge reservoir, *in* McIlreath, I.A., and Harrison, R.S., eds., Geology of selected carbonate oil, gas and lead-zinc reservoirs in Western Canada: Canadian Society of Petroleum Geologists, Core Conference, p. 23-45.

DIX, G.R., 1990, Stages of platform development in the Upper Devonian (Frasnian) Leduc Formation, Peace River Arch, Alberta. Bulletin of Canadian Petroleum Geology, v. 38A, p. 66-94.

DIXON, R.J., STOAKES, F.A., and CAMPBELL, C.V., 1991, Exploration for Nisku Formation isolate reefs of the Wood River area: a stratigraphic play-type in a structural world: Canadian Society of Petroleum Geologists Conference Program and Abstracts, p. 55.

EXPLORATION STAFF, CHEVRON STANDARD LIMITED, 1979, The geology, geophysics and significance of the Nisku reef discoveries, West Pembina area, Alberta, Canada: Bulletin of Canadian Petroleum Geology, v. 27, p. 326-359.

GILHOOLY, M. G., 1987, Sedimentology and geologic history of the Upper Devonian (Frasnian) uppermost Ireton and Nisku formations, Bashaw area Alberta: Unpublished M.Sc. thesis, University of Calgary, Calgary, Alberta, 203 p.

MACHEL, H.G., 1983, Facies and diagenesis of some Nisku buildups and associated strata, Upper Devonian, Alberta, Canada, *in* Harris, P.M., ed. ,Carbonate buildups - a core workshop: Society of Economic Paleontologists and Mineralogists, Core Workshop No. 4, p. 144-181.

MACHEL, H.G., 1984, Facies and dolomitization of the Upper Devonian Nisku Formation in the Brazeau, Pembina and Bigoray areas, Alberta, Canada: Canadian Society of Petroleum Geologists Core Conference 1984: Carbonates in subsurface and outcrop, p. 191-224.

MACHEL, H.G., 1985, Facies and diagenesis of the Upper Devonian Nisku Formation in the subsurface of Alberta: Unpublished PhD thesis, McGill University, Montreal, Quebec, 392 p.

SLINGSBY, A., and AUKES, P.G., 1989, Geology and Reservoir Heterogeneity of the Enchant Arcs "F" and "G" pools, *in* Canadian Society of Petroleum Geologists Core Conference, p. 4-1 to 4-27.

STABB, G.A., 1990, Nisku Sedimentology from Meekwap to Sturgeon Lake, Alberta, *in* Basin Perspectives, 1990 Canadian Society of Petroleum Geologists Convention Program and Abstracts, p. 135.

STOAKES, F.A., 1987, Evolution of the Upper Devonian of Western Canada, *in* Principles and concepts for exploration and exploitation of reefs in the Western Canada Basin. Short Course Notes, Canadian Reef Inventory project.

STOAKES, F.A., and WENDTE, J.C., 1987, The Woodbend Group, *in* Krause, F.F., and Burrowes, O.G., eds., Devonian lithofacies and reservoir styles in Alberta: 13th CSPG Core Conference and Second International Symposium on the Devonian System, p. 153-170.

WATTS, N.R., 1987, Carbonate sedimentology and depositional history of the Nisku Formation in south-central Alberta, *in* Krause, F.F., and Burrowes, O.G., eds., Devonian lithofacies and reservoir styles in Alberta: 13th CSPG Core Conference and Second International Symposium on the Devonian System, p.87-152.

CHAPTER 10

# WABAMUN MEGASEQUENCE

F.A. STOAKES

## INTRODUCTION

This sequence comprises a thick succession of shallow-water ramp carbonates assigned to the Wabamun Group of Famennian age (Figure 1). In southern Alberta the section becomes evaporitic and is subdivided into the Stettler and Big Valley formations. In this account the open-marine Big Valley Formation is grouped within the overlying Banff cycle of sedimentation and as such is detailed in the next chapter. To the north and west, in northeast British Columbia, the Wabamun interval has equivalents into the Tetcho and Kotcho formations which shale out into the basinal Besa River Formation.

The Wabamun cycle marks a reflooding of the WCSB following the Winterburn cycle of basin fill (Figure 2). At the initiation of this cycle the underlying Winterburn succession had infilled practically all of the existing topography throughout most of Alberta and northeast British Columbia. Consequently, the lack of any significant precursor topography, meant that the Wabamun megasequence preserves a rather monotonous succession of low angle, mud-dominated ramps. The exception to this is around the Peace River Arch which existed, at least initially, as an "island" in the Wabamun sea. Here a number of shallow, fringing platforms existed. The superposition of muddy ramp cycles makes discerning depositional patterns and lateral shifts of facies belts in the Wabamun relatively difficult other than at its distal prolongation in northeast British Columbia or its landward transition into evaporites in southeast Alberta (Figure 4). Examination of these areas reveals that the bulk of the Wabamun section comprises an initial transgressive phase and a subsequent regressive phase.

## STRATIGRAPHIC NOMENCLATURE

The name Wabamun Formation was first proposed by the Geological Staff, Imperial Oil Limited (1950) for a sequence of limestones and dolomites in central Alberta (Figure 1). Belyea (1955) recommended elevation to group status based on the presence of regionally correlatable units within it. Subdivision of the Wabamun Group into the lower, evaporitic Stettler and upper, open-marine Big Valley formations was proposed by Wonfor and Andrichuk (1956). Operators in the Crossfield area of Alberta coined the name Crossfield Member for a porous dolomite unit, up to 35 m in thickness, within the Stettler Formation.

In north-central Alberta the Wabamun was previously undivided until Halbertsma and Meijer-Drees (1987) proposed subdivision into six shallowing-upward sequences. They proposed that the lower five

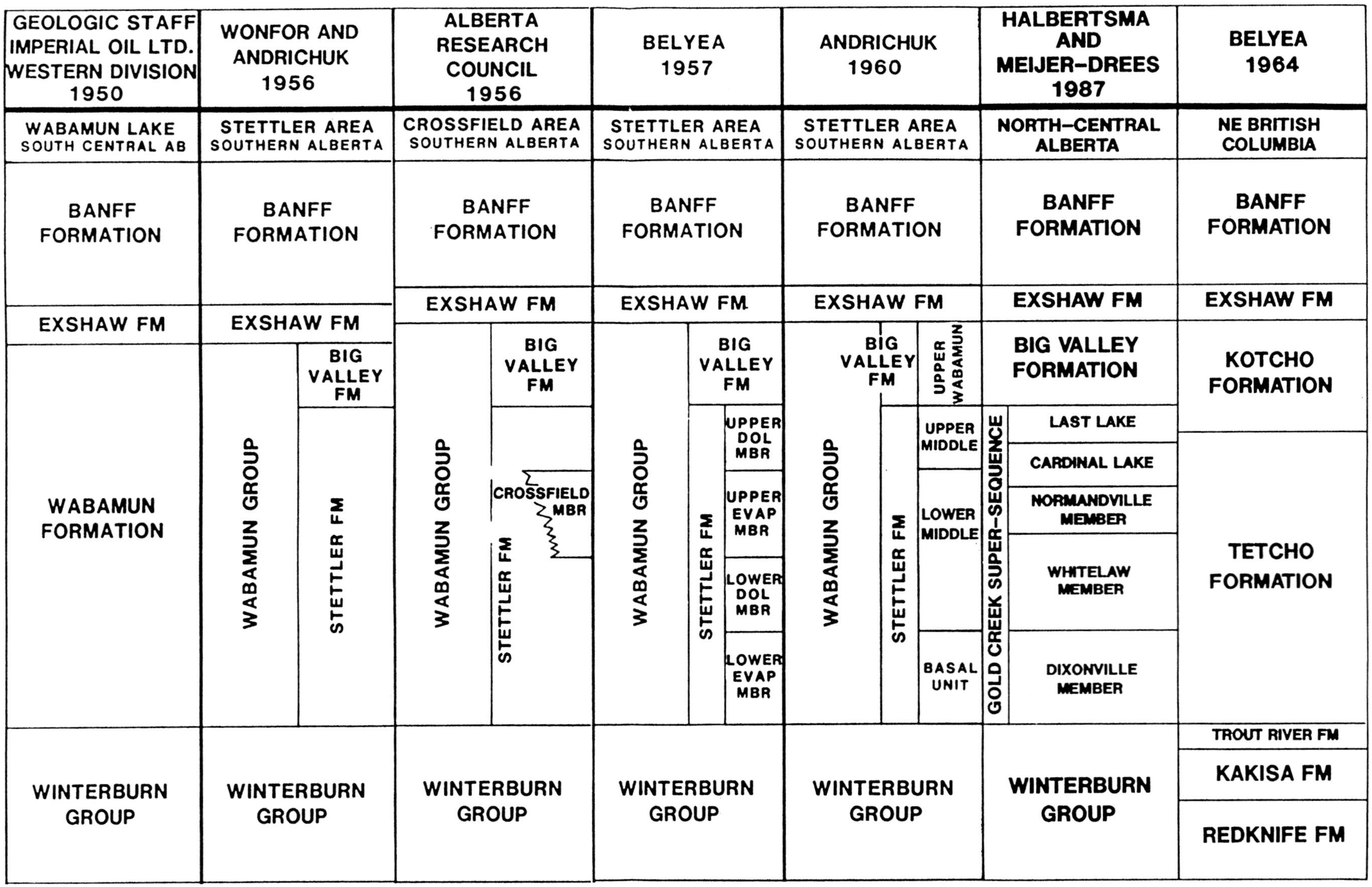


FIGURE 1. Stratigraphic Nomenclature of the Wabamun Group

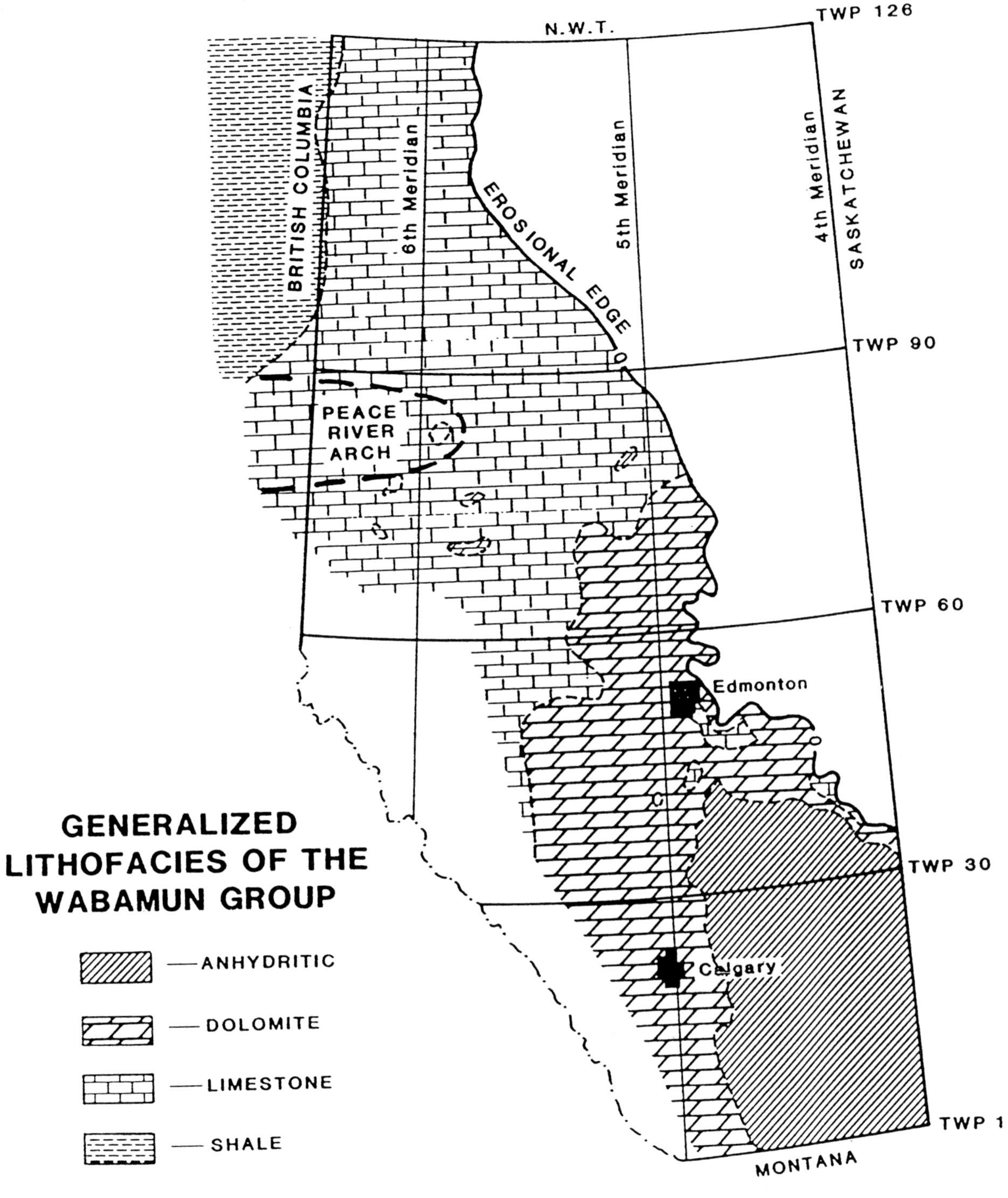


FIGURE 2. Lithofacies Distribution of the Wabamun Group in the Western Canada Basin

sequences, which they named the "Gold Creek" super-sequence, equated to the Stettler Formation of southern Alberta. Stoakes and Foellmer (1987) refer to seven informal but correlatable units characterizing the Wabamun in the Peace River Arch area (Units 1-7). One zone that has been consistently recognized over the years is the "Normandville zone" (Nishida, 1987), a stromatoporoid-bearing unit in the vicinity of the Peace River Arch.

In outcrop in the Alberta-British Columbia Front Ranges, the Wabamun equates to the Palliser Formation (Belyea and McLaren, 1956). Geldsetzer (1982) was able to subdivide the Palliser Formation into upper and lower units based on outcrop sections in northeast British Columbia. Further south it appears as a monotonous unit with few readily apparent subdivisions (Beales, 1956) although Halbertsma and Meijer-Drees (1987) state that a characteristic black band, dividing the upper from lower unit, can be traced as far south in outcrop as Fernie in British Columbia.

In northeast British Columbia the Wabamun shales out into a number of carbonate ramps assigned to the Tetcho and Kotcho formations and finally into the basinal Besa River Formation.

## GENERAL MODEL OF WABAMUN SEDIMENTATION

At the end of Winterburn "time" (see Chapter 9) the WCSB throughout Alberta was almost completely infilled with only the Peace River landmass remaining as a significant topographic feature. The increase in rate of sea-level rise that initiated the Wabamun cycle of sedimentation on the shelf, resulted in marine inundation of most of Alberta. Initial deposits show an overall transgressive or backstepping geometry as the seas invaded further southwards (Stoakes, 1987(a); Halbertsma and Meijer-Drees, 1987).

The Wabamun preserves a sequence of superposed low angle, mud-dominated ramp units which in large part reflect the lack of pre-existing topography following the Winterburn cycle of sedimentation. Around the Peace River landmass overall deepening initially led to the establishment of a number of fringing shallow-water platforms (Stoakes and Foellmer, 1987). These platforms backstep to higher levels on the Arch in response to rising sea level (Figure 9). As the coastline became stabilized in southern Alberta widespread evaporitic deposits of the Stettler Formation developed. In some cases extreme evaporation led to the deposition of halite in this area (Figure 3). As relative sea-level stabilized, regressive conditions became dominant. Commencing with the Normandville sequence ramp successions gradually prograded northwards filling the basin and downlapping onto previously drowned Wabamun platforms around the Arch. This progradation is best reflected as the distal prolongation of equivalent strata (Tetcho Formation) basinwards in northeast British Columbia.

At this point a major transgression took place and basinal conditions (basal Kotcho Formation) were established on the distal ramp succession of the Tetcho Formation (Figure 4). This transgression is reflected in the Peace River Arch area by a number of thin deeper-water distal units (Units 5 to 7 of Stoakes and Foellmer, 1987, Figure 9; Big Valley Member "sequence" of Halbertsma and Meijer-Drees, 1987) on top of the prograding Wabamun

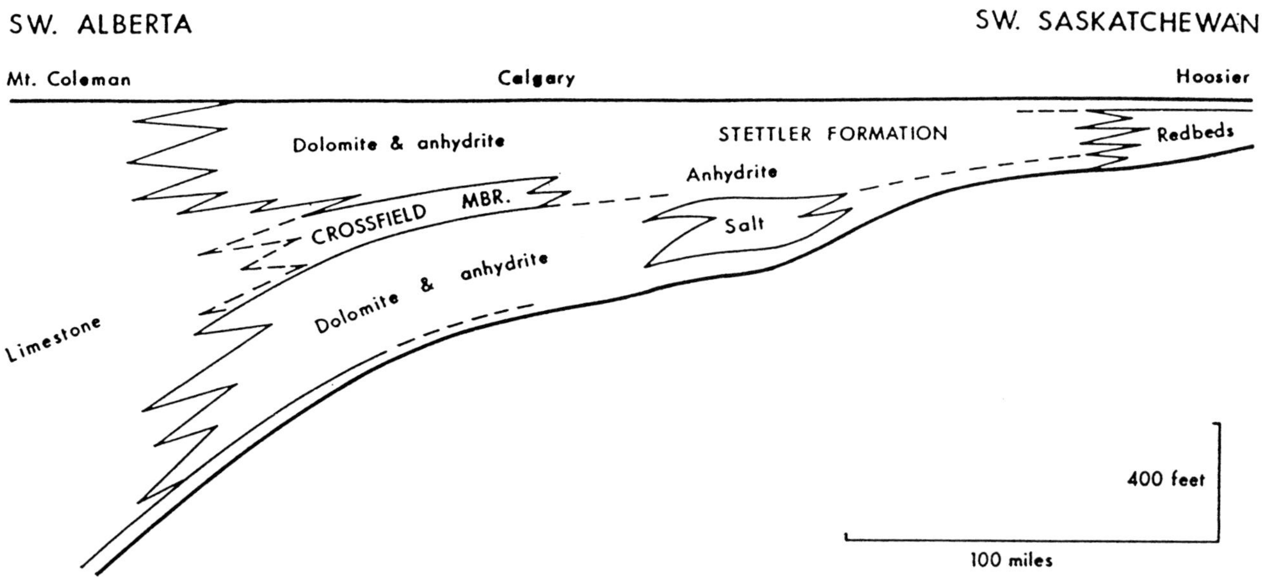


FIGURE 3. West-East Stratigraphic Cross Section of the Stettler Formation (from Fuller and Porter, 1969)

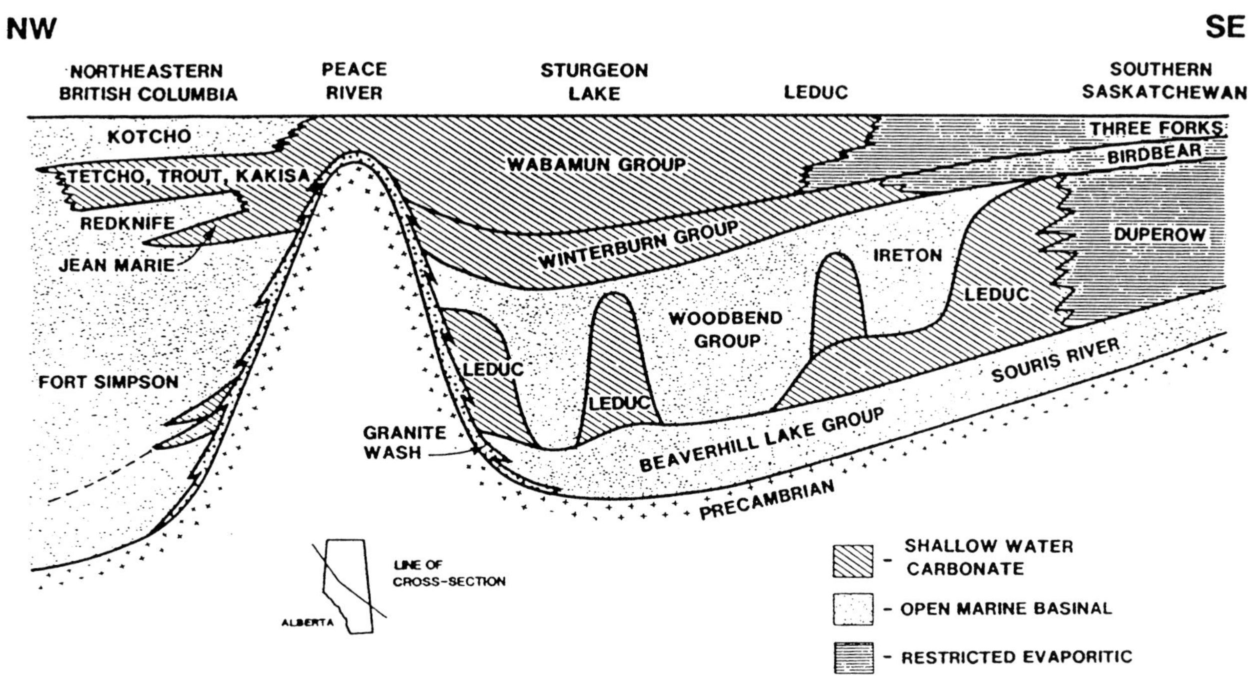


FIGURE 4. Schematic Cross Section of the Upper Devonian

ramps. This transgression also placed the open-marine Big Valley Formation onto the evaporitic Stettler Formation in southern Alberta. Technically these overlying strata are assigned to the transgressive deposits of the ensuing (Banff) sequence (Stoakes and Foellmer, 1987; Stoakes, 1987(a)) and are dealt with in the next chapter (Chapter 11).

The following sections describe the stratigraphy, sedimentology and evolution of the Wabamun megacycle in the various portions of the basin in more detail.

## SOUTHERN ALBERTA

In southern Alberta the Wabamun comprises a widespread deposit up to 150 m in thickness. It is comprised of a succession of dolomites and anhydrites which thin eastwards and intertongue with a red bed succession in Saskatchewan. Thinning also results from the truncation of strata at the sub-Cretaceous unconformity to the northeast.

Andrichuk (1960) divided the Wabamun into four units (Figure 5). The lower part of the Stettler Formation is comprised of dolomites and anhydrites which pass westwards into limestones and eastwards into halite. Fuller and Porter (1969) depict this relationship equating the succession to cyclic sabkha deposition (Figure 3). Locally the halite is removed by later solution and the Wabamun shows abrupt thinning with compensation in overlying Upper Cretaceous shales (Oliver and Cowper, 1983). This package of sediments appears to represent the transgressive, backstepping lower part of the Wabamun cycle resulting in maximum transgression at Crossfield Member "time" (Figure 6).

Workman and Metherell (1969) and Ostrom (1985) studied the lithofacies of the Crossfield Member, a biostromal unit that trends north-south through the Calgary-Foothills belt (Figure 6). It contains a number of oil and gas fields at its updip facies change to evaporitic sediments and marks the break-in-slope on the ramp at the point of maximum transgression. The unit grades vertically into more restricted dolomites and anhydrites of the regressive upper Stettler Formation.

Eliuk (1984) depicts a facies model for the Crossfield Member with stromatoporoid biostromes developed at the seaward limit of sabkha-salina sediments (Figure 6). Further out on the ramp profile sediments are dominantly bioturbated mudstones.

To the west the Wabamun-Palliser thickens to between 200 and 300 m and takes on an overall more open marine aspect. The upper Stettler package, which is dominantly evaporitic in the Crossfield area, undergoes a facies change to open-marine limestone (Eliuk and Hunter, 1987). This regressive cycle is terminated by transgression of the more open-marine limestones of the Big Valley Formation. A number of authors refer to the presence of two major regressive-transgressive cycles, reflected by the lower Stettler-Crossfield Member and upper Stettler Formation evaporites-Big Valley Formation limestones (Andrichuk, 1960; Stynan, 1984). This account differs only in assigning the lower Stettler-Crossfield-upper Stettler to one transgressive-regressive cycle and grouping the Big Valley Formation within the transgressive phase of the

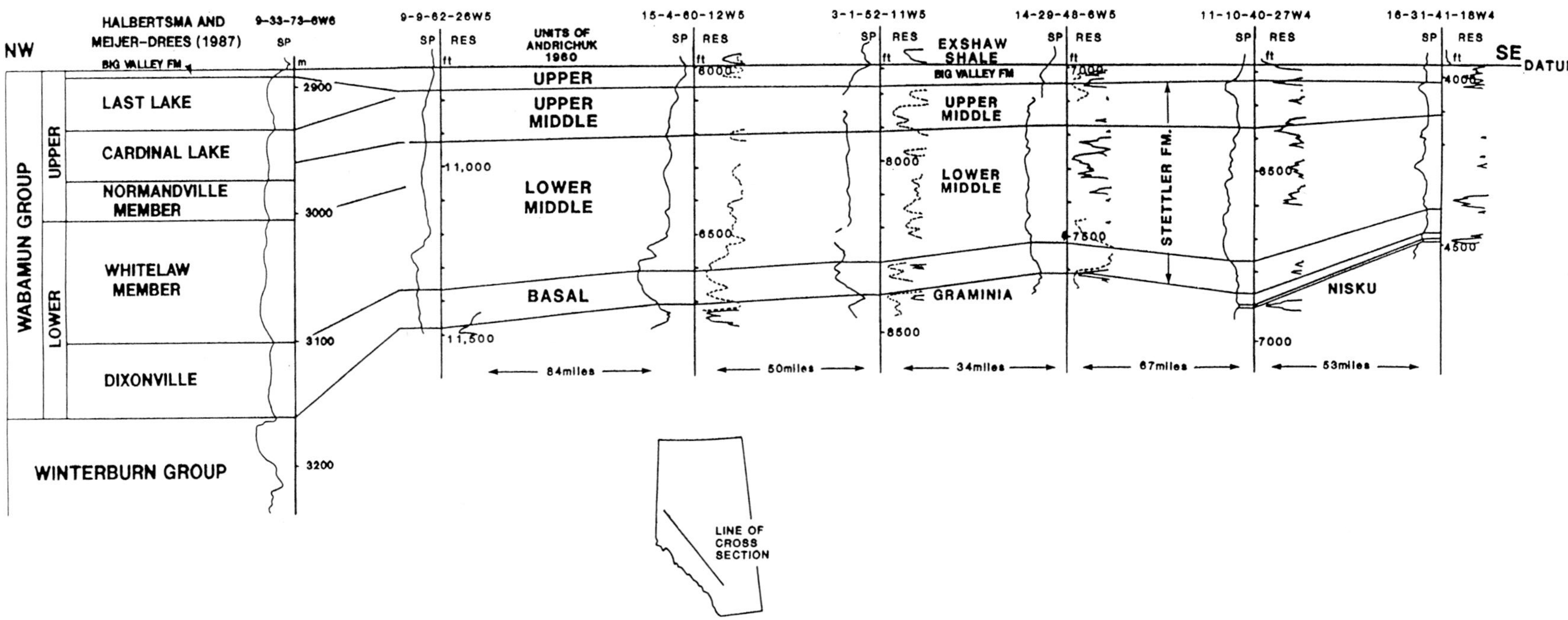


FIGURE 5. Well Log Cross Section Showing Regional Correlations in the Wabamun Group of Alberta

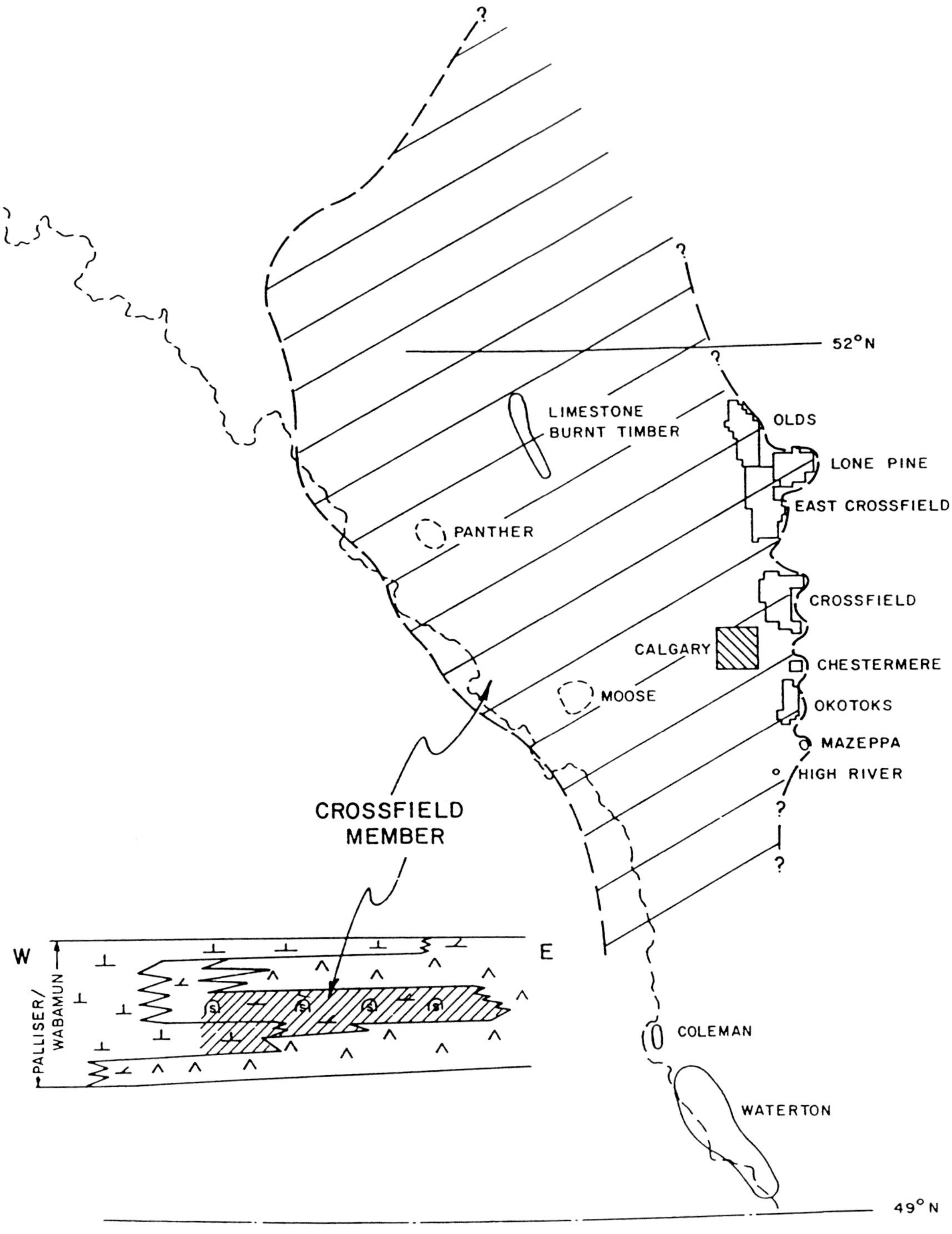


FIGURE 6. Extent of Crossfield Member in Southern Alberta from Eliuk (1984)

ensuing Banff cycle.

## NORTH-CENTRAL ALBERTA

In north-central Alberta the Wabamun is a widespread unit up to 260 m in thickness. It thins as a result of onlap onto the Peace River Arch landmass (Figure 4) and is indeed absent over the highest portions of the Arch (Stoakes, 1987(b) his Figure 3; Dix, 1990 his Figure 5).

The Wabamun is divisible into a number of correlatable ramp sequences (Halbertsma and Meijer-Drees, 1987; Stoakes and Foellmer, 1987) which comprise an overall transgressive-regressive cycle of sedimentation (Stoakes, 1987(a)). The point of maximum transgression equates to a widespread argillaceous unit at the base of the Normandville zone called the "Belcourt" transgression (Halbertsma and Meijer-Drees, 1987) after nearby outcrops. This tends to infer that the Normandville Member of north-central Alberta equates to the Crossfield Member of southern Alberta. Nishida (1987) and Halim-Dihardja and Mountjoy (1989) present facies models for the stromatoporoid buildups and associated facies of the Normandville zone. Like the labechiid stromatoporoid buildups of the Crossfield Member, it appears to mark the break-in-slope on the ramp between shallow, almost tidal-flat conditions, and more offshore deposition. Regional paleogeographic considerations suggest that the Normandville Member represents seaward progradation of this facies from the point of maximum incursion of the seas in southern Alberta (ie. the Crossfield Member shoreline). As such it should slightly postdate the Crossfield Member and this does appear to be substantiated by biostratigraphic studies (Stearn pers. comm. cited in Nishida, 1987).

Finally the upper part of the Wabamun (Cardinal Lake and Last Lake sequences) show further progradation basinwards up to the point of the Big Valley transgression.

The transgressive-regressive nature of the Wabamun in the Peace River Arch area is also expressed in the establishment of two separate depositional systems (Stoakes 1987(a); Stoakes and Foellmer, 1987):

1) A shallow water platform system fringing the Arch.
2) A regional ramp system (see Figure 7).

The geometry and nature of the regional ramp system has been detailed in the preceding section so attention will be focused here on the fringing platforms. The rise in sea level that flooded the pre-existing Winterburn surface isolated the still emergent Peace River landmass as an island within a shallow epeiric sea. This allowed a fringing platform system to become established. Shallow-water carbonate platforms, developed around the Arch, are recognized in the subsurface by their "clean" blocky gamma-ray response, as opposed to the gradually "shoaling" response of adjacent ramp units (Figure 8). The different nature of the facies of these platform units is also readily evident in core. These

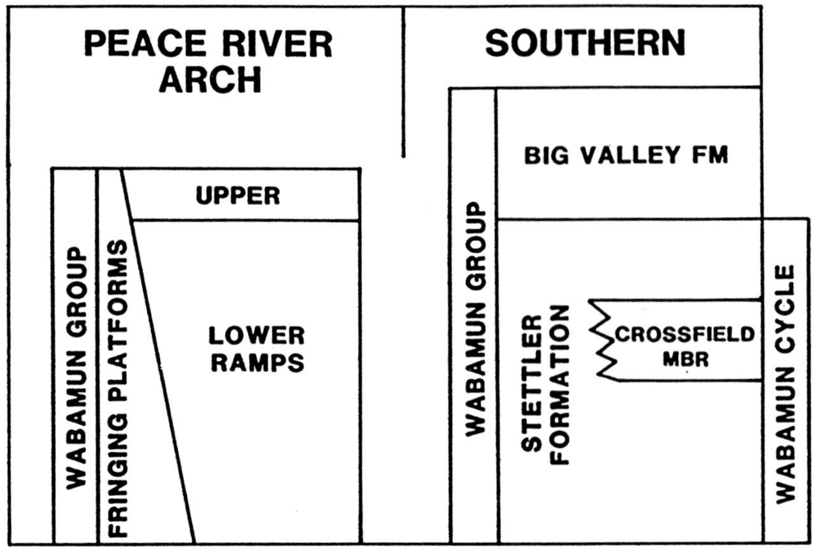


FIGURE 7. Stratigraphy of the Wabamun Cycle (from Stoakes 1987)

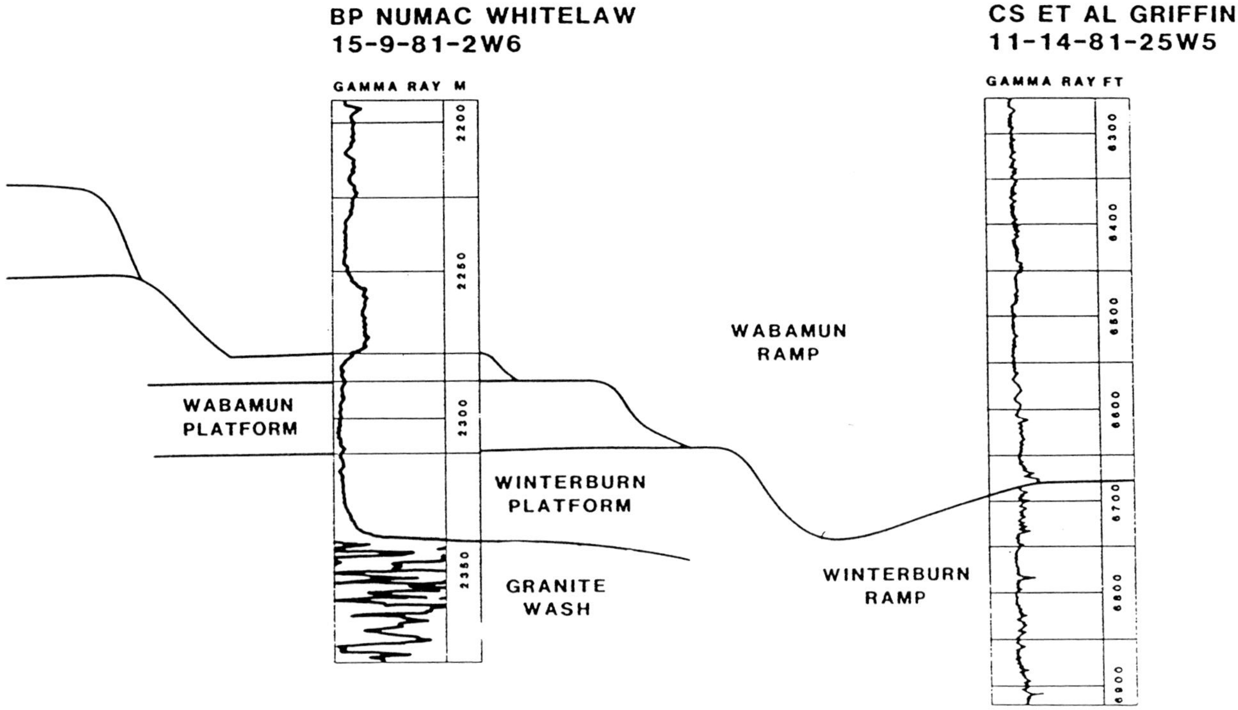


FIGURE 8. Log Response of Platform Versus Ramp Sediments

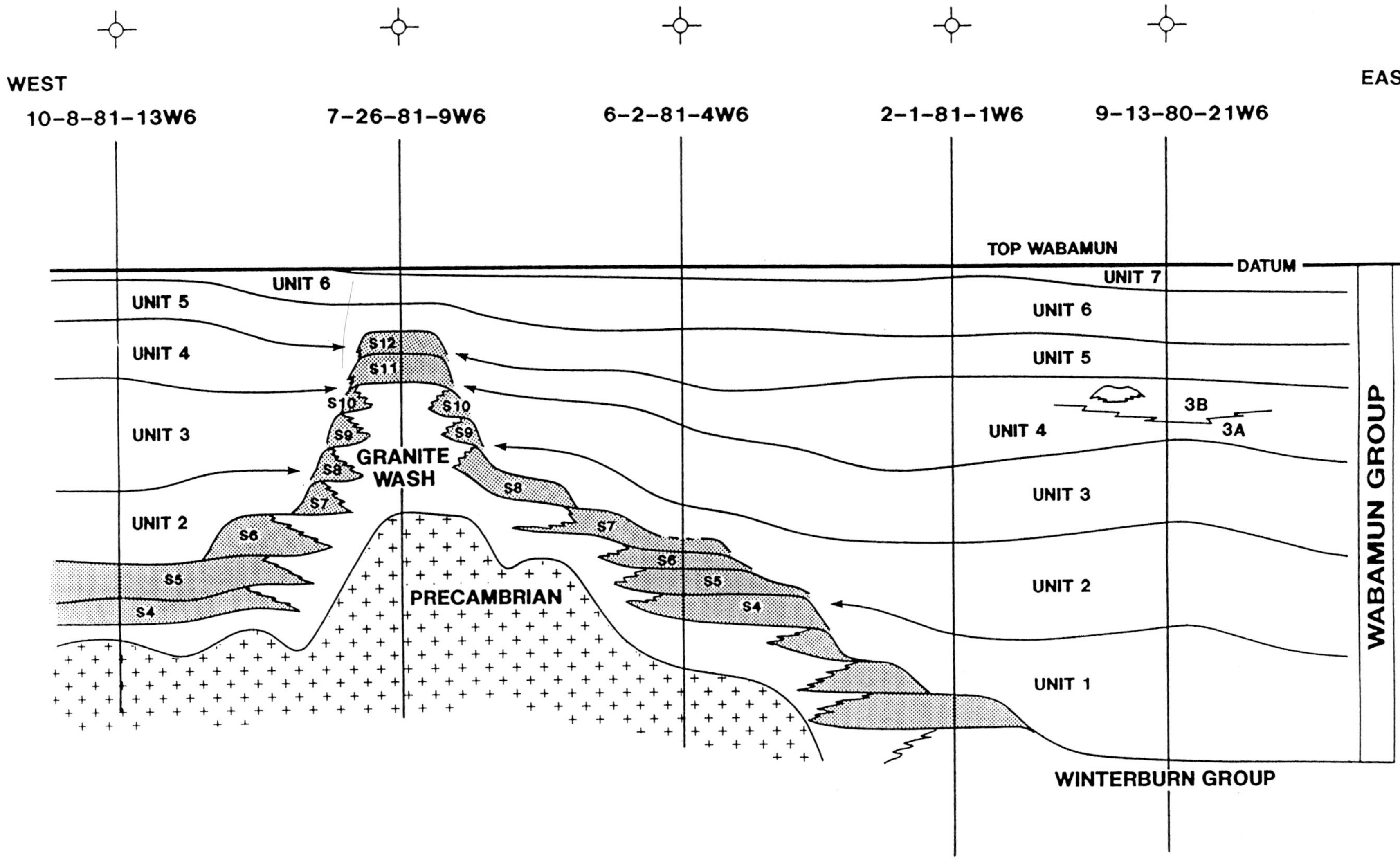


FIGURE 9. West-East Section Showing Fringing Wabamun Platforms and Regional Ramp Units

units are frequently more fossiliferous and grainier in aspect, dolomitized and often interbedded with siliciclastics shed from further up on the Arch. Correlations reveal a total of at least 13 separate cycles each tending to backstep to progressively higher levels on the Arch (Figure 9). These units are progressively downlapped and finally engulfed by the regional regressive ramp system (Units 1-5).

## NORTHEAST BRITISH COLUMBIA

When traced north and westward from the Peace River Arch the Wabamun Group thins and changes facies along its distal limits to more of an open-marine deeper water limestone. This is evident in cores from the Parkland field just across the Alberta-British Columbia border. The mid-Wabamun point of maximum deepening ("Belcourt" transgression) is evident in the Parkland field area as a laminated deep-water shale unit in the middle of the Wabamun package. Further to the north the Wabamun has equivalents to the Tetcho and Kotcho formation ramps. Halbertsma (1990) equates his Dixonville Member with the Tetcho Formation and the overlying five members with the Kotcho Formation.

## SUMMARY

Deposition of the Wabamun sequence records a mega transgressive-regressive cycle of sedimentation. The transgression took place over a surface of very low relief and the majority of Wabamun deposition conforms to a low angle, mud-dominated ramp. Overall, the Wabamun has more of an open-marine transgressive aspect when compared to the underlying Winterburn sequence. For this reason, it is interpreted as the initial deposit in a latest Devonian-Early Mississippian phase or major cycle of deposition.

In the initial transgressive phase ramp sequences backstepped into southern Alberta, an area that became dominated by stacked carbonate-evaporite cycles. The maximum cratonward advance of marine conditions corresponds to the development of stromatoporoid bioherms of the Crossfield Member in southern Alberta. Basinwards, this level equates to the "Belcourt" transgression of Halbertsma and Meijer-Drees (1987) and the mid-Wabamun "shale break" in northeast British Columbia. A regressive depositional style characterizes the remainder of Wabamun sedimentation. Sabkhas prograded basinward in southern Alberta and shallow distal ramp units built basinward in northeast British Columbia. The regional ramp system eventually inundated the Peace River landmass downlapping previously deposited shallow-water platform carbonates that fringed the Peace River Arch landmass, all but covering the very highest parts.

The uppermost part of the Wabamun Group (Big Valley Formation) records the transgressive pulse that initiates the ensuing Banff cycle discussed in the next chapter (Chapter 11).

## REFERENCES

ANDRICHUK, J.M., 1960, Facies analysis of the Upper Devonian Wabamun Group in west-central Alberta, Canada: Bulletin of American Association of Petroleum Geologists, v. 44, p. 1651 - 1681.

BEALES, F.W., 1956, Conditions of deposition of Palliser (Devonian) limestone of southwestern Alberta: Bulletin of American Association of Petroleum Geologists, v. 40, p. 848-870.

BELYEA, H.R., 1955, Cross sections through the Devonian System of the Alberta plains: Geological Survey of Canada, Paper 55-3.

BELYEA, H.R., 1957, Correlation of Devonian subsurface formations, southern Alberta: Geological Survey of Canada, Paper 55-38.

BELYEA, H.R., 1964, Upper Devonian, *in* Geological History of Western Canada, McCrossan, R.G., and Glaister, R.P., eds., Alberta Society of Petroleum Geologists, p. 66-88.

BELYEA, H.R., and MCLAREN, D.J., 1956, Devonian sediments of Bow Valley and adjacent areas: Alberta Society of Petroleum Geologists, Sixth Annual Field Conference Guidebook, p. 66-91.

DIX, G.R., 1990, Stages of platform development in the upper Devonian (Frasnian) Leduc Formation, Peace River Arch, Alberta: Bulletin of Canadian Petroleum Geology, v. 38A, p. 66-92.

ELIUK, L.S., 1984, A hypothesis for the origin of hydrogen sulfide in Devonian Crossfield Member dolomite, Wabamun Formation, Alberta, Canada, *in* Eliuk, L.S., ed., Carbonates in subsurface and outcrop: Canadian Society of Petroleum Geologists, Core Conference, p. 245-289.

ELIUK, L.S., and HUNTER, D.F., 1987, Wabamun Group structural thrust fault fields: The Limestone-Burnt Timber example: Canadian Society of Petroleum Geologists, 13th Core Conference and Display, 2nd International Symposium on the Devonian System, 1987, p.39-62.

FULLER, J.G.C.M., and PORTER, J.W., 1969, Evaporite formations with petroleum reservoirs in Devonian and Mississippian of Alberta, Saskatchewan and North Dakota: Bulletin of American Association of Petroleum Geologists, v. 53, p. 909-926.

GELDSETZER, H.H.J., 1982, Depositional history of the Devonian succession in the Rocky Mountains southwest of the Peace River Arch: Geological Survey of Canada, Current Research, Part C, Paper 82-1C, p.55-64.

GEOLOGICAL STAFF, IMPERIAL OIL LIMITED, WESTERN DIVISION, 1950, Devonian nomenclature in Edmonton Area, Alberta, Canada: Bulletin of American Association of Petroleum Geologists, v. 34, p. 1807-1825.

HALBERTSMA, H.L., 1990, Wabamun/Palliser strata of the Western Canada Sedimentary Basin: Canadian Society of Petroleum Geologists convention, Program and Abstracts, p.62.

HALBERTSMA, H.L., and MEIJER-DREES, N.C., 1987, Wabamun limestone sequences in north-central Alberta: Canadian Society of Petroleum Geologists, 13th Core Conference and Display, 2nd International Symposium on the Devonian System, 1987.

HALIM-DIHARDJA, M.K., and MOUNTJOY, E.W., 1989, A stromatoporoid patch reef in the Upper Devonian Wabamun Group, Normandville Field, North-Central Alberta, *in* Geldsetzer, H.H.J., James, N.P. and Tebbutt, G.E., eds., Reefs Canada and adjacent areas: Canadian Society of Petroleum Geologists, Memoir 13, p. 448-453.

NISHIDA, D.K., 1987, A Famennian stromatoporoid patch reef in the Wabamun Group, west-central Alberta, Canada: Canadian Society of Petroleum Geologists, 13th Core Conference and Display, 2nd International Symposium on the Devonian System, 1987.

OLIVER, T.A., and COWPER, N.W., 1983, Wabamun salt removal and shale compaction effects, Rumsey Area, Alberta: Bulletin of Canadian Petroleum Geology, v. 31, p. 161-168.

OSTROM, J.N., 1985, Lithofacies analysis of the Upper Devonian Crossfield Member, Stettler Formation, Olds area, southwestern Alberta: Unpublished B.Sc. Thesis, University of British Columbia.

STOAKES, F.A., 1987(a), Evolution of the Upper Devonian of western Canada, *in* Principles and concepts for exploration and exploitation of reefs in the Western Canada Basin, Short Course Notes, Canadian Reef Inventory Project.

STOAKES, F.A., 1987(b), Fault controlled dolomitization of the Wabamun Group, Tangent Field, Peace River Arch,Alberta: Canadian Society of Petroleum Geologists, 13th Core Conference and Display, 2nd International Symposium on the Devonian System, 1987, p. 73-85.

STOAKES, F.A., and FOELLMER, K.E.H., 1987, The Wabamun Group of the Peace River Arch area, West Central Alberta: Recognition of Fringing Platforms, a new exploration play-type (Abs.), *in* 2nd International Symposium on the Devonian System, Program and Abstracts, p. 213.

STYAN, W., 1984, Palliser Formation in the Canmore area, *in* Richards, B., Styan, W. and Harrison, R., (field trip leaders, Paleozoic Carbonate Reservoir Units and Facies Distribution: Canadian Society of Petroleum Geologists and Canadian Society of Exploration Geophysicists, Exploration Update '84, Field Trip Bl Guidebook, p. 1-14.

WONFOR, J.S., and ANDRICHUK, J.M., 1953, Upper Devonian in the Stettler area, Alberta: Canada News Bulletin, Alberta Society of Petroleum Geologists v. 1, no. 9, p. 3-6.

WONFOR, J.S., and ANDRICHUK, J.M., 1956, The Wabamun Group in the Stettler area, Alberta: Journal of the Alberta Society of Petroleum Geologists, v. 4, p.99-111.

WORKMAN, L.E., and METHERELL, R.G., 1969, Geology of Crossfield East and Lone Pine Creek gas fields, Alberta: Bulletin of Canadian Petroleum Geology, v. 17, p. 92-108.

CHAPTER 11

# EARLY MISSISSIPPIAN MEGASEQUENCES

F.A. STOAKES

## INTRODUCTION

Early Mississippian sediments of the Western Canada Sedimentary Basin comprise a thick sequence of basinal shales, shallow-water carbonates and isolate Waulsortian reefs. The sediments described in this section are assigned to the uppermost Wabamun Group and the Exshaw Formation of largely Famennian age and the Banff, Pekisko and Shunda formations of Tournaisian age (Figure 1). This section comprises at least two transgressive-regressive cycles of sedimentation. The lowermost cycle, termed the Banff cycle by Stoakes, (1987a), actually commences in the uppermost Wabamun (Figure 2). It terminates with progradation of ramp deposits of the lower part of the Pekisko Formation. The second cycle commences with backstepping of the Pekisko ramps and initiation, in downslope positions, of isolate Waulsortian reefs. The cycle terminates with progradation and basin-filling of the Shunda Formation. Although the overall picture is still not that clear it would appear that each Mississippian cycle represents increased reflooding of the WCSB and as such is part of a much large depositional phase. The area where these patterns are perhaps best known is in the area of the Peace River Arch where an anomalously thick sedimentary succession of Carboniferous age is preserved (O'Connell, 1990).

## STRATIGRAPHIC NOMENCLATURE

The stratigraphic definition of the Wabamun Group was dealt with in the previous chapter (Chapter 10). It was Stoakes (1987a) that first recognized the distal upper ramp successions of the Wabamun Group in the Peace River Arch area and having equated them to the Big Valley Formation of southern Alberta assigned them to the transgressive phase of the Banff cycle.

The Exshaw Formation was originally named for a section of laminated black shales at Jura Creek, near Exshaw, Alberta. Later authors (Clark, 1949; Harker and McLaren, 1958; Macauley, 1958; and MacQueen and Sandberg, 1970) have included a sequence of silty limestones and siltstones above the shales within the Exshaw Formation. It is now generally accepted, following Richards and Higgins (1988) that the Exshaw Formation comprises a lower anomalously radioactive shale member and an upper siltstone member. Conodonts within the black shale member indicate that the Devonian/Carboniferous boundary lies within that unit.

The Banff Formation is a shallowing-upward succession of argillaceous shales and carbonates that form a prograding sequence of ramp to basin deposits (Beauchamp and others, 1986). O'Connell

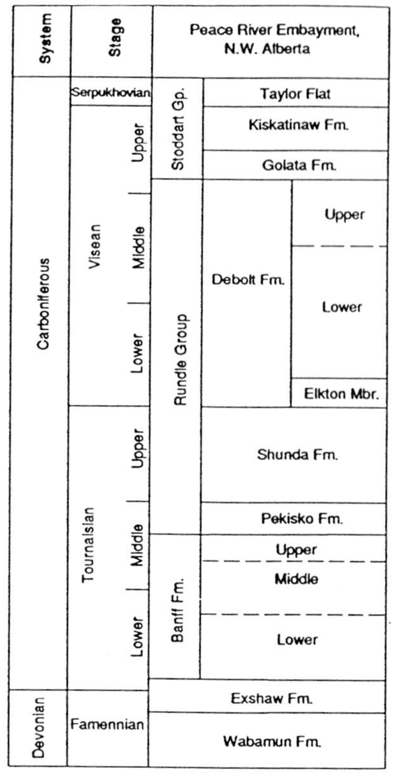


Figure 1. Stratigraphic nomenclature of the Peace River Arch region (from Bamber and others, 1980).

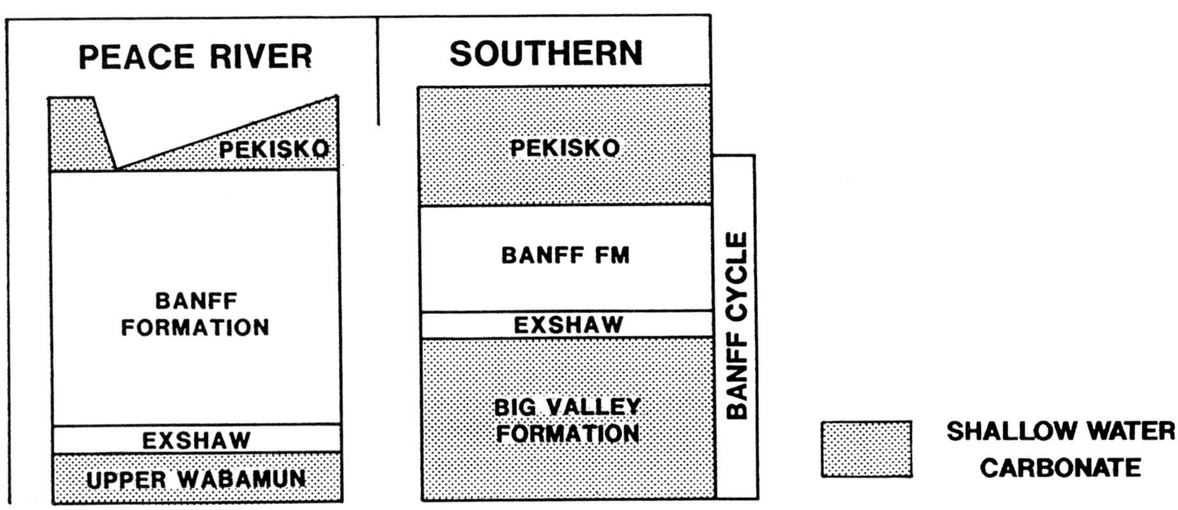


Figure 2. Stratigraphy of the Banff cycle (from Stoakes, 1987a).

(1990) subdivided the Banff Formation of the Peace River Embayment into upper, middle and lower, based on lithological variations. This basin-fill unit constitutes the regressive portion of the Banff cycle.

The overlying Pekisko-Shunda cycle commences with transgresive backstepping ramp deposits of the Pekisko Formation. In basinward areas isolate Waulsortian reefs comprised of crinoids and bryozoans became established (Davies and others, 1988). Stoakes and others, (1989) proposed a number of informal member level subdivisions to better portray the depositional complexity in this interval (Figure 3). To a certain extent their Cranberry shale member and a portion of the Heart River Member ramp belong to the regressive phase that infilled the depositional relief of the basin. This basin-filling continued with deposition of the Shunda Formation shales.

## GENERAL MODEL OF EARLY MISSISSIPPIAN SEDIMENTATION

At the end of the Wabamun megacycle (Chapter 10) shallow ramp carbonate deposition existed across most of Alberta with the Peace River landmass remaining only as a series of small islands (Stoakes, 1987b his Figure 3). Increased rate of sea-level rise led to renewed inundation of the shelf and deposition of open-marine limestones of the Big Valley Formation in southern Alberta. In the Peace River Arch area this equates to the deep-water distal ramps that comprise the uppermost Wabamun (Units 5 to 7 of Stoakes and Foellmer, 1987; Big Valley Member "sequence" of Halbertsma and Meijer-Drees, 1987). This transgressive deepening culminated with deposition of organic laminites of the Exshaw Formation under anoxic bottom conditions (Creaney and Stoakes, 1987). The Banff shales are generally regarded as a westerly prograding system (Beauchamp and others, 1986). However this assumption is often hampered by a lack of detectable time-based stratigraphic markers or horizons (O'Connell, 1990). Overall the Banff interval thins through onlap onto the Peace River Arch and stratigraphic cross sections by O'Connell (1990) reveal a westerly and southerly aspect to the progradation at least within his Lower Banff section. The ensuing Pekisko-Shunda cycle is initiated by renewed relative sea-level rise. This led to backstepping and then aggradation of a series of crinoidal ramp deposits in the form of widespread carbonate shelves (Macauley and others, 1964). Coeval with this rise, Waulsortian reefs became established along the basin margin in much the same way as disphyllid coral reefs characterize this position in the Winterburn sequence of the West Pembina area (Chapter 9). Finally prograding shales, this time of the Shunda Formation, complete the regressive phase of this cycle.

## GENERAL OBSERVATIONS

The general model of cyclicity outlined here represents an early stage in our understanding of these sequences. Fortunately more attention is being focussed on this interval which should lead to a better understanding of it's depositional patterns. However a number of questions remain, primarily prograding patterns in the Banff Formation of the Peace River Embayment are not reflected in

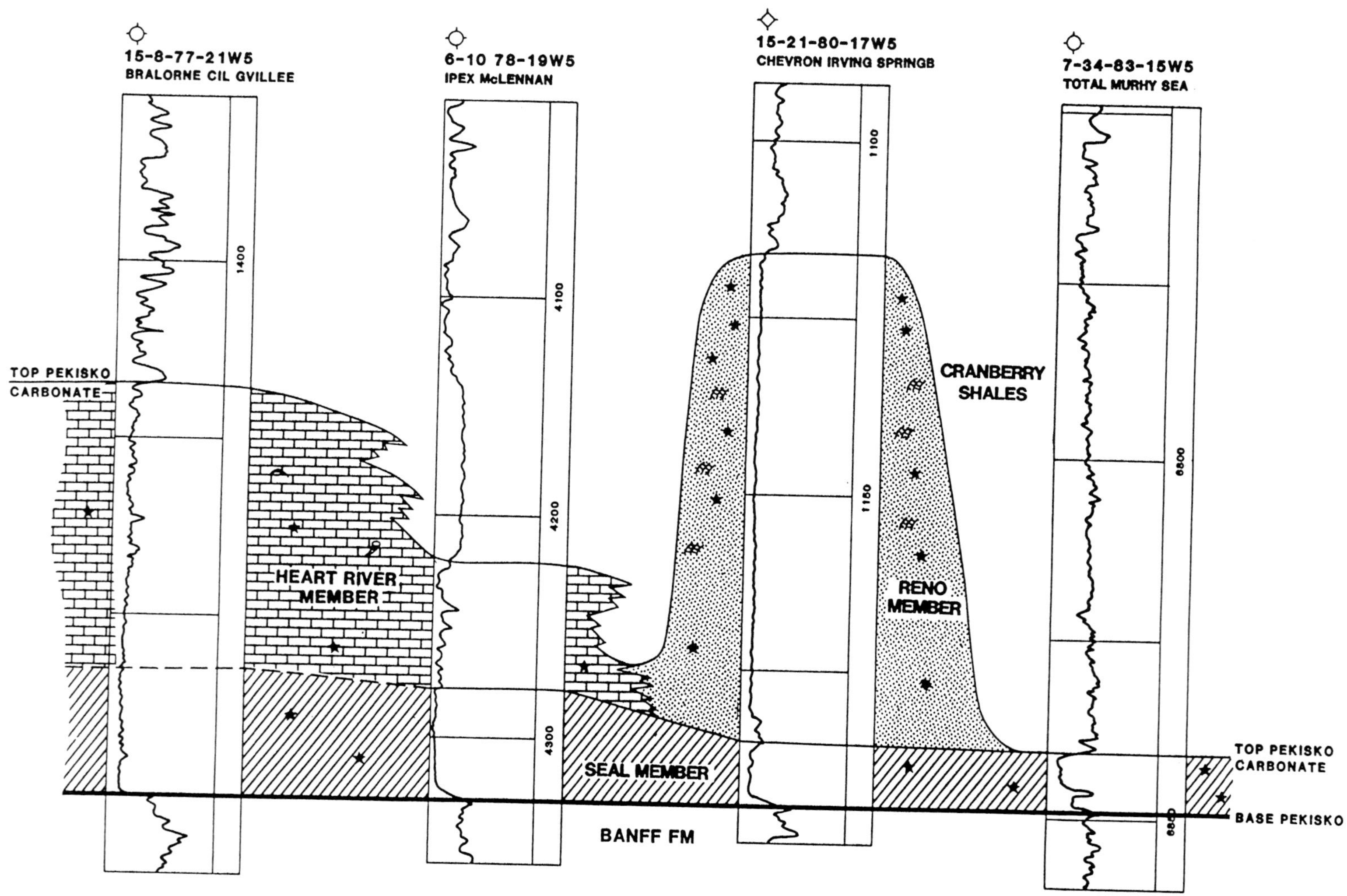


Figure 3. Gamma ray log cross section showing member level stratigraphy proposed for the Pekisko of the Peace River Arch area (from Stoakes and others, 1989).

the paleogeography of the ensuing Pekisko. Tectonic movement of the Arch is a possibility but O'Connell (1990) states that during deposition of both the Banff and Pekisko Formations there was no pronounced differential subsidence of the underlying Arch. What then gave rise to the east-west trending embayment (Figure 4) of the Pekisko? The Peace River Embayment is generally portrayed as a westward facing system (Figure 5) but in detail it appears as more of an eastward facing one, which narrows to the west and apparently opens to the east (Stoakes and others, 1989; O'Connell, 1990). This point has significant economic importance as it's geometry relates to mapping of the potential reef fairway. Presently known Waulsortian mounds lack any hydrocarbons (with only one exception - heavy biodegraded oil in the Seal area, Edwards, 1987). They have good porosity and permeability but analysis of source laminites adjacent to the pinnacle reveals them to be thermally immature (Stoakes and others, 1989). Using the analogy of the West Pembina pinnacles of the Winterburn sequence (Chapter 9) it should be possible to extrapolate this trend into areas of greater burial and hence greater prospectivity. It is critical therefore to fully understand the paleogeography.

This also brings up another consideration. If pinnacle reef development characterizes the distally deepened ramp of basal transgressive phases, as indeed it does in the Winterburn cycle and the Pekisko-Shunda cycle, where are the pinnacle reefs of the transgressive phase of the preceding Wabamun and Banff cycles?

## SUMMARY

Latest Devonian-Early Mississippian successions belong to part of a longer term depositional phase that represents reflooding of the Western Canadian Sedimentary Basin following Winterburn basin filling. This reflooding commences with the Wabamun cycle (Chapter 10) and continues into the Banff and Pekisko-Shunda cycles outlined in this chapter. In seems likely that this depositional phase ends with progradation of shallow-water carbonates of the Debolt Formation.

Although our understanding of depositional patterns characterizing the latest Devonian-Early Mississippian is somewhat limited in the geographic area of the Peace River Arch it appears to comprise two major cycles of deposition. The first or Banff cycle commences with transgressive deepening over the progradational Wabamun carbonates resulting in deposition of a number of backstepping ramps of the Big Valley Formation. A thin source interval, the Exshaw Formation, developed under deeper water anoxic bottom conditions prior to regressive carbonate and shale deposition of the Banff Formation.

The second or Pekisko-Shunda cycle is initiated by transgressive backstepping of Pekisko ramp carbonates. Along the basin margin a number of isolate crinoid-bryozoan Waulsortian mounds became established and grew concomitant with aggradation of the regional carbonate shelf. The regressive phase of the cycle equates to carbonate-shale interbeds of the Pekisko and then Shunda formations. The present stratigraphic nomenclature appears inadequate to properly portray the depositional interrelationships

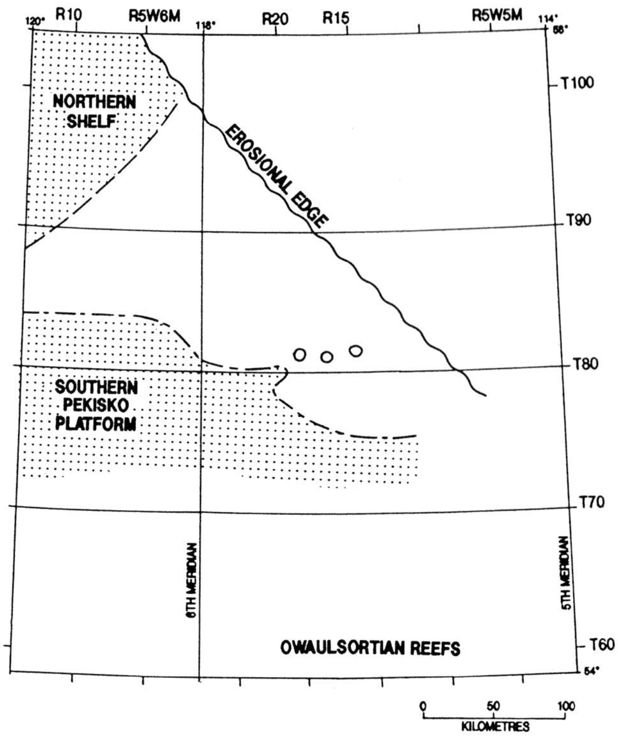


Figure 4. Paleogeography of the Pekisko embayment in the vicinity of the Peace River Arch (modified after O'Connell, 1990).

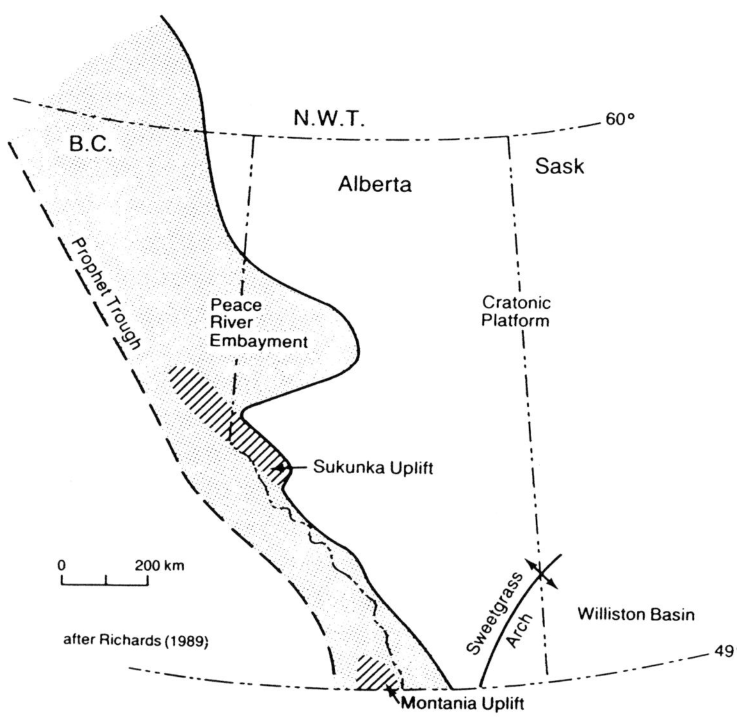


Figure 5. Generalized paleogeography and tectonic elements in Alberta and British Columbia during the Early Carboniferous (from Richards, 1989).

in this area although Stoakes and others, (1989) have proposed a seemingly workable informal terminology. An analogous situation was the undivided Nisku Formation prior to the discovery of hydrocarbon-bearing pinnacle reefs at West Pembina in the mid-1970's. The nomenclature of the Pekisko-Shunda cycle awaits a commercial discovery!

## REFERENCES

BAMBER, E.W., MacQUEEN, R.W., AND RICHARDS, B.C., 1980, Facies relationships at the Mississippian carbonate margin, Western Canada: Geological Survey of Canada, Open File Report 674, p. 18.

BEAUCHAMP, B., RICHARDS, B.C., BAMBER, E.W., AND MAMET, B.L., 1986, Lower Carboniferous lithostratigraphy and carbonate facies, upper Banff Formation and Rundle Group, east-central British Columbia, *in* Current Research, Part A: Geological Survey of Canada, Paper 86-1A, p. 627-644.

CLARK, L.M., 1949, Geology of Rocky Mountain Front Ranges near Bow River, Alberta: American Association of Petroleum Geologists Bulletin, v. 33, p. 585-609.

CREANEY, S., AND STOAKES, F.A., 1987, The Exshaw Formation laminites in the Peace River area: pattern of maturation and migration with exploration implications for the Wabamun Group: Canadian Society of Petroleum Geologists, Second International Symposium on the Devonian System, Program and Abstracts, p. 64.

DAVIES, G.R., EDWARDS, D.E., AND FLACH, P.D., 1988, Lower Carboniferous (Mississippian) Waulsortian reefs in the Seal area of north-central Alberta., *in* Geldsetzer, H.H.J., James, N.P., and Tebbutt, G.E., eds., Reefs Canada and Adjacent Areas: Canadian Society of Petroleum Geologists, Memoir 13, p. 643-648.

EDWARDS, D., 1987, Lower Carboniferous carbonate mounds in the Seal area, *in* Bloy, R., and Hopkins, J.C., organizers and compilers, Principles and Concepts for Exploration and Exploitation of Reefs in the Western Canada Basin: Canadian Society of Petroleum Geologists, Canadian Reef Inventory Project - Short Course, 16 p.

HALBERTSMA, H.L., AND MEIJER-DREES, N.C., 1987, Wabamun limestone sequences in north-central Alberta, *in* Krause, F.F., and Burrowes, O.G., eds., Devonian Lithofacies and Reservoir Styles in Alberta: Canadian Society of Petroleum Geologists, 13th Core Conference and Display and Second International Symposium on the Devonian System, p. 21-38.

HARKER, P., AND McLAREN, D.J., 1958, The Devonian-Mississippian boundary in the Alberta Rocky Mountains, *in* Goodman, A.J., ed., Jurassic and Carboniferous of Western Canada: American Association of Petroleum Geologists, John Andrew Allan Memorial Volume, p. 244-259.

MACAULEY, G., 1958, Late Paleozoic of Peace River area, Alberta, *in* Goodman, A.J., ed., Jurassic and Carboniferous of Western Canada: American Association of Petroleum Geologists, John Andrew Allan Memorial Volume, p. 289-308.

MACAULEY, G., PENNER, D.G., PROCTER, R.M., AND TISDALL, W.H., 1964, Carboniferous, *in* McCrossan, R.G., and Glaister, R.P., eds., Geological History of Western Canada: Alberta Society of Petroleum Geologists, p. 89-102.

MacQUEEN, R.W., AND SANDBERG, C.A., 1970, Stratigraphy, age and interregional correlation of the Exshaw Formation, Alberta Rocky Mountains: Bulletin of Canadian Petroleum Geology, v. 18, p. 32-66.

MARTIN, H.L., 1969, The Banff and Exshaw formations in the Wabamun Lake area, Alberta (83G): Geological Survey of Canada, Paper 69-7, 24 p.

O'CONNELL, S.C., 1990, The development of the Lower Carboniferous Peace River Embayment as determined from Banff and Pekisko formations depositional patterns: Bulletin of Canadian Petroleum Geology, v. 38A, p. 93-114.

RICHARDS, B.C., 1989, Upper Kaskaskia Sequence: uppermost Devonian and lower Carboniferous, *in* Ricketts, B.D., ed., Western Canada Sedimentary Basin - A Case History, Canadian Society of Petroleum Geologists, Special Publication no. 30, p. 164-201.

STOAKES, F.A., 1987a, Evolution of the Upper Devonian of Western Canada, *in* Bloy, R., and Hopkins, J.C., organizers and compilers, Principles and Concepts for Exploration and Exploitation of Reefs in the Western Canada Basin: Canadian Society of Petroleum Geologists, Canadian Reef Inventory Project - Short Course, 36 p.

STOAKES, F.A., 1987b, Fault controlled dolomitization of the Wabamun Group, Tangent Field, Peace River Arch, Alberta: Canadian Society of Petroleum Geologists, 13th Core Conference and Display and Second International Symposium on the Devonian System, p. 73-85.

STOAKES, F.A., AND FOELLMER, K.E.H., 1987, The Wabamun Group of the Peace River Arch area, west central Alberta: Recognition of Fringing Platforms, a new exploration play-type: Canadian Society of Petroleum Geologists, Second International Symposium on the Devonian System, Program and Abstracts, p.213

STOAKES, F.A., HASSLER, G.T., AND FOELLMER, K.E.H., 1989, Stratigraphy and hydrocarbon potential of the Pekisko Waulsortian Mound play, Peace River Arch, Alberta, *in* Exploration Update `89 - Integration of Technologies: Canadian Society of Petroleum Geologists, Programs and Abstracts, p. 172.

# CHAPTER 12

## SUMMARY

F.A. STOAKES
J.C. WENDTE
C.V. CAMPBELL

This section summarizes the nature of cyclicity, controls on depositional patterns and the occurrence of hydrocarbons in Devonian and early Mississippian strata in the WCSB. These strata comprise a thick succession of shallow-water carbonates and basin-filling shales, limestones and evaporites. The foregoing account provides a sequence stratigraphic framework for these deposits. This framework was established through the recognition of cyclicity at a number of different levels. The first-order cycles approximate the group level stratigraphy of the basin. At a finer level, the use of the depositional sequences defined in this manual provides a higher resolution of correlation than can be attained from either biostratigraphy or seismic stratigraphy. The recognition of cyclicity allows for predicting depositional patterns at many different levels. The understanding of facies distributions in time and space is critical to successfull exploration. This is especially important in the Devonian and early Mississippian of the WCSB where the majority of hydrocarbon traps are stratigraphic in nature.

### CYCLICITY

The succession of Devonian and early Mississippian strata in the WCSB record the nature of cyclicity at many levels. Each level of cyclicity is characterized by an initial deepening or transgressive phase and a later regressive phase of deposition.

Within the context of the scope of this manual, the longest depositional phases are estimated to be on the order of 40 to 50 million years in duration. Stratigraphic successions described in this manual comprise two of these long-term phases: from the onset of Devonian deposition to the top of the Winterburn succession (Gedinian to earliest Famennian) and from the uppermost Devonian (Wabamun) sequence into the Mississippian (Famennian to at least the Tournaisian).

The initial transgressive successions of the first major phase correspond to evaporitic sequences of the Lower Elk Point interval. These deposits are restricted in occurrence to sub-basins in northern and central Alberta. Ensuing transgressive successions correspond to times of more open-marine circulation and the areal expansion of deep-water basins. The Upper Elk Point interval, the first interval described in detail in this manual, marks the initial open-marine inundation of the Alberta Basin. Deposits in the regressive phase of this megasequence, however, remain evaporitic and fill in basin topography created during the lower transgressive phase of this megasequence. The deposition of the Beaverhill Lake and the initial part of the Woodbend succession correspond to open-marine phases that mark continued expansion and deepening of the Alberta Basin. The culmination of this transgressive phase occurs just prior to the end of middle Leduc deposition.

The regressive portion of this first major stratigraphic succession corresponds to the infilling of the basin created in the preceding Beaverhill Lake and early Woodbend intervals. Infilling occurred during episodes of Woodbend (Ireton) and Winterburn shale progradation. The culmination of this phase is reached by progradation of coastal plain deposits of the top of the Winterburn succession across most of Alberta.

The Wabamun and early Mississippian intervals described in this manual are part of a second major phase of deposition. The Wabamun marks the initial transgressive part of this major phase. Further inundation is recorded by backstepping of the latest Devonian Big Valley and early Mississippian Banff through Pekisko formations. The time of maximum inundation corresponds to the initiation and partial development of the downslope mounds in the Pekisko Formation. The onset of the regressive phase is characterized by progradation of Shunda clinothem successions and probably culminates in shallow-water prograding sediments of the Debolt Formation.

Both of these major successions are divided into a number of "first-order" cycles of variable duration estimated to be on the order of a few million years. These first-order cycles reflect the major shifts of facies belts in the WCSB. The transgressive phase of these cycles corresponds to backstepping of carbonate platforms and ramps, initiation of isolate reef growth and creation of deep-water basins. The regressive phase of these cycles records forestepping of ramp-clinothem successions into the basin, termination of growth on some reefs and infilling of the basin.

These first-order cycles are comprised of many second-order cycles, defined by significant shifts or backsteps of facies belts. Smaller third-order cycles correspond to the level of lagoonal and peritidal successions in platform and isolate reef interiors.

Some of the first-order cycles contain "mid-megasequence" cycle divisions that appear to be on a level between these first and second-order cycles. Examples of these include the pre-Zama carbonate level in the Upper Elk Point megasequence, the pre-top middle Leduc in the Woodbend succession and the Blue Ridge base in the Winterburn succession.

### *Controls On Depositional Patterns*

The evolution of Devonian and early Mississippian sequences record the interaction of sea-level, depositional and tectonic controls.

Sequence evolution is primarily related to changes in sea level. In this manual, no attempt has been made to differentiate between the effects of eustasy and regional subsidence. Accordingly, variations in sea level refer to relative changes encompassing both eustatic and subsidence components. Backstepping of carbonate platforms and ramps and the initiation of isolate reefs during transgressive phases are attributed to relatively rapid rates of sea-level rise. Carbonate forestepping and infilling of the basin during regressive phases are generally interpreted to reflect slower rates of sea-level rise.

Our knowledge of Devonian and early Mississippian sea-level change is by no means complete. Many questions remain unanswered. Foremost amongst these are how the regressive phase of one first-order cycle switches over to the transgressive phase of the next first-order cycle. The Beaverhill Lake megacycle appears to be terminated by a lowering of sea level. This lowering resulted in exposure of the upslope portion of the Beaverhill Lake (Mildred) ramp, while the "correlative conformity" in downslope positions corresponds to further shoaling. The nature of the change from the Woodbend to Winterburn and from the Winterburn to Wabamun megasequences is not known. What is clear, however, is that the vast majority of basinal sediments have equivalents on the surrounding platform or upslope portions of ramps. As an example, the increment of basinal strata corresponding to the interval of emergence at the end of the Beaverhill Lake (or start of the Woodbend succession) is very thin. Deposition of most basinal strata, therefore, represents "highstand" conditions. This style of basin-filling contrasts markedly with contemporary modern models of terrigenous clastic deposition alluded to in Chapter 2.

Depositional controls exert a major influence on stacking patterns of first-order sequences in Middle and Late Devonian time. During this interval, east to west water circulation was established as a result of the easterly trade winds. This pattern of water circulation resulted in the progradation and forestepping of ramp-clinothem successions in east-central Alberta into the basin to the west. This style of asymmetric forestepping characterizes the upper regressive phases of the Beaverhill Lake, Woodbend and Winterburn successions.

The availability of extrabasinally derived terrigenous sediment also influenced depositional patterns. Even though the postulated source of the terrigenous sediment was distant, the Ellesmerian Fold Belt in the Canadian Arctic Archipelago, the delivery of these sediments impacted the style of basin filling and modified stacking patterns in the WCSB. This delivery of terrigenous sediment was especially important during the evolution of the Woodbend and Winterburn successions. Although most of the terrigenous sediment was fed laterally to the east side of the Alberta Basin, some lobes depart from this configuration. The role of these lobes in modifying stacking patterns is perhaps best illustrated by comparison of the succession of Nisku strata in the Meekwap and West Pembina areas of Alberta. In early Nisku "time", the basin in central Alberta was open to both the north and to the west. In the Meekwap area, a post-Z Marker lobe of terrigenous sediment prograded into the area from the northeast, infilling this portion of the basin and cutting it off from the open seaway to the north. The upper Nisku Meekwap shelf formed along the southern side of this lobe. In the West Pembina area further to the southeast, the lower Nisku (pre-Z Marker) was developed on the upslope portion of an Ireton clinoform. Significant terrigenous sediment only reached this area in late Nisku "time", corresponding to the deposition of the Cynthia shale. The preceding upper Lobstick, Bigoray and lower Cynthia ramp units progressively backstep away from the basin, displaying a style much different from the regressive aspect in the Meekwap area.

Tectonic elements such as the Peace River Arch, the Tathlina Arch and the Western Alberta Arch were positive features during at least some of Middle and Late Devonian time. Topographic relief on these features was sufficient to localize carbonate platforms and isolate reefs. Significantly though, the greater rates of subsidence away from these features do not appear to have had the same impact on stacking patterns as did sea-level and depositional controls. Subsidence creates accommodation, which in itself, favours a transgressive style of sedimentation. The area of greatest subsidence in late Middle Devonian time was along the eastern flank of the Beaverhill Lake Basin. Beaverhill Lake ramp units, however, forestep toward the west side of the basin. Obviously, high rates of sediment production and the westerly dispersal of sediments overrode the effects of subsidence.

Other tectonic controls remain more enigmatic. For example, the coincidence of a portion of the Leduc reef chain with the Precambrian "Rimbey Arc" is obvious. Just how this basement feature acted to localize Leduc reefing is, however, not understood.

### *Occurrence of Hydrocarbons*

The Devonian of the WCSB is one of the most prolific hydrocarbon producing intervals in North America. This occurrence reflects the coincidence of many factors:

1. The presence of porous and permeable rocks in stratigraphic trapping configurations. Reservoirs include isolate reefs on carbonate platforms and ramps, and up-dip parts of carbonate platforms.
2. The occurrence of thermally mature, organic-rich source intervals. These intervals mainly correspond to deep-water, laminated limestones and shales that occur in basin-centre positions.
3. The presence of permeable carrier beds, allowing for the secondary migration of hydrocarbons from thermally mature source intervals to potential reservoir facies in trapping configurations further updip.
4. The occurrence of impermeable, potential sealing facies overlying and juxtaposed to the reservoir beds. Clinoforming shales and limestones cap many of the isolate reef and platform-margin reservoirs.

The cyclic context of Devonian strata in the WCSB provides a framework useful in all stages of hydrocarbon exploration and exploitation. An understanding of the first-order context of cycles provides a general model that can be used to highgrade, evaluate and predict potential reservoir, source and seal facies. Stacking patterns expressed at the second-order level constitute the architecture of the individual exploration play types. An appreciation of facies patterns at both the second and third-order cycle level is necessary to determining the internal continuity of carbonate reservoirs.